5
Topics in Organometallic Chemistry

Topics in Organometallic Chemistry

Forthcoming volumes:

Organometallics in Process Chemistry
Volume Editor: R. Larsen

CVD Precursors
Volume Editor: R. Fischer

Metallocenes in Stereoselective Synthesis
Volume Editor: T. Takahashi

New Aspects of Transition Metal Arene Complexes in Organic Synthesis
Volume Editor: P. Kündig

New Aspects of Zirconium Containing Organic Compounds
Volume Editor. I. Marek

Ruthenium Catalysts and Fine Chemistry
Volume Editors: C. Bruneau, P. Dixneuf

Previous volumes:

Vol. 4 (1999) Organometallic Bonding and Reactivity – Fundamental Studies;
Vol. Eds.: J. M. Brown, P. Hofmann; ISBN 3-540-64253-6
Vol. 3 (1999) Activation of Unreactive Bonds and Organic Synthesis;
Vol. Ed.: S. Murai; ISBN 3-540-64862-3
Vol. 2 (1999) Lanthanides: Chemistry and Use in Organic Synthesis;
Vol. Ed.: S. Kobayashi; ISBN 3-540-64526-8
Vol. 1 (1998) Alkene Metathesis in Organic Synthesis;
Vol. Ed.: A. Fürstner; ISBN 3-540-64254-4

Springer-Verlag Berlin Heidelberg GmbH

Organolithiums in Enantioselective Synthesis

Volume Editor: D.M. Hodgson

With contributions by
P. Beak, M. Brüggemann, J. Clayden, B. Goldfuss,
E. Gras, D.M. Hodgson, D. Hoppe, M. Iguchi,
T.A. Johnson, D.D. Kim, S.H. Lim, F. Marr,
S. Nakamura, J.F. Normant, M.A.H. Stent,
K. Tomioka, K. Tomooka, T. Toru, K. Yamada

Springer

The series *Topics in Organometallic Chemistry* presents critical overviews of research results in organometallic chemistry, where new developments are having a significant influence on such diverse areas as organic synthesis, pharmaceutical research, biology, polymer research and materials science. Thus the scope of coverage includes a broad range of topics of pure and applied organometallic chemistry. Coverage is designed for a broad academic and industrial scientific readership starting at the graduate level, who want to be informed about new developments of progress and trends in this increasingly interdisciplinary field. Where appropriate, theoretical and mechanistic aspects are included in order to help the reader understand the underlying principles involved.
The individual volumes are thematic and the contributions are invited by the volumes editors.
In references Topics in Organometallic Chemistry is abbreviated
Top. Organomet. Chem. and is cited as a journal

Springer WWW homepage: http://www.springer.de

Visit the TOMC home page at
http://link.springer.de/series/tomc
http://link.springer-ny.com/series/tomc

ISSN 1436-6002

DOI 10.1007/978-3-540-36117-6

Cataloging-in-Publication Data applied for

Bibliographic information published by Die Deutsche Bibliothek
Die Deutsche Bibliothek lists this publication in the Deutsche Nationalbibliographie; detailed bibliographic data is available in the Internet at <http:// dnb.ddb.de>.

Typesetting: Medio Technologies AG, Berlin
Production editor: Christiane Messerschmidt, Rheinau
Cover: Medio Technologies AG, Berlin

Printed on acid-free paper 02/3020 – 5 4 3 2 1 0
www.springer.com/mycopy

Volume Editor

Dr. David M. Hodgson
Dyson Perrins Laboratory
Department of Chemistry
University of Oxford
South Parks Road
Oxford, OX1 3QY
U.K.
E-mail: david.hodgson@chem.ox.ac.uk

Editorial Board

Topics in Organometallic Chemistry
Also Available Electronically

For all customers with a standing order for Topics in Organometallic Chemistry we offer the electronic form via SpringerLink free of charge. Please contact your librarian who can receive a password for free access to the full articles by registration at:
http://link.springer.de/orders/index.htm
If you do not have a standing order you can nevertheless browse through the table of contents of the volumes and the abstracts of each article at:
http://link.springer.de/series/tomc
http://link.springer-ny.com/series/ tomc
There you will find also information about the

- Editorial Board
- Aims and Scope
- Instructions for Authors

Preface

Organolithiums are versatile, widely used reagents and intermediates in organic synthesis. As part of the accelerating interest in developing methods for asymmetric synthesis, the use of organolithiums in enantioselective synthesis has witnessed spectacular advances in the last dozen years. This volume constitutes the first comprehensive treatise of this exciting area.

Carving up the field of enantioselective organolithium chemistry into discrete chapters based on the type of process involved is of course to some extent an artificial process and could, no doubt, have been carried out in other ways. Nevertheless, it is believed that the eight main chapters of this volume constitute a reasonable division into the major areas where significant progress has been made. Also, in a multi-author volume of this nature, it is inevitable, and at times desirable (for the completeness of an individual chapter) that there is the occasional slight overlap of content between chapters. It must also be accepted that the level of detail and style of chapters will differ, as each contribution brings a flavour of the author's own knowledgeable views to the precise area under discussion. For example, several chapters include appropriate background on diastereoselective (especially chiral auxiliary) based approaches to the use of organolithiums in synthesis; this material serves to put the enantioselective (ligand-assisted) contributions into proper perspective.

I am very grateful to my fellow chemists who generously put their expertise into this project: P. Beak, M. Brüggemann, J. Clayden, B. Goldfuss, E. Gras, D. Hoppe, M. Iguchi, T. A. Johnson, D. D. Kim, S. H. Lim, F. Marr, S. Nakamura, J. F. Normant, M. A. H. Stent, K. Tomioka, K. Tomooka, T. Toru, and K. Yamada. I am also grateful to J. Brown (Oxford) for encouragement and C. Moisa (Springer) for help with the preparation of this volume.

Oxford, January 2003 David Hodgson

Contents

Topics Organomet Chem (2003) 5: 1–20
DOI 10.1007/b10341

Overview of Organolithium-Ligand Combinations and Lithium Amides for Enantioselective Processes

David M. Hodgson, Matthew A. H. Stent

Dyson Perrins Laboratory, Department of Chemistry, University of Oxford, South Parks Road, Oxford, OX1 3QY, U.K. *E-mail: david.hodgson@chem.ox.ac.uk*

The first reports of the use of additives to induce enantioselectivity in reactions of organolithiums appeared in the late 1960s, with widespread interest developing from the late 1980s onwards mirroring the heightened interest in asymmetric synthesis. This chapter provides an overview of the use of external chiral ligands as the source of enantioinduction in organolithium processes, mainly in the areas of asymmetric additions and enantioselective deprotonations. Key selected developments are presented, with emphasis on important ligand classes investigated, the source of these ligands and the effect of ligand structure on the success of a process. Chiral lithium amides/alkoxides have been extensively investigated in a number of valuable asymmetric processes, however, in this chapter only those examples of the use of lithium amides that result in the formation of a new enantioenriched organolithium species are discussed.

Keywords. Chiral ligands, Enantioselective processes, Lithium amides, Organolithiums

1
Organolithium-Ligand Combinations

1.1
Historical Perspective

Organolithium chemistry pervades organic chemistry, and aspects of organolithium chemistry have been reviewed on several occasions [1–5]. Its beginnings can be traced to the early part of the last century: in 1917 Schlenk and Holtz reported the first organolithium compound (by transmetallation of dimethylmercury with lithium) [6]; the earliest reported example of metallation by an organolithium compound (the reaction of ethyllithium with fluorene to give 9-fluorenyllithium and ethane) appeared in 1928 [7]; shortly thereafter preparations of organolithium reagents from organic halides and lithium metal appeared by Ziegler [8], Wittig [9], and Gilman [10]. As the utility of organolithium reagents in synthesis developed, it became clear that the solvent(s) used, as well as the addition of small quantities of Lewis bases such as amines, could have important affects on reactivity. Key ligand developments occurred around 1960, with investigations (mainly industrial) into the use of chelating diamines to enhance anionic polymerisation of unsaturated hydrocarbons such as ethylene and butadiene using organolithiums [11]. Langer, Eberhardt and Eastham reported independently that the reactivities of organolithium compounds were particularly enhanced by chelating ditertiary amines, such as TMEDA **1** or (–)-sparteine **2** (Fig. 1) [11].

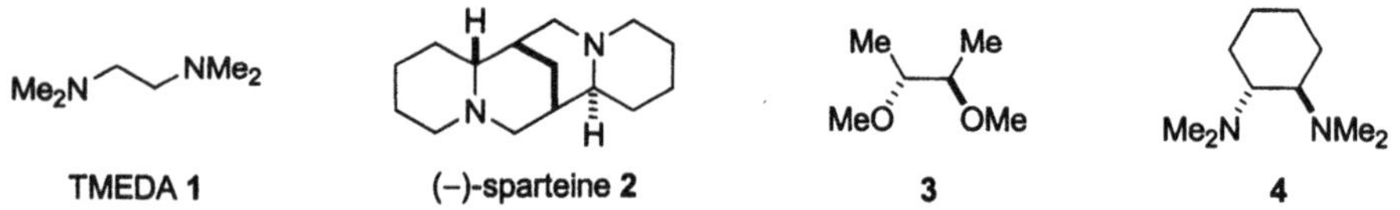

Me_2N NMe_2 TMEDA **1** (–)-sparteine **2** **3** **4**

Fig. 1.

Studies on enantioselective transformations of Grignard reactions were reported as early as 1953 [using (2*R*,3*R*)-dimethoxybutane **3** as solvent] [12]. The seminal work on organolithiums in enantioselective synthesis, albeit with low levels of enantioinduction, was reported by Nozaki and coworkers in the late 1960s using (–)-sparteine **2** as an added ligand in the asymmetric lithiations of isopropylferrocene, ethylbenzene (e.g., Scheme 1, op=optical purity) and in allene generation from cyclopropyl carbenoids [13].

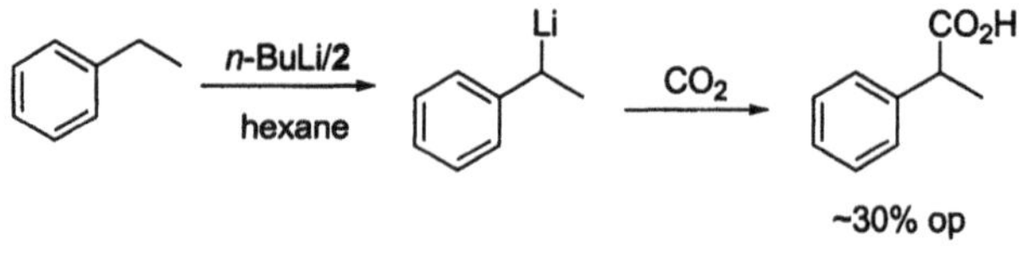

n-BuLi/**2**, hexane; CO_2; ~30% op

Scheme 1.

Enantioselective reactions using organolithiums with various additives were studied sporadically in the 1970s and with increasing frequency in the 1980s, mirroring the heightened interest in asymmetric synthesis [14]. Tomioka reviewed the developing field of asymmetric synthesis utilising external chiral ligands in 1990 [15], and work by Hoppe at the end of the 1980s concerning lithiation-electrophile trapping of carbamate derivatives in very high ees led to a surge of interest in enantioselective processes using organolithiums. Hoppe and Hense reviewed enantioselective synthesis with lithium/sparteine carbanion pairs in 1997 [16], and a recent review focused on the important topic of configurational stability and transfer of stereochemical information in the reactions of enantioenriched organolithium reagents [17]. However, the current volume is the first dedicated to a comprehensive coverage of organolithiums in enantioselective synthesis.

1.2
Scope of Processes

In this section an overview is provided of the use of external chiral ligands as the source of enantioinduction in organolithium processes, mainly in the areas of asymmetric additions and enantioselective deprotonations, however a few other significant examples will be discussed. Only selected key developments are presented, with emphasis on important ligand classes investigated, the source of these ligands (often an important criteria in utility of the chemistry) and the effect of ligand structure on the success of a process. The aim is to provide a "feel" for the type of reactions which have been investigated, the level of success currently achieved and the range of ligands which have been explored to date; the reader is referred to subsequent chapters for detailed analyses.

1.2.1
Asymmetric Additions

1.2.1.1
Additions to C=O

Nozaki, in his pioneering studies on ligand-assisted enantioselective lithium transformations, also examined the addition of BuLi to benzaldehyde (6% optical purity) using sparteine [13]. Early studies, with marginally higher optical purities of alcohols, were also carried out by Langer and Whitney using 1,2-bis(*N*,*N*-dimethylamino)cyclohexane **4** (Fig. 1), a chiral analogue of TMEDA **1** [11]. This ligand is available as either enantiomer from *trans*-1,2-diaminocyclohexane via resolution using tartaric acid. Around 1980 the groups of Mukaiyama [18] and Cram [19] both developed chiral ligands for the addition of organolithiums to benzaldehyde in high ee (Scheme 2, see also Goldfuss, in this volume); the use of very low temperatures is essential to obtain high levels of asymmetric induction.

Mukaiyama's tridentate diamino alcohol **6** is derived from the coupling of two L-proline units followed by reduction of the ester/amide carbonyl groups; *ent*-**6**

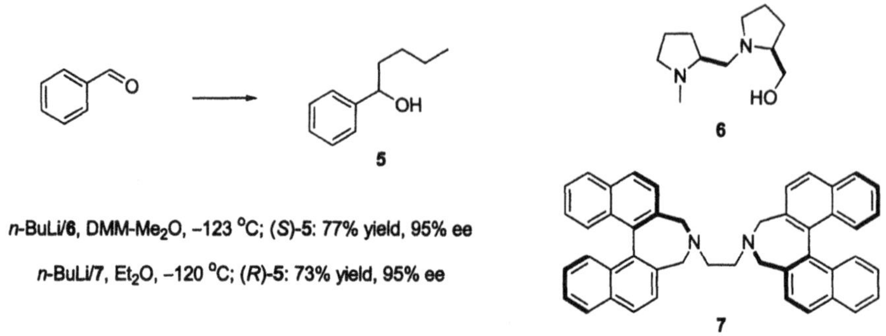

Scheme 2.

and analogues are therefore available. The reaction is quite specific for n-BuLi, other primary organolithiums (methyl, ethyl, propyl) gave only moderate levels of ee.

Bis(binaphthylamine) ligand **7**, developed by Cram, is synthesised by the coupling of two equivalents of 2,2'-bis(bromomethyl)-1,1'-dinaphthyl with 1,2-diaminoethane, a subsequent (–)-dibenzoyl tartrate resolution affording both enantiomers of **7**. As with Mukaiyama's system, the use of other organolithiums in the addition gave only moderate levels of ee.

More recently, the focus has been the development of ligands capable of inducing good levels of asymmetric induction in this type of addition under less rigorous experimental conditions (i.e., at –78 °C), a considerable challenge for a process with a rapid reaction rate in the absence of ligand.

Naef has introduced 1-amino-1,2-diphenylethanols (both enantiomers of the corresponding free amino alcohol being commercially available) as ligands for additions of this type; the reaction of n-BuLi with benzaldehyde in the presence of **8** (Fig. 2) (THF, –78 °C, 1 h) affording (S)-**5** (Scheme 2) in 78% ee [20]. C_2-Symmetric diamino diols have been used as chiral ligands to induce asymmetric addition of lithium acetylides to carbonyl groups [21]. The enantiomeric excesses observed depended on the structure of the acetylene, but were up to 99% ee for the addition of the TBDMS ether of propargyl alcohol to benzaldehyde using ligand **9** (Fig. 2).

Fort and coworkers have reported the use of BuLi-(S)-N-methyl-2-pyrrolidine methoxide **10** for the chemo, regio- and enantioselective reaction of pyridine derivatives with aldehydes [22] (Scheme 3); in this chemistry the amine alkoxide plays a dual role in first preventing classical nucleophilic addition of BuLi onto the heteroaromatic ring, and secondly mediating the enantioselective addition.

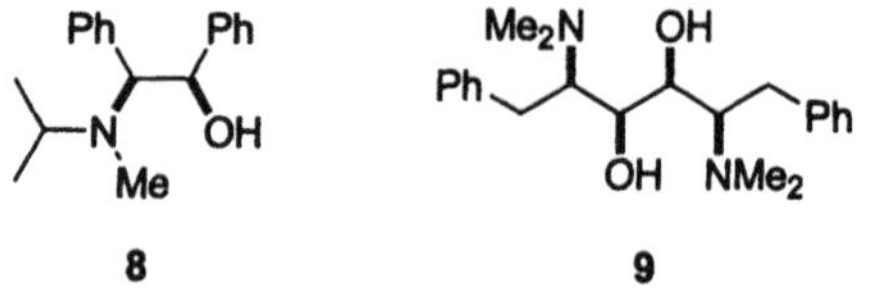

Fig. 2.

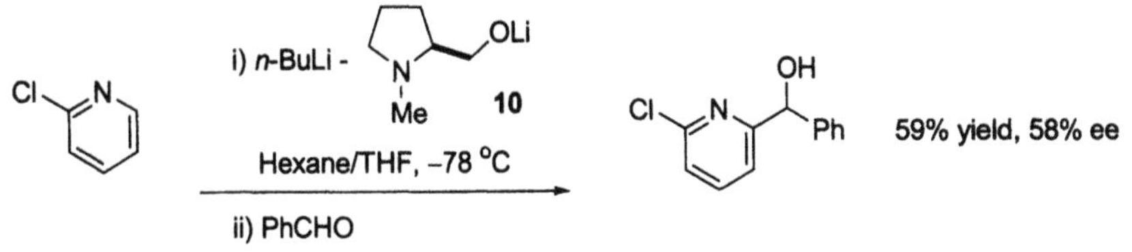

Scheme 3.

1.2.1.2
Conjugate Additions

Regioselectivity for 1,4-addition of organolithiums to α,β-unsaturated esters and imines can be promoted by a bulky ester/imine substitutent. Tomioka and coworkers initially reported asymmetric conjugate addition to α,β-unsaturated aldimines, where 1,2-addition is prevented by an *N*-cyclohexyl substituent (Tomioka, in this volume). Several C_2-symmetric diether and diamino ligands were investigated in the addition of *n*-BuLi and PhLi to imine **11** (Scheme 4), diether **3** (Fig. 1) giving promising levels of asymmetric induction (53% ee), whilst **12** and **13** afforded material with very low ee (<15%) [23]. More recently, Tomioka described the rational design of diether **14** (Scheme 4), which mediates the addition of either MeLi or *n*-BuLi to imine **11** in >90% ee [24]. Diethers **3** and **12** are derived from the corresponding diols, either enantiomer being commercially available in both cases, similarly tetramethyldiamine **13** is prepared from the commercially available diamine by methylation. Diether **14** is derived from the corresponding *trans*-stilbene via asymmetric dihydroxylation.

Tomioka has also extensively investigated conjugate 1,4-additions to α,β-unsaturated BHA (2,6-di-*tert*-butyl-4-methoxyphenyl) esters; stilbene-derived di-

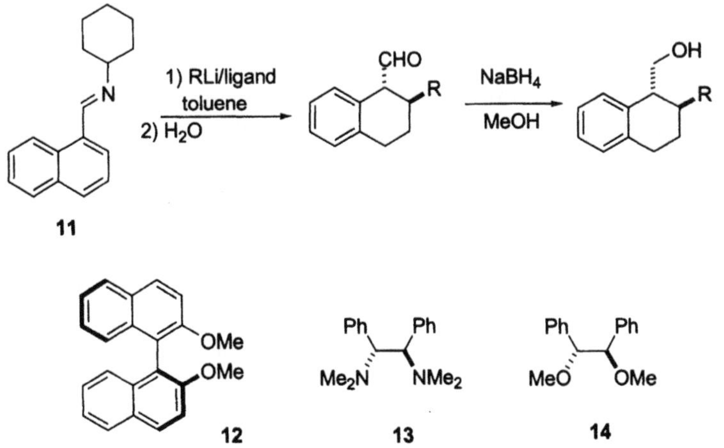

Scheme 4.

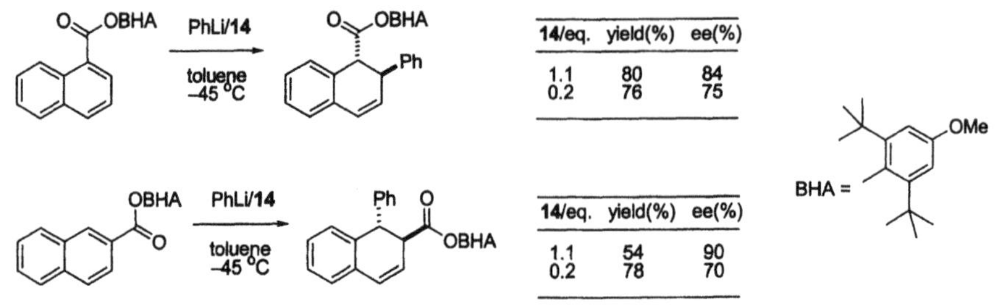

14/eq.	yield(%)	ee(%)
1.1	80	84
0.2	76	75

14/eq.	yield(%)	ee(%)
1.1	54	90
0.2	78	70

Scheme 5.

ether **14** again mediating the asymmetric addition of PhLi in high ee. In a further development, reduction of the amount of ligand used to substoichiometric quantities continued to afford good levels of asymmetric induction (Scheme 5) [25].

Comparison of diether **14** and (–)-sparteine **2** reveal that they are complementary to each other, with **14** giving the best levels of asymmetric induction (up to 93% ee) for phenyl- and vinyllithium, whilst (–)-sparteine gives the best results for butyl- and ethyllithium [26].

1.2.1.3
Additions to Imines

The first report of an external chiral ligand-mediated enantioselective addition of an organolithium reagent to an imine appeared in 1990 (Tomioka, in this volume) [27]. Tomioka and coworkers found that 1,2-addition of MeLi, *n*-BuLi and vinyllithium to the 4-methoxyphenylimine of benzaldehyde **15** gave, in the presence of tridentate aminoether **16**, the corresponding tertiary amines **17** in excellent yields and good ees (71–77%) (Scheme 6).

Ligand **16**, derived from L-phenylalanine, is recovered quantitatively from the reaction following chromatography, and can also be used in substoichiometric quantities: ees of 40–64% being observed when 5–30 mol% of **16** was utilised [28].

In 1994 Denmark and coworkers reported a major advance in the asymmetric addition of organolithium reagents to imines using bisoxazolines as ligands (e.g., **19**, Scheme 7); excellent yields and high levels of ee were observed [29, 30]. C_2-Symmetric bisoxazolines have been widely used as ligands in asymmetric

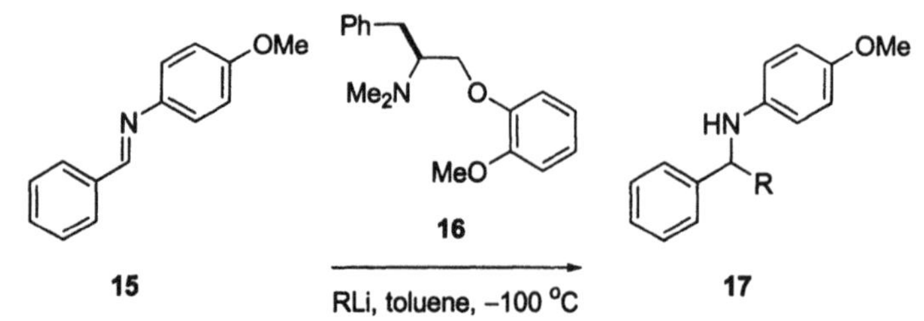

Scheme 6.

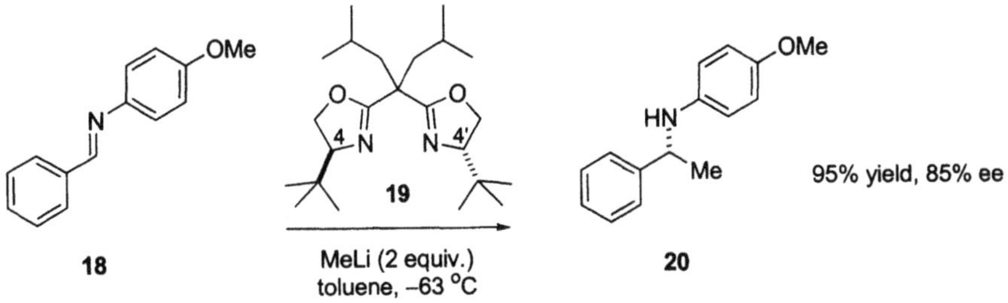

18 → 20: 19, MeLi (2 equiv.), toluene, –63 °C; 95% yield, 85% ee

Scheme 7.

synthesis (although Denmark was the first to demonstrate their utility in enantioselective organolithium chemistry), and their C4/4' and linking substituents are easily varied depending on the precursor amino acid/alcohol and substituted malonic acid used; thus both enantiomeric forms and numerous analogues are available. In Denmark's work, the ligand substitution pattern is partly designed to avoid addition of organolithium to the oxazoline C=N bond.

Whilst the addition of both *n*-BuLi and PhLi was achieved in only moderate levels of enantiomeric excess, the reaction of both MeLi and vinyllithium afforded the corresponding amines in excellent yield and high ee. The use of (–)-sparteine **2** was also investigated as a comparison in this reaction, high levels of enantiomeric excess (72–82%) being observed with MeLi, *n*-BuLi and PhLi. As the non-ligand-mediated background reaction between imine **18** and MeLi is very slow in toluene (6% yield of **20** after 4h at –78°C) the potential for catalytic use of the bisoxazolines in this ligand-accelerated process was investigated. The addition of MeLi, *n*-BuLi and vinyllithium to various imines was found to proceed in only slightly reduced levels of asymmetric induction when only 20 mol % of the bisoxazoline ligands were used compared to when stoichiometric quantities were used. However, the use of catalytic quantities of (–)-sparteine resulted in significantly lower levels of enantioinduction.

More recently, Andersson and Tanner have developed C_2-symmetric bisaziridine ligands for the enantioselective addition of organolithium reagents to imines. The synthesis of these ligands is not trivial, requiring at least nine steps from commercial materials (the chirality being introduced by asymmetric dihydroxylation). A number of these ligands (e.g., **22**) were examined in the addition of MeLi, *n*-BuLi and vinyllithium to imine **21**, the resulting amines **23** being obtained in 6–89% ee (Scheme 8) [31].

1.2.1.4 Other Additions

In 1992 Tomioka and coworkers reported catalytic enantioselective nucleophilic aromatic substitution giving binaphthyls in high ee. Chiral diether **14** was used to mediate the reaction between naphthyllithium and naphthylimines, the resulting binaphthyls being obtained in up to 90% ee (Scheme 9) [32].

The asymmetric ring opening of oxabicylic compounds has been extensively investigated, principally by Lautens, the processes proceeding with a variety of

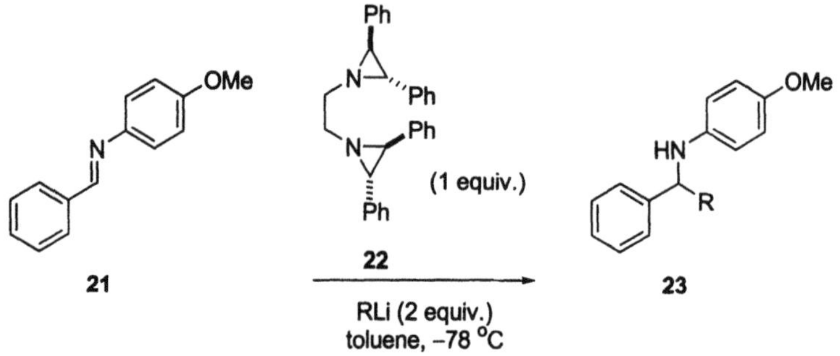

Ph OMe N Ph Ph N OMe N HN (1 equiv.) Ph R 21 22 23 RLi (2 equiv.) toluene, –78 °C

Scheme 8.

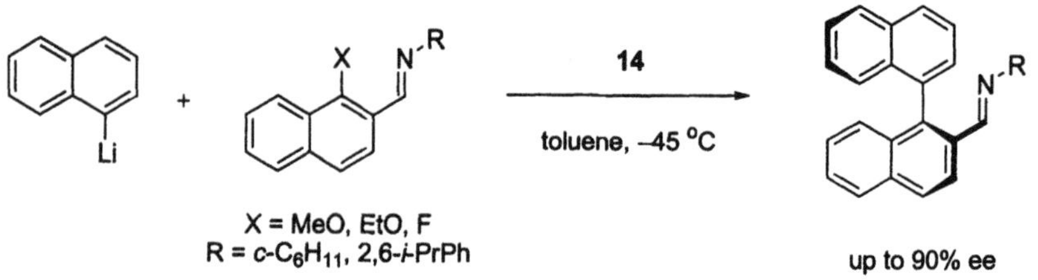

X N R 14 N R + Li toluene, –45 °C X = MeO, EtO, F R = c-C_6H_{11}, 2,6-i-PrPh up to 90% ee

Scheme 9.

organometallic species in the presence of transition metal complexes. In 1993 the use of an organolithium reagent in the presence of a chiral ligand was used to successfully perform such an enantioselective ring-opening reaction (Scheme 10). Of the ligands investigated in this reaction, catalytic quantities (15 mol %) of (–)-sparteine gave the highest levels of induction (52% ee), whereas diamines **13** and **24** afforded cycloheptenol in only 4% and 20% ee, respectively [33].

Enantioselective carbolithiation of simple alkenes using sparteine, has been developed following the seminal disclosure in 1995 by Normant and coworkers (Normant, in this volume) [34]. The efficiency of the catalytic process, using as little as 1% (–)-sparteine **2**, is noteworthy for an organolithium transformation (Scheme 11).

Kündig and co-workers have reported the regio- and enantioselective addition of alkyl-, vinyl- and aryllithium reagents to a prochiral arene-$Cr(CO)_3$ complex in the presence of chiral ligands [35]. (–)-Sparteine **2** and three di(methyl ether) ligands (the source of ligands **3** and **14** was discussed above, **26** is derived

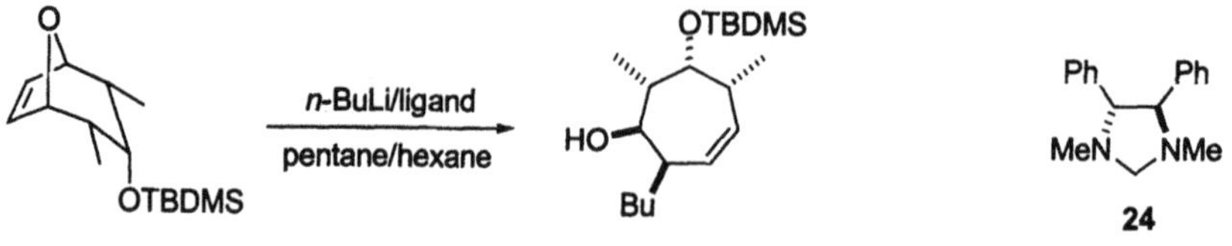

O OTBDMS n-BuLi/ligand Ph Ph pentane/hexane HO MeN NMe OTBDMS Bu 24

Scheme 10.

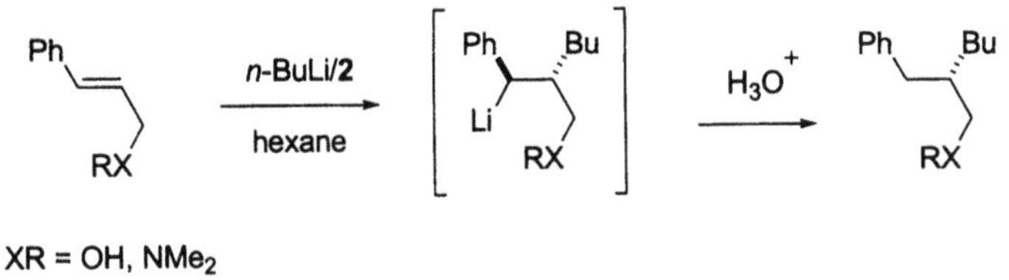

$XR = OH, NMe_2$

Scheme 11.

from *trans*-cyclopentane diol – commercially available as either enantiomer) were screened in the addition of MeLi, *n*-BuLi, vinyllithium and PhLi to arene complex **25** (Scheme 12). (–)-Sparteine **2** gave only moderate levels of ee (34–54%), **3** only afforded high levels of asymmetric induction in conjunction with PhLi (81% ee), cyclic diether **26** gave >80% ee for the addition of both MeLi and PhLi, whilst stilbene-derived **14** gave the best overall results (65–90% ee).

Enantioselective additions of organolithiums to epoxides, oxetanes and cyclic acetals, have been reported by Tomioka, Alexakis and Müller; during these studies the compatibility of $BF_3{\cdot}Et_2O$, to assist ring-openings, with sparteine was demonstrated [36].

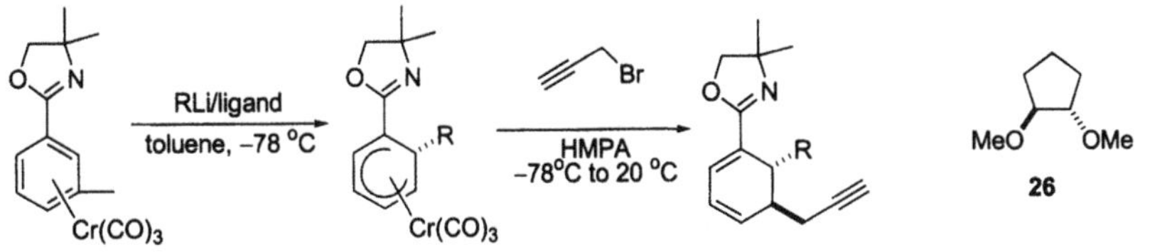

Scheme 12.

1.2.2
Enantioselective Transformations via Deprotonation

In contrast to the range of ligands successfully influencing asymmetric additions of organolithium reagents, the use of organolithium/ligand combinations to effect enantioselective transformations via deprotonation has mainly focused on the use of (–)-sparteine **2** [16]; bisoxazoline ligands have also been effective in some specific reaction classes.

The origin of asymmetric induction may be either (a) an initial enantioselective deprotonation to give a configurationally stable organolithium (often a "saturated" carbanion α to N or O) which reacts stereospecifically with an electrophile, or (b) formation of a configurationally unstable organolithium (often benzylic or allylic) which, if interconversion is rapid or slow relative to electrophile trapping, can generate enantioenriched product via dynamic kinetic resolution, or via dynamic thermodynamic resolution, respectively [17]. These issues are discussed in more detail in the following chapters, especially that of Toru and Nakamura, in this volume.

1.2.2.1
Carbanions α to O

Enantioselective synthesis by lithiation adjacent to O and electrophile incorporation (Hoppe, in this volume) has been dominated by the studies of Hoppe using carbamates (e.g., Scheme 13).

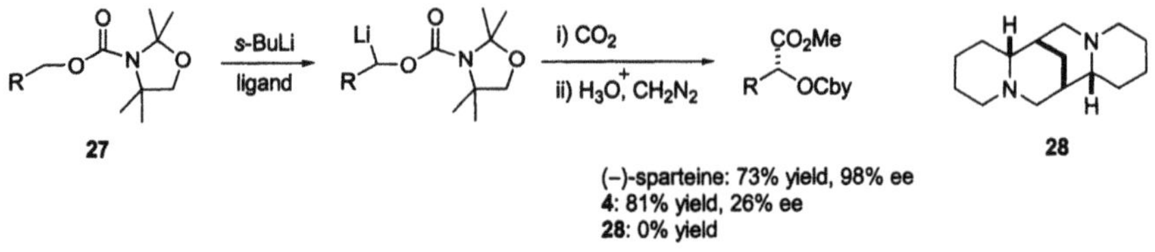

s-BuLi, ligand; i) CO_2, ii) H_3O^+, CH_2N_2

27 28

(–)-sparteine: 73% yield, 98% ee
4: 81% yield, 26% ee
28: 0% yield

Scheme 13.

Hoppe has concentrated almost exclusively on the use of (–)-sparteine **2** as ligand in these transformations (generally high enantioinduction is observed). However, the C_2-symmetric lupine alkaloid (-)-α-isosparteine **28** was briefly investigated by Zschage and Hoppe in 1992 as an alternative to (-)-sparteine as a chiral ligand in the BuLi-mediated deprotonation of an allylic carbamate where, following transmetallation with $Ti(O\textit{i}\text{-}Pr)_4$ and reaction with butanal, it induced 16% ee [(-)-sparteine **2** gave 31% ee] in the resultant homoallylic alcohol [37]. (-)-α-Isosparteine **28** occurs naturally and is available commercially, albeit in milligram quantities [38]. It can be prepared by $AlCl_3$-promoted isomerisation of (-)-sparteine **2** [39] and may additionally be dried as a solution in ether over CaH_2 prior to use [40], or alternatively by the use of an additional equivalent of RLi on the monohydrate (which binds water much more tightly than sparteine **2**) in situ [41].

More recently, Hoppe has also compared (–)-sparteine **2** with (–)-α-isosparteine **28** and *N,N,N',N'*-tetramethyldiamine **4** in the enantioselective deprotonation of alkyl carbamates. It was found that α-isosparteine **28** does not support the deprotonation of alkyl carbamates at all, whereas **4** generally gave lower levels of ee (Scheme 13) compared with sparteine [42]. However, the "slimmer" ligand **4** facilitated enantioselective deprotonation of hindered carbamate **27** (R = *t*-Bu) in 79% ee, whereas (–)-sparteine **2** failed to effect deprotonation; a delicate balance in the steric demand of the *CH*-acid (alkyl carbamate) and the inducing diamine therefore determines the success of the deprotonation. Quantum-chemical calculations (PM3, ab initio methods) on several models for the competing diastereomeric transition states of the deprotonation under the influence of **2** were found to reflect well the sense and the magnitude of the experimentally observed chiral induction.

As discussed more fully in Hodgson et al, in this volume, enantioselective Wittig rearrangements initiated by deprotonation α to O using organolithiums were first investigated by Mazaleyrat in 1983, who reported the kinetic resolution of a racemic binaphthyl either via [1,2]-Wittig rearrangement using sparteine **2**, albeit in modest ee [43]; more recently bisoxazoline ligands (e.g., **19**, Scheme 7) have been used with success in [1,2]-Wittig rearrangements (up to 62% ee), including catalytic processes [44]. Organolithiums in enantioselective

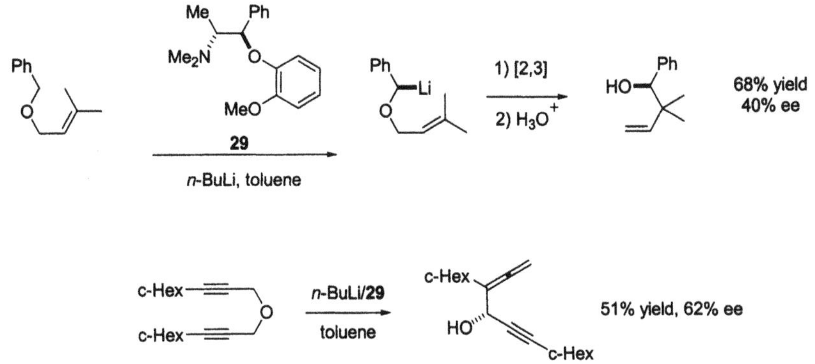

Scheme 14.

[2,3]-Wittig rearrangements were first examined by Kang and coworkers in 1994 using sparteine **2**, and α-isosparteine **28** [40]. A rare example of the use of ligands other than sparteines and bisoxazolines in enantioselective deprotonation is found in the [2,3]-sigmatropic rearrangement work of Manabe using the tridentate pseudonorephedrine-derived ligand **29** (Scheme 14) [45].

More recently, bisoxazolines were studied as ligands in [2,3]-Wittig rearrangements (up to 89% ee) [46]; these ligands were also effective in a catalytic manner [47]. Bisoxazoline **30** provides remarkably high enantioselectivity in the benzylic lithiation-electrophile trapping reactions of benzyl methyl ether, isochroman and phthalan (up to 97% ee) (e.g., Scheme 15) [48].

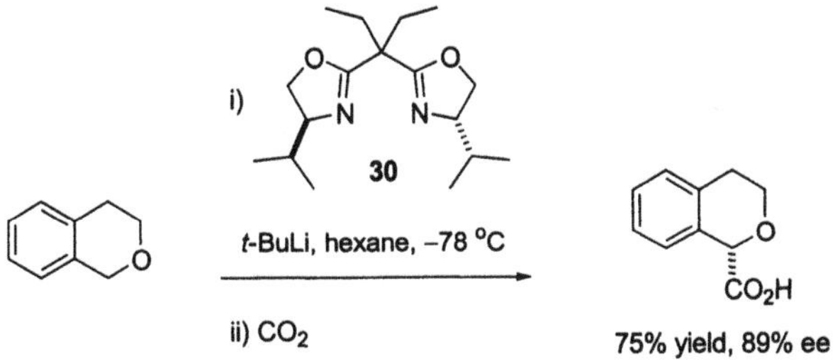

Scheme 15.

As discussed in Hodgson et al, in this volume, transformations of achiral epoxides based on enantioselective α-deprotonation have been extensively studied by Hodgson and coworkers since 1996 [49, 50]; ligands used were (-)-sparteine **2** [49], (−)-α-isosparteine **28** and (usually less effectively) bisoxazolines (e.g., **30**), with the bisoxazoline study [51] being the first report of such ligands in an enantioselective deprotonation.

1.2.2.2
Carbanions α to N

Beak and coworkers carried out the seminal (and much of the subsequent) studies on the use of organolithiums in enantioselective synthesis via lithiocarbanions α to N (Beak, in this volume) [16, 17, 52, 53]. The first example of this type was reported in 1991: the enantioselective deprotonation of Boc-pyrrolidine **31** by *s*-BuLi-(–)-sparteine **2** was found to proceed at –78 °C in ether with a high degree of selectivity (94% ee) for removal of the pro-*S* hydrogen of **31** to give **32**, which could be trapped with a range of electrophiles (e.g., Scheme 16).

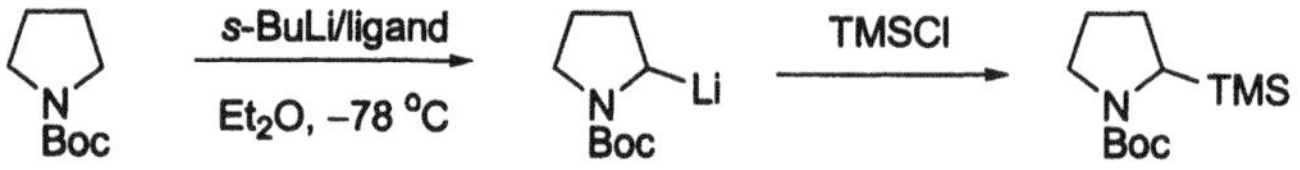

s-BuLi/ligand; Et_2O, –78 °C; TMSCl

N Boc **31** → N Boc Li **32** → N Boc TMS **33**

Scheme 16.

The enantioselective deprotonation of Boc-pyrrolidine **31** is probably the most studied example of an organolithium/ligand enantioselective deprotonation. As Beak has nicely summarised [54]: "The kinetics of the deprotonation are consistent with formation of a thermodynamically favourable three-component complex of alkyllithium, sparteine **2**, and **31** prior to a rate-determining transfer of the pro-*S* H to the complexed *i*-PrLi. Computational results indicate that the origin of the enantioselectivity observed for the deprotonation is largely a steric phenomenon [55]. The most stable complex, that leading to transfer of the pro-*S* hydrogen of **31**, also has the lowest activation energy for lithiation; an analogous but less stable complex, leading to transfer of the pro-*R* hydrogen of **31**, is more congested and a number of the steric interactions responsible for the higher ground-state energy of this complex persist in the transition state for proton transfer. The relatively large difference in the transition state energies calculated for removal of the pro-*R* and pro-*S* hydrogens of **31** ($\Delta\Delta H\ddagger = 4.5$ kcal/mol, $\Delta\Delta G\ddagger = 3.2$ kcal/mol) are fully consonant with the highly enantioselective character of the process."[1]

A recent example of enantioselective synthesis via lithiocarbanions α to N is found in the work of Voyer et al [56], which allows access to *N*-Boc protected phenylglycines (Scheme 17).

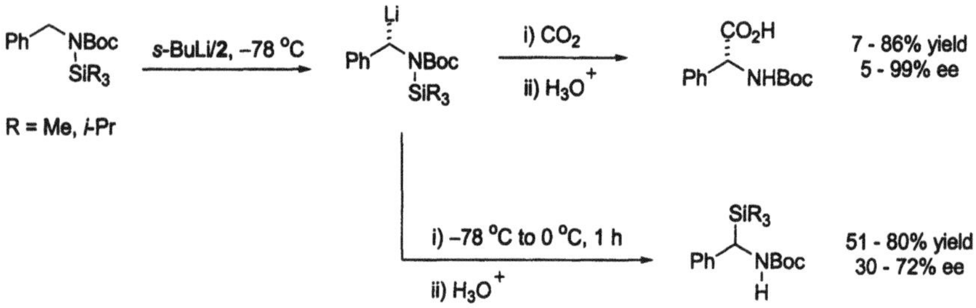

Ph NBoc SiR_3 —(s-BuLi/**2**, –78 °C)→ Ph(Li) NBoc SiR_3 —(i) CO_2; ii) H_3O^+)→ Ph(CO_2H) NHBoc: 7 - 86% yield, 5 - 99% ee

R = Me, *i*-Pr

—(i) –78 °C to 0 °C, 1 h; ii) H_3O^+)→ Ph(SiR_3) NBoc H: 51 - 80% yield, 30 - 72% ee

Scheme 17.

1 Excepted with permission from J. Am. Chem. Soc. 2002, 124, 1889–1896. Copyright 2002 American Chemical Society.

1.2.2.3
Carbanions α to S or P

Organolithium-induced deprotonation α to sulfur (Toru, in this volume) and subsequent formation of enantioenriched products in an enantioselective manner has utilised sparteine **2** but, notably, bisoxazolines were found by Toru to be significantly more effective in asymmetric substitution reactions of primary α-sulfenyl carbanions (up to 99% ee) (Scheme 18) [57, 58].

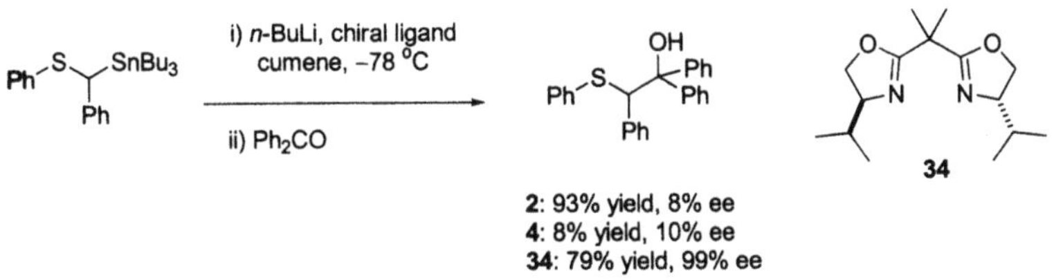

i) *n*-BuLi, chiral ligand cumene, –78 °C
ii) Ph_2CO

2: 93% yield, 8% ee
4: 8% yield, 10% ee
34: 79% yield, 99% ee

Scheme 18.

Discrimination between enantiotopic alkyl groups in a phosphine by deprotonation α to phosphorous using organolithium reagents was probably first achieved by Raston and White with $PhPMe_2$ and BuLi/sparteine, since X-ray analysis of the so-formed lithiated phosphine/sparteine complex (40%) indicated that it was composed of homochiral dimers; electrophile trapping was unsuccessful, however [59]. Activation at phosphorus, for example by oxidation or complexation, assists α-deprotonation. $PhMe_2P=O$ underwent reaction with BuLi/sparteine followed by addition of EtI to give PhMePrP=O in 14% optical purity [59]. Efficient enantioselection between alkyl groups in phosphine-borane complexes (Toru, in this volume) was first reported by Evans in 1995 [60]; BuLi-sparteine is the favoured base-ligand combination (up to 99% ee).

1.2.2.4
Planar Chiral Compounds

Although early studies by Nozaki examined (-)-sparteine **2** in the asymmetric lithiation of isopropylferrocene (as noted in Sect. 1.1 above), the first enantioselective generation of planar chirality in good ees using an organolithium (Claydon, in this volume) was reported by Uemura in 1994 in the lithiation of tricarbonyl(η^6-phenyl carbamate)chromium complexes using chiral diamines. After quenching with electrophiles, enantioenriched (*o*-substituted phenyl carbamate)chromium complexes were obtained in up to 82% ee (Scheme 19) [61].

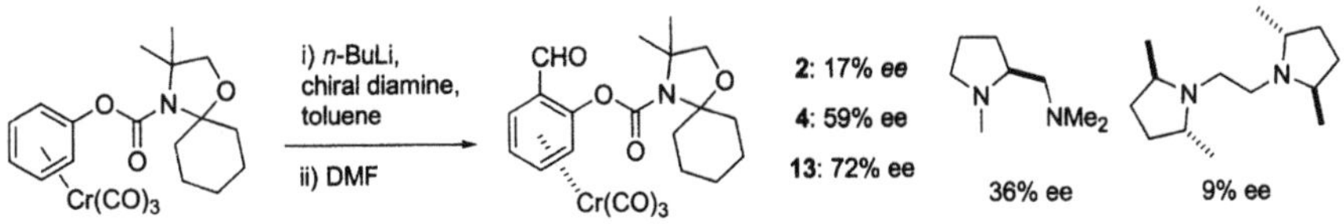

i) *n*-BuLi, chiral diamine, toluene
ii) DMF

2: 17% ee
4: 59% ee
13: 72% ee

36% ee
9% ee

Scheme 19.

Related chemistry has been reported by Snieckus using ferrocene-carboxamides with sparteine as additive, giving planar chiral adducts in excellent ees (up to 99%) [62].

1.3 Sparteine and Other Diamine Ligand Availability

(-)-Sparteine **2** is commercially available in large quantities by extraction from certain papilionaceous plants (e.g., Scotch broom), and usually can be readily recycled for further reactions by acid-base extraction. These beneficial factors all contribute to (–)-sparteine's popularity as a chiral diamine ligand in asymmetric synthesis. From the examples discussed above (–)-sparteine can clearly be seen to have been the most widely used, and generally most successful (especially in deprotonations) external chiral ligand in combination with organolithiums. Indeed, it is remarkable that this alkaloid effects such a wide range of enantioselective transformations using organolithiums and also other organometallics, such as Grignards [63] and in organopalladium chemistry [64, 65].

Nevertheless, there are two important issues that can arise following the use of sparteine in any particular transformation. Firstly, although (+)-sparteine also occurs naturally it is much less easily obtained and so for a specific application if (–)-sparteine leads to the undesired enantiomer, then it is currently difficult, even despite an impressive recent total synthesis of (+)-sparteine [66], to find a substitute for (–)-sparteine that will allow access to the desired enantiomer. A recent example is found in work by Clarke and Travers towards insecticidal 4-alkynyloxazolines, which used a novel sparteine-mediated enantioselective deprotonation-alkylation of propargylic amide systems (Scheme 20), and unfortunately did not lead to the biologically relevant enantiomer [67].

It should be noted that there are a few specific examples where (–)-sparteine can be used to generate either product-type enantiomer; for example via transmetallation [17, 53], the use of electrophiles with different leaving groups [68], double lithiation in the generation of planar chirality [69], or a solvent switch [70].

Opposite enantioselectivity was also observed between *N,N*-diisopropyl-*o*-allyloxymethylbenzamide and *N,N*-diethyl-*o*-allyloxymethylbenzamide in (–)-sparteine-mediated enantioselective [2,3]-Wittig rearrangements [71]. A related

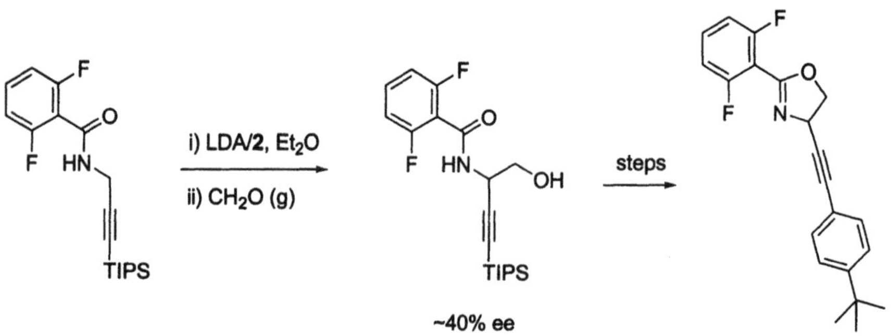

i) LDA/**2**, Et_2O
ii) CH_2O (g)
steps
~40% ee

Scheme 20.

example is found in the formal synthesis of both enantiomers of curcuphenol, where high and opposite enantiodiscriminations were observed between tertiary amides and secondary amides in the sparteine-mediated lateral lithiation-allylation of 2-ethyl-*m*-toluamide derivatives (Scheme 21) [72].

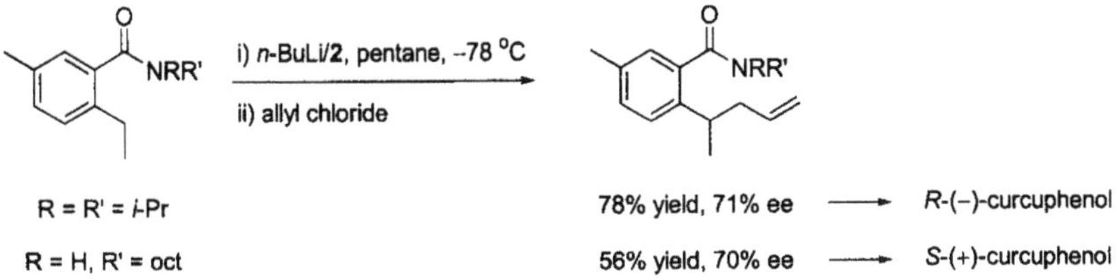

Scheme 21.

A second key issue with the use of sparteine concerns modification/simplification of the sparteine skeleton to improve ees (if required), and/or to attempt to evaluate the factors which influence enantioselectivity. An increasing number of studies have sought to address these important issues and have often used the deprotonation of Boc-pyrrolidine as a test case. This case is also useful for assaying the utility of ligands, since almost no lithiation occurs in the absence of a diamine. Beak conducted a screen of numerous other ligands in combination with *s*-BuLi in the asymmetric deprotonation of Boc-pyrrolidine **31** (Scheme 16), however only those compounds closely analogous to (–)-sparteine gave anything close to the same levels of asymmetric induction [73]. (-)-α-Isosparteine **28** was found to be a poor ligand using *s*-BuLi in ether [10% yield and 61% ee, compared with 87% yield and 96% ee using (-)-sparteine **2**]. In contrast, *s*-BuLi/*N*,*N*,*N'*,*N'*-tetramethylcyclohexanediamine **4**, gave excellent conversion (~90%) in the deprotonation /electrophilic trapping (TMSCl) of Boc-pyrrolidine, but the product was racemic [73]. Bispidines (e.g., **35**, Fig. 3) containing the core 3,7-diazabicyclo[3.3.1]nonane ring system of (–)-sparteine proved to be reasonable sparteine mimics, (*R*)-**33** being obtained in up to 75% ee (following addition of TMSCl to the 2-lithiated intermediate **32**) [73]. The bispidines examined by Beak were prepared via a double Mannich reaction of *N*-methyl-4-piperidone with enantiomerically pure amines and a subsequent Wolff-Kishner reduction.

Kozlowski has studied lithiation of *N*-Boc pyrrolidine (cf. Scheme 16) using racemic 1,5-diaza-*cis*-decalins (e.g., **36** and **37**, Fig. 3), and found that for **36** the ligands with the smallest most electron rich R groups (Me>Et>CH_2t-Bu>CH_2CF_3≈Bn) were most effective [74, 75]. The ready availability of both

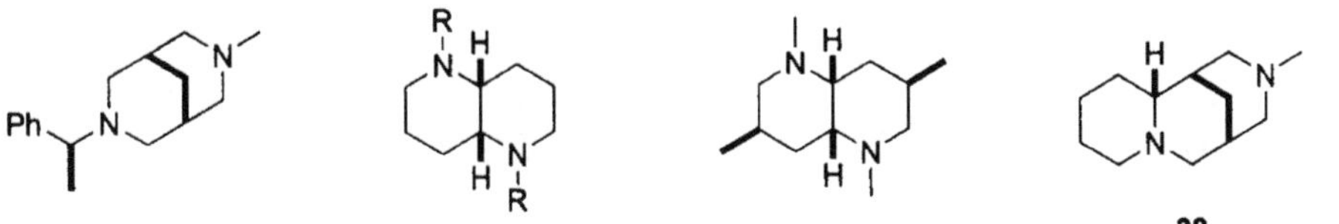

Fig. 3.

enantiomers of the 1,5-diaza-*cis*-decalins and the ability to tune steric and electronic properties renders these compounds an attractive new class of diamine ligands. Although, the use of resolved forms of the optimal ligands (*R*,*R*)-**36** (R= Me) and 3,7-dimethyl derivative **37** in this reaction led to disappointingly low selectivity (12% and 18% ee, respectively), higher ees (45 and 60%, respectively) were observed in lateral lithiations.

O'Brien has also developed bispidine ligands and examined them in this reaction, his approach involving building up the racemic core piperidone via the double Mannich reaction and then attempting to resolve the final ligand. Bispidine **38** (Fig. 3), prepared in 55% ee by partial resolution, gave silylated pyrrolidine **33** (Scheme 16) in 53% ee, but in only 41% yield [76]. However, if a ligand of this type can be prepared as either pure enantiomer, and there is a linear correlation of product and ligand ee, then this ligand class may provide a long-sought solution to some of the above issues.

2 Lithium Amides

Chiral lithium amides/alkoxides have been extensively investigated in a number of valuable asymmetric processes (e.g., diastereoselective additions, chiral enolate formation, enantioselective eliminations) [77–80]. However, since the current volume is concerned with organolithium chemistry, only representative examples of the (much more restricted) use of lithium amides that result in the formation of enantioenriched organolithiums leading to enantioenriched products are highlighted here. Because of their weaker thermodynamic basicity compared with alkyllithiums, the use of lithium amides to generate organolithiums is usually restricted to carbon acids which have pKa values in the upper 30s or better.

For those carbon acids having pKa values in the upper 30s, unfavourable equilibria in the deprotonation imply that if reaction is to proceed satisfactorily in the forward direction, the resultant organolithium must either undergo a rearrangement, or be trapped out by an external electrophile present in situ (the electrophile therefore requiring compatibility with the lithium amide on the reaction timescale).

The first examples of chiral lithium amides to be used to generate a chiral organolithium species were reported by Marshall in 1987 [81] in his pioneering work into [2,3]-sigmatropic rearrangements (Hodgson et al, in this volume) for which the lithium amide **39** (Fig. 4) was used (up to 60% ee); the amine precursors to both **39** and *ent*-**39** are commercially available.

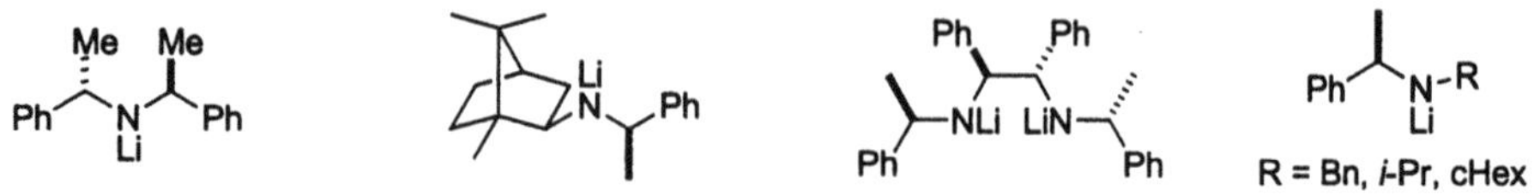

R = Bn, *i*-Pr, cHex

39 **40** **41** **42**

Fig. 4.

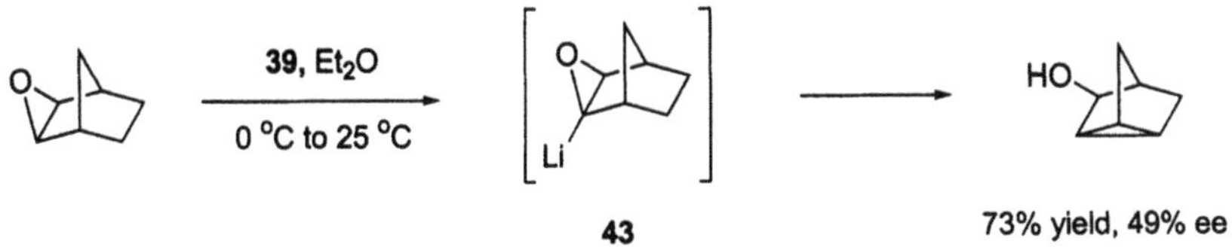

39, Et_2O
0 °C to 25 °C
43
73% yield, 49% ee

Scheme 22.

A more recent example of chiral lithium amide-induced enantioselective deprotonation-rearrangement is the conversion of *exo*-norbornene oxide to nortricyclanol which proceeds via the lithiated epoxide **43** (Scheme 22) [82].

α-Sulfonyl and sulfinyl carbanions have been generated using chiral bases **40** and **41** and reacted with electrophiles in up to 64 and 88% ee, respectively (Toru, in this volume) [80, 83–85]. *meso*-Phospholene oxides undergo deprotonation α to phosphorus with chiral lithium amide **39**, followed by electrophile trapping, in up to 92% ee [86].

$Cr(CO)_3$-complexed benzylic ethers and sulfides are acidic enough to undergo lithium amide deprotonation, α to the heteroatom, followed by electrophile trapping; the more elaborate base **41** [prepared via the addition of PhMgCl to the enantiomerically pure bis-imine derived from glyoxal and (*R*)-α-methylbenzylamine] usually giving the best ees (up to 97%), e.g., Scheme 23 [87].

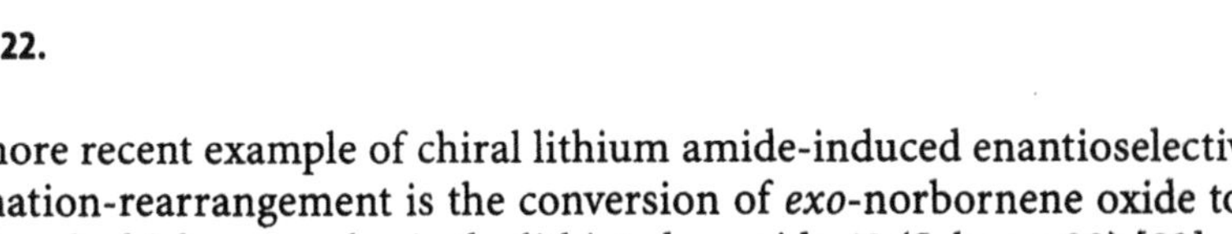

1. **41**, LiCl, THF, –78 °C
2. Electrophile
X = O, S
R = Me, Bn
E = PhS, Me

Scheme 23.

The synthesis of an enantiomerically enriched chromium complex via asymmetric lithiation of a prochiral tricarbonyl(η^6-arene)chromium complex using a chiral lithium amide base was first demonstrated in 1994 by Simpkins [88]. Arene complex **44** was treated with C_2-symmetric chiral base *ent*-**39** in the presence of TMSCl as an internal quench and silylated complex **45** was obtained in 84% ee (Scheme 24).

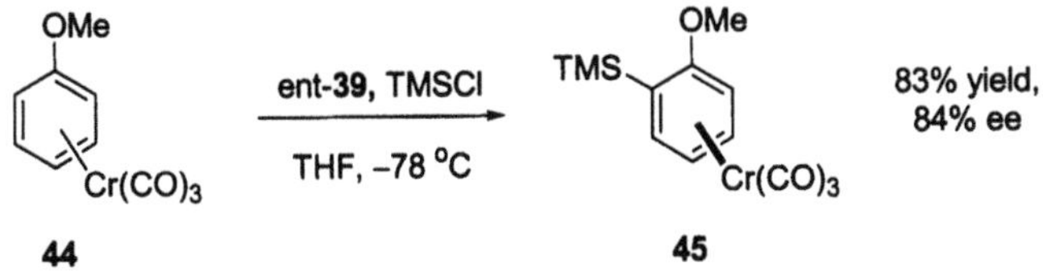

ent-**39**, TMSCl
THF, –78 °C
44
45
83% yield, 84% ee

Scheme 24.

Investigation of other lithium amides in this process (**40**, **42**; Fig. 4) gave only moderate (<60% ee) levels of enantioselectivity [89].

Using camphor-derived chiral lithium amide **40**, the tricarbonylchromium complexes of phenyl carbamates could be *ortho*-deprotonated with good levels

1. **40**, THF, –78 °C
2. E^+

Scheme 25.

of enantioselectivity, a range of external electrophiles being successful in the reaction (Scheme 25, up to 81% ee) [90].

Kündig has examined a number of chiral lithium amides (Fig. 5) in this process (quenching with TMSCl), with no improvement in enantioselectivity observed compared with using **40** [91].

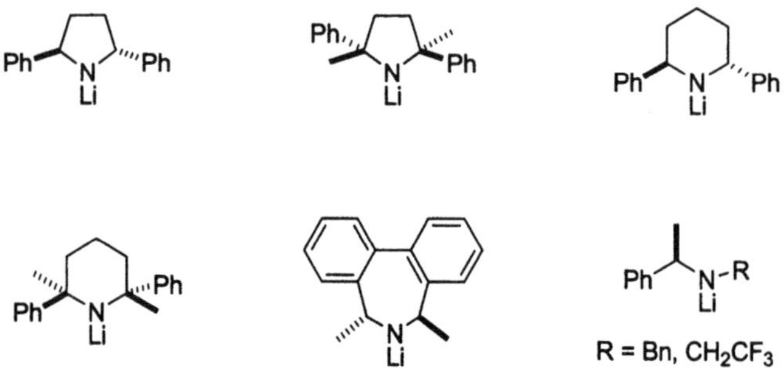

Fig. 5.

Note added in proof. O'Brien has now obtained bispidine ligand **38** (Fig. 3) from (–)-cytisine in 3 steps and demonstrated that it function as a readily accessible (+)-sparteine surrogate; in most processes, the stereoselectivity of reactions with ligand **38** are comparable to those with (–)-sparteine as a ligand [92]. Imino Diels-Alder reactions have been investigated as a new route to sparteine analogs, and diamines **38** and *epi*-**38** prepared by this methodology have been examined in the lithiation on *N*-Boc-pyrrolidine [93].

References

1. Clayden J (2002) Organolithiums: selectivity for synthesis. Pergamon, Oxford
2. Schlosser M (2002) In: Schlosser M (ed) Organometallics in synthesis. Wiley, Chichester, p 1
3. Gray M, Tinkl M, Snieckus V (1995) In: Abel EW, Stone FGA, Wilkinson G (eds) Comprehensive organometallic chemistry II, vol 11. Pergamon, New York, p 1
4. Sapse A-M, Schleyer PvR (1995) Lithium chemistry: A theoretical and experimental overview. John Wiley, New York
5. Wakefield BJ (1988) Organolithium methods. Academic Press, London
6. Schlenk W, Holtz J (1917) Ber dtsch chem Ges 50:262
7. Schlenk W, Bergmann E (1928) Annalen 463:98
8. Ziegler K, Colonius H (1930) Annalen 479:135
9. Wittig G, Leo M (1931) Ber dtsch chem Ges 64:2395
10. Gilman H, Zoellner EA, Selby WM (1932) J Am Chem Soc 54:1957

11. Langer AW, Jr. (ed) (1974) Polyamine-chelated alkali metal compounds. American Chemical Society, Washington, D.C.
12. Morrison JD, Mosher HS (1971) Asymmetric organic reactions. Prentice-Hall, Englewood Cliffs, N.J.
13. Nozaki H, Aratani T, Toraya T, Noyori R (1971) Tetrahedron 27:905
14. Morrison JD (ed) (1983) Asymmetric synthesis, vol 2. Academic Press, New York
15. Tomioka K (1990) Synthesis 541
16. Hoppe D, Hense T (1997) Angew Chem Int Ed Engl 36:2282
17. Basu A, Thayumanavan S (2002) Angew Chem Int Ed 41:716
18. Mukaiyama T, Soai K, Sato T, Shimizu H, Suzuki K (1979) J Am Chem Soc 101:1455
19. Mazaleyrat J-P, Cram DJ (1981) J Am Chem Soc 103:4585
20. Schön M, Naef R (1999) Tetrahedron: Asymmetry 10:169
21. Jiang B, Feng Y (2002) Tetrahedron Lett 43:2975
22. (a) Fort Y, Gros P, Rodriguez AL (2001) Tetrahedron: Asymmetry 12:2631; (b) Gros P, Fort Y (2002) Eur J Org Chem 3375
23. Tomioka K, Shinda M, Koga K (1989) J Am Chem Soc 111:8266
24. Shindo M, Koga K, Tomioka K (1998) J Org Chem 63:9351
25. Tomioka K, Shinda M, Koga K (1993) Tetrahedron Lett 34:681
26. Asano Y, Iida A, Tomioka K (1997) Tetrahedron Lett 38:8973
27. Tomioka K, Inoue I, Shindo M, Koga K (1990) Tetrahedron Lett 31:6681
28. Tomioka K, Inoue I, Shindo M, Koga K (1991) Tetrahedron Lett 32:3095
29. Denmark SE, Nakajima N, Nicaise OJ-C (1994) J Am Chem Soc 116:8797
30. Denmark SE, Nicaise OJ-C (1996) Chem Commun 999
31. Andersson PG, Johansson F, Tanner D (1998) Tetrahedron 54:11549
32. Shindo M, Koga K, Tomioka K (1992) J Am Chem Soc 114:8732
33. Lautens M, Gajda C, Chiu P (1993) J Chem Soc Chem Commun 1193
34. Klein S, Marek I, Poisson J-F, Normant J-F (1995) J Am Chem Soc 117:8853
35. Amurrio D, Khan K, Kündig EP (1996) J Org Chem 61:2258
36. Müller P, Nury P, Bernardinelli G (2001) Eur J Org Chem 4137
37. Zschage O, Hoppe D (1992) Tetrahedron 48:5657
38. Apin Chemicals, Milton Park, Abingdon; <url>http://www.apinchemicals.com</url>
39. Galinovsky G, Knoth P, Fischer W (1955) Monatsh Chem 86:1014
40. Kang J, Cho WO, Cho HG, Oh HJ (1994) Bull Korean Chem Soc 15:732
41. Hodgson DM, Cameron ID, Christlieb M, Green R, Lee GP, Robinson LA (2001) J Chem Soc Perkin Trans 1 2161
42. Würthwein E-U, Behrens K, Hoppe D (1999) Chem Eur J 5:3459
43. Mazaleyrat JP, Welvart Z (1983) Nouv J Chim 7:491
44. Tomooka K, Yamamoto K, Nakai T (1999) Angew Chem Int Ed Engl 38:3741
45. Manabe S (1997) Chem Commun 737
46. Tomooka K, Komine N, Nakai T (1998) Tetrahedron Lett 39:5513
47. Tomooka K, Komine N, Nakai T (2000) Chirality 12:505
48. Tomooka K, Wang L-F, Okazaki F, Nakai T (2000) Tetrahedron Lett 41:6121
49. Hodgson DM, Lee GP (1996) Chem Commun 1015
50. Hodgson DM, Gras E (2002) Synthesis 1625
51. Hodgson DM, Lee GP (1997) Tetrahedron: Asymmetry 8:2303
52. Beak P, Basu A, Gallagher DJ, Sun Park Y, Thayumanavan S (1996) Acc Chem Res 29:552
53. Beak P, Anderson DR, Curtis MD, Laumer JM, Pippel DJ, Weisenburger GA (2000) Acc Chem Res 33:715
54. Bailey WF, Beak P, Kerrick ST, Ma S, Wiberg KB (2002) J Am Chem Soc 124:1889
55. Wiberg KB, Bailey WF (2001) J Am Chem Soc 123:8231
56. Barberis C, Voyer N, Roby J, Chénard S, Tremblay M, Labrie P (2001) Tetrahedron 57:2965
57. Nakamura S, Nakagawa R, Watanabe Y, Toru T (2000) Angew Chem Int Ed Engl 39:353
58. Nakamura S, Nakagawa R, Watanabe Y, Toru T (2000) J Am Chem Soc 122:11340
59. Byrne LT, Engelhardt LM, Jacobsen GE, Leung WP, Papasergio RI, Raston CL, Skelton BW, Twiss P, White AH (1989) J Chem Soc Dalton Trans 105

60. Muci AR, Campos KR, Evans DA (1995) J Am Chem Soc 117:9075
61. Uemura M, Hayashi Y, Hayashi Y (1994) Tetrahedron: Asymmetry 5:1427
62. Tsukazaki M, Tinkl M, Roglans A, Chapell BJ, Taylor NJ, Snieckus V (1996) J Am Chem Soc 118:685
63. Shintani R, Fu GC (2002) Angew Chem Int Ed Engl 41:1057
64. (a) Jensen DR, Pugsley JS, Sigman MS (2001) J Am Chem Soc 123:7475; (b) Jensen DR, Sigman MS (2003) Org Lett 5:63
65. Ferreira EM, Stoltz BM (2001) J Am Chem Soc 123:7725
66. Smith BT, Wendt JA, Aubé A (2002) Org Lett 4:2577
67. Clark D, Travis DA (2001) Bio Med Chem 9:2857
68. Thayumanavan S, Lee S, Liu C, Beak P (1994) J Am Chem Soc 116:9755
69. Tan Y-L, White AJP, Widdowson DA, Wilhelm R, Williams DJ (2001) J Chem Soc Perkin Trans 1 3269
70. Schlosser M, Limat D (1995) J Am Chem Soc 117:12342
71. Kawasaki T, Kimachi T (1999) Tetrahedron 55:6847
72. Kimachi T, Takemoto Y (2001) J Org Chem 66:2700
73. Gallagher DJ, Wu S, Nikolic NA, Beak P (1995) J Org Chem 60:8148
74. Li X, Schenkel LB, Kozlowski MC (2000) Org Lett 2:875
75. Xu Z, Kozlowski MC (2002) J Org Chem 67:3072
76. Harrison JR, O'Brien P, Porter DW, Smith NM (2001) Chem Commun 1202
77. Bull SD, Davies SG, Roberts PM, Savory ED, Smith AD (2002) Tetrahedron 58:4629
78. Plaquevent J-C, Perrard T, Cahard D (2002) Chem Eur J 8:3300
79. O'Brien P (1998) J Chem Soc Perkin Trans 1 1439
80. Cox PJ, Simpkins NS (1991) Tetrahedron: Asymmetry 2:1
81. Marshall JA, Lebreton J (1987) Tetrahedron Lett 28:3323
82. Hodgson DM, Lee GP, Marriott RE, Thompson AJ, Wisedale R, Witherington J (1998) J Chem Soc Perkin Trans 1 2151
83. Simpkins NS (1988) Chem Ind 387
84. Armer R, Begley M, Cox PJ, Persad A, Simpkins NS (1993) J Chem Soc Perkin Trans 1 3099
85. Blake AJ, Westaway SM, Simpkins NS (1997) Synlett 919
86. Hume SC, Simpkins NS (1998) J Org Chem 63:912
87. Gibson (née Thomas) SE, Reddington EG (2000) Chem Commun 989
88. Price DA, Simpkins NS, MacLeod AM, Watt AP (1994) J Org Chem 59:1961
89. Ewin RA, MacLeod AM, Price DA, Simpkins NS, Watt AP (1997) J Chem Soc Perkin Trans 1 401
90. Kündig EP, Quattropani A (1994) Tetrahedron Lett 35:3497
91. Pache S, Botuha C, Franz R, Kündig EP, Einhorn J (2000) Helv Chim Acta 83:2436
92. Dearden MJ, Firkin CR, Hermet JR, O'Brien P (2002) J Am Chem Soc 124:11870
93. Danieli B, Lesma G, Passarella D, Piacenti P, Sacchetti A, Silvani A, Virdis A Tetrahedron Lett (2002) 43:7155

Topics Organomet Chem (2003) 5: 21–36
DOI 10.1007/b10340

Enantioselective Addition of Organolithiums to C=O Groups and Ethers

Bernd Goldfuss

University of Cologne, Institute for Organic Chemistry, Greinsraße 4, 50939 Köln.
E-mail: Goldfuss@uni-koeln.de; www.uni-koeln.de/goldfuss

Chiral, non-racemic ligands and additives implement enantioselective additions of organolithiums to C=O functions of carbonyl compounds (i.e., aldehydes and ketones) or enantioselective cleavages of C-O units in strained ethers (i.e., epoxides and oxetanes) as well as in acetals. The achievement of high enantioselectivities is desirable and hence factors controlling these enantioselectivities are subjects of intensive research. Beneficial external influences on enantioselectivity are given by low temperature, suitable solvents (e.g., dimethoxymethane, dimethyl ether) and the exclusion of salt impurities. The formation of mixed anionic aggregates from in situ lithiated ligands and the organolithium reagents enables especially efficient chiral modifications. In some cases such mixed aggregates can be identified and studied structurally. The improved understanding of the nature of chiral organolithium aggregates provides the way to a more rational design of new enantioselective organolithium reagents.

Keywords. Enantioselectivity, Organolithiums, Chiral ligands, Carbonyl compounds, Strained ethers, Acetals

1 Introduction

In this section we will cover the chiral ligand-mediated enantioselective cleavages of π- or σ-C-O bonds, which proceed via additions of organolithiums to carbonyl compounds, i.e., aldehydes and ketones, to strained cyclic ethers, i.e., epoxides and oxetanes, as well as to acetals. In one addition step, the formation of new C-C bonds *and* new stereocenters are established, which makes these reactions very attractive synthetically [1–3]

Due to the electropositive character of lithium, organolithiums [4] are sometimes regarded as "carbanions". It should be considered however that they are "free carbanions" in only a few cases in the condensed phase (e.g., solvent separated ion pairs in highly polar media, strong lithium ion coordination by crown ethers or cryptands) [5]. Since the lithium counterion plays also an eminent role for the reactivity and selectivity, organolithiums should be regarded as (highly) polar organometallics rather than as "carbanions" [6, 7].

To achieve high enantioselectivities, the organolithium reagent should be tightly coordinated by the chiral ligand during the enantioselective step of the C-C bond formation. Competing background additions without enantiopure additives, yielding racemic products, should be avoided. In some cases good enantioselectivities are observed even with sub-stoichiometric or catalytic amounts of chiral additives. For such catalytic enantioselective applications,

(a) the ligand modified system apparently exhibits a higher reactivity than the ligand free system,
(b) this chiral additive should be readily detached from the product and reenter the catalytic cycle, and
(c) the ligand structure should efficiently differentiate the two diastereomeric transition states [8, 9].

Prior to organolithiums, Grignard reagents were employed in enantioselective additions to aldehydes and ketones [10], and the influence of chiral solvents such as 2,3-dimethoxybutane [11, 12] or 2-methyltetrahydrofuran [13] was studied. Nowadays, organolithium chemistry shows very dynamic growth, and organolithiums are well-established precious reagents for organic syntheses [14–16].

2 Additions to Aldehydes

Enantioselective additions of organolithiums to aldehydes are employed to synthesize chiral secondary alcohols, but these reactions are also quite frequently used for testing new chiral ligands and additives. The latter is usually performed with the standard addition of organolithium reagent {RLi, in many cases *n*-butyllithium, *n*-BuLi) to benzaldehyde [Eq. (1)], L* denotes a chiral ligand} [9, 15, 17–19].

$$\mathrm{PhCHO} \xrightarrow[\mathrm{RLi}]{\mathrm{L^*}} \mathrm{PhC^*H(OH)R} \tag{1}$$

Enantioselective additions of lithium enolates to aldehydes forming aldols (β-hydroxyaldehydes) are synthetically well established and have been reviewed elsewhere [20]. A catalytic variant, the Mukaiyma aldol reaction, i.e., the addition of silyl enol ethers to aldehydes, is usually mediated by chiral Lewis acids [21, 22].

In 1968, Nozaki et al. performed the first enantioselective addition with *n*-BuLi according to Eq. (1) [23, 24]. To an equimolar mixture of *n*-BuLi and (–)-

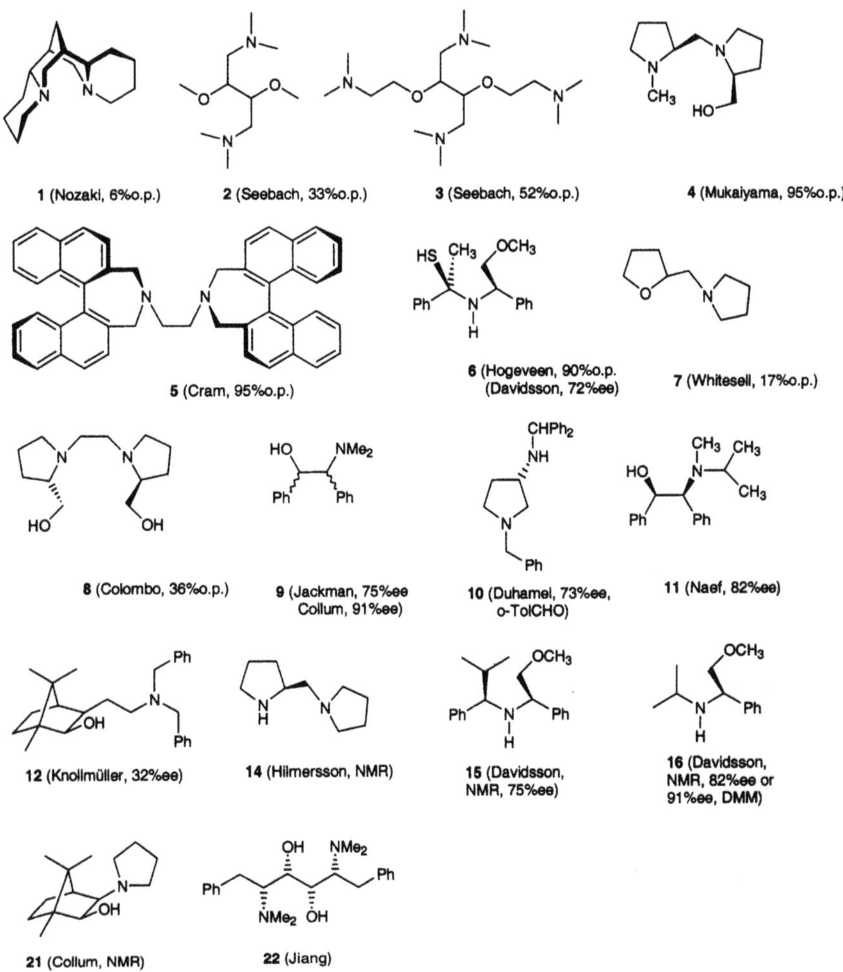

Scheme 1. Chiral ligands, the enantioselectivities refer to Eq. (1), R=*n*-Bu

sparteine (**1**, Scheme 1), benzaldehyde was added in hexane at –70°C. (+)-1-Phenyl-1-pentanol was isolated from this reaction in 6% optical purity (op).

In 1969, Seebach et al. employed mixtures of *n*-pentane and the tartrate derivative 1,4-bis(dimethylamino)-2,3-dimethoxybutane (**2**) as a chiral co-solvent in *n*-BuLi additions to PhCHO [25]. They obtained 1-phenyl-1-pentanol in up to 33% optical purity at –130°C and achieved higher enantioselectivities than with hexane/sparteine or hexane/2,3-dimethoxybutane mixtures under similar conditions. After further systematic studies [26–28], Seebach et al. showed that two equivalents of the polydentate amino ether ligand **3** yield an optical purity of 52% (–78°C, pentane) while a 1:1 ratio gives only 38% ee [29].

In 1978 Mukaiyama et al. reported 72% optical purity of (–)-(*S*)-1-phenyl-1-pentanol obtained from addition of *n*-BuLi to PhCHO using the (*S*)-proline-derived bispyrrolidine alcohol **4** [30]. The reaction was performed with a ratio of protic chiral additive **4**:*n*-BuLi:PhCHO of 4.05:6.75:1.0 in benzene at –123°C. As the *O*-methylated derivative of **4** gave a lower optical purity (14%) than **4**, it was suggested that the free hydroxy function and the resulting in situ formed lithium alkoxide unit are essential for high enantioselectivity. The optical purity for this reaction was even improved to 95% by employing a 1:1 mixture of dimethoxymethane ($MeOCH_2OMe$) and dimethyl ether (Me_2O) as solvent at –123°C [31, 32]. The addition of lithium trimethylsilylacetylide to aliphatic aldehydes was performed with **4** in good enantioselectivities (40–80 % optical purity) [33]. Johnson et al. employed Mukaiyama's ligand **4** for an enantioselective step in a corticoid synthesis in 90% ee by adding a lithium acetylide to an aliphatic aldehyde [34].

Mazaleyrat and Cram achieved 95% optical purity of (+)-(*R*)-1-phenyl-1-pentanol using the (*R*,*R*)-binaphthylamine **5** with *n*-BuLi and PhCHO at a reagent ratio of 5.0:4.4:1.2 at –120°C in diethyl ether [35]. For a ratio of 3.6:3.4: 1.2 they observed only 59% optical purity of the formed (*R*)-1-phenyl-1-pentanol.

Eleveld and Hogeveen employed the benzylamine **6** (respectively its *N*-lithiated derivative) with *n*-BuLi and PhCHO in a ratio of 4.0:2.7:1.0. The (–)-(*S*)-1-phenyl-1-pentanol product was obtained in 90% optical purity using a 1:1 mixture of dimethyoxymethane and dimethyl ether at –120°C [36]. Toluene as solvent decreased the optical purity of the product drastically to 18%. However, reinvestigations of this experiment with chiral gas chromatographic analyses rather than optical rotation analyses gave only 72% ee [37].

Whitesell's aprotic pyrrolidine ligand (–)-**7** was found to yield with *n*-BuLi and PhCHO in a ratio of 1.6:1.3:1.2 only 17% optical purity of (+)-(*R*)-1-phenyl-1-pentanol (inpentane at –78°C) [38]. Similar enantioselectivities (19% op) were obtained with methyllithium, while the Grignard reagent MeMgCl gave no optical induction.

Colombo et al. tested the hydroxypyrrolidine **8** as well as its *O*-methylated derivative with *n*-BuLi and PhCHO in a ratio of 3:2:1 [39]. In analogy to Mukaiyama's results, protic **8** gave in dimethoxymethane a higher optical purity [36% of (*R*)-1-phenyl-1-pentanol] than its aprotic dimethyl ether derivative in hexane [15% of (*S*)-1-phenyl-1-pentanol]. Salt impurities (e.g., LiI, $LiClO_4$) were found to decrease the enantioselectivities.

Alberts and Wynberg studied the addition of ethyllithium to PhCHO, mediated by the lithiated product (*R*)-1-phenyl-1-propanol-d_1 [40]. The (*R*)-1-phenyl-1-propanol enantiomer was formed in 17% enantiomeric excess in the reaction. For this asymmetric induction rendered by the product, the authors coined the expression "enantioselective autoinduction". It was suggested that the formation of mixed aggregates consisting of the labeled chiral mediator and the organolithium reagent "influences the stereochemistry of subsequent C-C bond formation" [41].

Jackman's group performed enantioselective additions of RLi (R=Me, *n*-Bu) to PhCHO, mediated by a series of chiral lithium alkoxides [42]. The highest enantioselectivities were achieved with the lithiated ephedrine derivative (*1R*,*2S*)-**9** with *n*-BuLi and PhCHO in a ratio of 0.3:0.15:0.1 with 75% ee (GC) of (*S*)-1-phenyl-1-pentanol in THF at –78°C. An analysis during the course of addition showed that no "enantioselective autoinduction" [40] by the chiral product occurred. After addition of LiCl or $LiClO_4$ to the reaction mixtures, a decrease of enantioselectivity was observed. This was pointing to the formation of achiral mixed anionic halide aggregates, which give rise to racemic products.

Kang et al. performed enantioselective additions of 2-lithio-1,3-dithiane to aldehydes mediated by stoichiometric amounts of isosparteine and achieved up to 70% ee for PhCHO as substrate in diethyl ether at −78 °C [43].

Duhamel's group employed 3-aminopyrrolidine derivatives and obtained with **10** enantioselectivities of up to 73% in *n*-BuLi additions to *o*-tolualdehyde [ratio: 1.5:2.5:1.0, in THF at −78 °C, (*R*)-configured product] [44]. Structural variations of the ligands were performed to elucidate the basis for the observed enantioselectivities [45].

Schön and Naef tested a variety of 2-aminoalcohols according to Eq. (1) and achieved the highest selectivities with the lithiated amino alcohol **11** [82% ee, (*S*)-1-phenyl-1-pentanol, 1:1:1 ratio] in THF at −78 °C [46]. More equivalents of **11** (e.g., a ratio of 4:1:1) had no significant influence on the enantioselectivity.

Knollmüller et al. tested seven camphor-derived 1,4-amino alcohols according to Eq. (1) and achieved with **12** up to 32% ee of (*R*)-1-phenyl-1-pentanol (5.2:9.2:1 ratio, in Et_2O at −78 °C) [47]. Diethyl ether was found to give for all ligands superior enantioselectivities than THF.

Aspinall et al. employed lithium lanthanide binaphtholates **13** (Scheme 2, THF molecules are omitted) in *n*-BuLi additions to PhCHO (1:2:1 ratio) with 67% ee [(*S*)-1-phenyl-1-pentanol, diethyl ether, −98 °C] [48]. With less equivalents of *n*-BuLi (1:1:1 ratio) only a 39% ee was observed under the same conditions.

From these and from other experiments in organolithium chemistry it can be concluded that higher enantioselectivities in alkyllithium additions to alde-

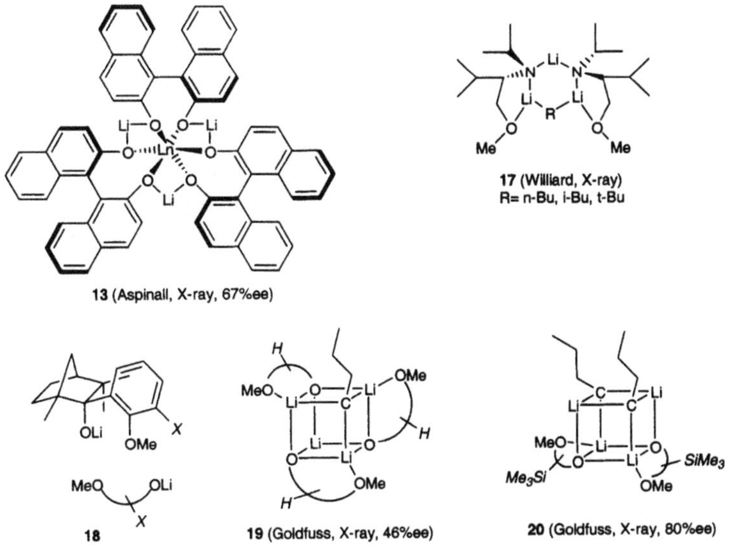

13 (Aspinall, X-ray, 67%ee)

17 (Williard, X-ray)
R= n-Bu, i-Bu, t-Bu

18

19 (Goldfuss, X-ray, 46%ee)

20 (Goldfuss, X-ray, 80%ee)

Scheme 2. Chiral complexes, enantioselectivities refer to Eq. (1), R=*n*-Bu

hydes are observed for higher quantities of chiral additives. It was stated "that lithium alkoxides influence the reactivity of *n*-BuLi only to a small extent"[49]. McGarrity et al. employed the "rapid-injection NMR" technique to study the *n*-BuLi addition to PhCHO in THF at −85 °C [50]. Dimeric *n*-BuLi was found to be only 10 times more reactive than the tetrameric species. This slightly increased reactivity was also found for mixed *n*-BuLi/BuOLi aggregates: $Li_4(n\text{-}Bu)_2(OBu)_2$ was measured to have a similar reactivity as $(n\text{-}BuLi)_2$. The more supporting coordination in the transition state rather than in the pre-complex was found to be the origin of kinetically favored directed *ortho*-metallations of benzene derivatives with organolithiums [51–54]. Coordinating solvents (THF) or ligands (TMEDA) de-aggregate organolithiums and hence give rise to higher reactivities than the oligomers [55], e.g., in enolate alkylations [56, 57]. However, due to the rather high reactivity of non-modified organolithiums towards aldehydes, catalytic procedures, e.g., with less than 5 mol% of chiral additive, seem to be hardly achievable with high enantioselectivities for generally applicable procedures. Ethereal solvents, low temperatures and protic ligands, which are prone to lithiation in situ and then form mixed aggregates with the organolithium reagent, are apparently the best choice for highly enantioselective aldehyde alkylations. Salt additives (e.g., LiCl, LiOH, $LiClO_4$) decrease enantioselectivities due to the formation of competing achiral mixed aggregates, which are frequently observed in solution [58–60] and in the solid state (see below) [61]. Organozinc or organotitanium catalysts [62–67] were found to be more suitable to achieve high enantioselectivities in aldehyde alkylations, especially with catalytic amounts of chiral additives. Due to the central role of mixed chiral organolithiums aggregates in alkylation reactions [e.g., Eq. (1)], the nature of such aggregates became the subject of intensive research.

Hilmersson and Davidsson studied by means of intensive NMR investigations a mixed 1:1 complex of *n*-BuLi and the lithiated methoxyamine **6** (Scheme 1) in diethyl ether at −80 °C [68]. A fluxional exchange between tetrameric and dimeric structures of the chiral lithiated pyrrolidine **14** and *n*-BuLi is apparent in diethyl ether solution [69]. With the lithiated amine **15**, a derivative of **6**, 75% ee of (*S*)-1-phenyl-1-pentanol was obtained in *n*-BuLi additions to PhCHO, using a ratio of 1.0:0.45:0.25 in diethyl ether at −116 °C [37]. The *N*-methyl derivative gave only 2% ee, while the *N*-isopropyl derivative **16** yielded 82% ee under the same conditions. This demonstrates the crucial role of the *N*-isopropyl substituent in **16**. Addition of dimethoxymethane increased the enantioselectivity of **16** to 91% ee of (*S*)-1-phenyl-1-pentanol. With lithiated **16**, enantioselectivities of up to 98.5% ee were achieved in butylations of aliphatic aldehydes [70]. Hilmersson demonstrated that the mixed lithium amide/*n*-BuLi aggregate alkylates aldehydes faster than the pure *n*-BuLi oligomers, Eq. (2) [71].

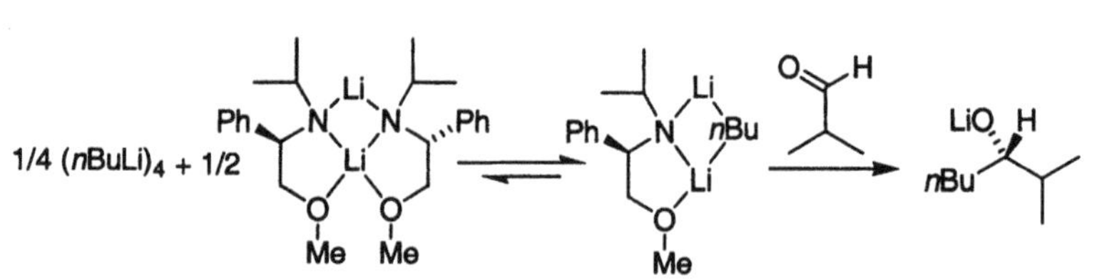

(2)

Despite the frequent employment of *n*-BuLi in enantioselective additions, only very few complexes of *n*-BuLi (or other organolithiums) and chiral, enantiopure ligands have been structurally analyzed. Williard et al. reported in 1997 X-ray crystal structures of mixed complexes of *n*-, *i*-, and *t*-butyllithium with lithium *N*-isopropyl-*O*-methylvalinolate (**17**, Scheme 2) [72]. The 2:1 stoichiometry of the mixed aggregates in the solid state was also found in solution with non-coordinating solvents such as hexane or toluene [73].

Goldfuss et al. developed a series of modular anisyl fencholates (e.g., **18**) [74, 75], which all form enantiopure mixed aggregates with *n*-BuLi (e.g., **19** and **20**) and which could be studied by X-ray analyses [76, 77]. Using the modular nature of these anisyl fencholates, the compositions of the mixed aggregates as well as enantioselectivities in benzaldehyde butylations can be controlled. While H as *ortho*-substituent (*X*) yields a 3:1 aggregate with *n*-BuLi (**19**), for X=$SiMe_3$ a 2:2 complex (**20**) was identified. Higher enantioselectivities in butylations of PhCHO were achieved with **20** [80% ee of (*R*)-1-phenyl-1-pentanol in toluene at -78 °C] than with **19** (46% ee). Similar mixed aggregates with a 2:2 composition are observed for *X*=*t*-Bu and $SiMe_2$*t*-Bu [78]. However, *X*=Me yields an unprecedented mixed aggregate, which is composed of two fencholates and four *n*-BuLi units and hence can be regarded as a enantiopure modification of hexameric *n*-BuLi [79].

The group of Collum, inspired by highly selective acetylide additions to ketones (see below), performed lithium phenylacetylide additions to PhCHO, mediated by the camphor-derived amino alkoxide **21** in THF at -78 °C [80]. Variations of the alkoxide:acetylide ratio revealed changes in the sense of enantioselectivity and the highest enantioselectivity (78% ee) was observed at a ratio of 1:3.

Collum's group also performed *n*-BuLi additions to PhCHO [Eq. (1)] mediated by 2-amino alcohols [81]. The highest enantioselectivity was found for the 1,2-diphenyldimethylaminoethanol (*1S,2R*)-**9**, it's enantiomer was used in Jackman's studies [42], with 91% ee of 1-phenyl-1-pentanol at –105 °C in THF/pentane (1.5:1:1 ratio). The corresponding pyrrolidine derivative gave 79% ee (–78 °C in THF/pentane, 1.5:1:1 ratio) [81]. Detailed NMR spectroscopic investigations showed that butylide additions to the aldehyde are much faster than structural exchanges of the alkylating mixed aggregates and that a 2:2 mixed cubic tetramer is most likely the reactive species.

Jiang et al. achieved recently in alkynylations of PhCHO up to 99% ee with the C_2 symmetric diaminodiol **22** [82]. The groups of Knochel and Carreira developed catalytic alkynylation procedures based on cesium and zinc species [83, 84].

3 Additions to Ketones

Enantioselective additions of organolithiums to prochiral ketones yield tertiary chiral alcohols, which are a synthetically highly attractive class of substances [85]. The formation of enolates via deprotonation (of α-hydrogens in the substrate) or reduction (via β-hydrogen atoms in the organometallic) are sometimes critical side reactions. The propensity for these side reactions is less pronounced for organolithiums than for Grignard reagents [86, 87].

The most prominent example of enantioselective organolithium additions to ketones is the development of the anti-AIDS drug efavirenz **23** (Scheme 3), a non-nucleoside reverse transcriptase inhibitor for a variety of HIV-1 mutant strains [88, 89].

The key step in the synthesis of **23** is the enantioselective generation of its quaternary carbon center. This can be accomplished by adding to the protected ketoaniline precursor **24** the lithium acetylide **25** [Eq. (3)] [90]. Due to the high medical relevance of the final product efavirenz (**23**), this enantioselective organolithium addition has been intensively studied. These detailed investigations provide intriguing information on the nature of reactive intermediates and the origins of enantioselectivity.

(3)

For this 1,2-addition Thompson et al. first tested lithiated quinine and quinidine additives, which were applied successfully in additions to imines [91, 92, 93], but for ketone **24** [Eq. (3)] they obtained only 50–60% ee [90]. Higher enantioselectivities were achieved with lithiated ephedrine derivatives with different *N*-substituents (e.g., NMe_2, NEt_2), among which the pyrrolidine substituent in **26** gave the highest enantioselectivity (82% ee) for the efavirenz precursor **27** [Eq. (3)]. An optimized protocol was developed, whereby the cyclopropylacetylene (two equivalents) and the chiral amino alcohol (two equivalents) were lithiated with *n*-BuLi at –15 °C. This mixture of **25** and **26** was then "aged" for 30 min at 0 °C. Addition of the ketoaniline **24** (one equivalent) to this mixture at –55 °C in THF affords **27** in up to 98% ee!

These high enantioselectivities were not observed with simple alkyllithiums and they are also sensitive to acetylide β-carbon substituents [88]. The *para*-methoxybenzyl protecting group in **24** was found to be crucial for high enantioselectivity, a free amino group decreases the enantioselectivity to 72% ee [89]. It was also observed that an excess of acetylide **25** relative to alkoxide **26** erodes the enantioselectivity. The equilibration of the mixed aggregates at higher temperatures ("aging" of the alkynylation mixture at 0 °C) proved to be essential for good enantioselectivites. Generation *and* addition of the alkoxide-acetylide mixture below –50 °C gave only up to 85% ee [89, 90].

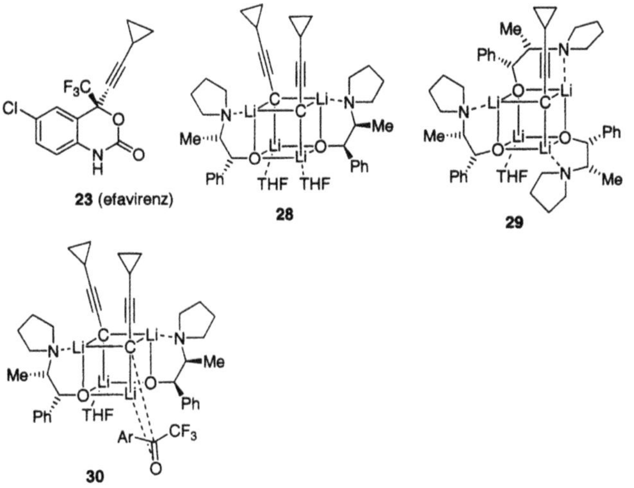

Scheme 3.

What are the origins of the two puzzling adjustments which are essential for the highly enantioselective synthesis of **27**,

(a) the necessary "aging" of the alkynylation mixture (**25** and **26**) at higher temperatures (0 °C) and

(b) the required 2:2:1 ratio of **25**, **26** and the ketone **24**?

Due to its central role, the nature of the mixed acetylide-alkoxide aggregate was studied intensively by ^{6}Li-, ^{15}N- and ^{13}C-NMR techniques in THF solution [88]. Generated at low temperatures (–78 °C), homonuclear oligomers of lithium acetylide **25** and of lithium alkoxide **26** as well as mixed acetylide-alkoxide aggregates were found. Above –40 °C, all the homonuclear oligomers of lithium acetylide **25** and of lithium alkoxide **26** convert to *one* stable, C_2-symmetrical 2:2 mixed aggregate **28** (Scheme 3). The high temperature necessary for this conversion points to an unusually slow mixed aggregate exchange [94]. The level of asymmetric amplification with scalemic mixtures of **26** (50% ee of **26** yields 77% ee of **27**) was also seen to be consistent with a cubic 2:2 tetramer like **28** [89].

Two equivalents of acetylide **25** and alkoxide **26** are required for full conversion of one equivalent of ketoaniline **24**. Addition of ketone **24** to an equimolar alkynylation mixture of **25** and **26** (1:1:1 ratio) proceeds rapidly, but only up to 50% conversion at low temperature (–90 °C). Further ketone addition at these temperatures gives rise to IR detectable carbonyl as well as NH functions of ketone **24**, but no C-H signal of a protonated acetylene from **25** could be detected. Hence, **24** then coexists with a rather unreactive acetylide-alkoxide species, which neither alkynylates the CO function nor deprotonates the NH group in **24** at low temperatures (–90 °C)! The completion of the reaction for this 1:1:1 stoichiometry proceeds at 0 °C, but then even requires several hours and also shows significantly lower enantioselectivity for the last than for the first 50% conversion [88].

A model for the rather unreactive species, which is formed after 50% conversion, is given by the 3:1 tetramer **29**, which forms, according to NMR investigations, from three equivalents of alkoxide **26** and one equivalent of acetylide **25** [88]. The solid state structure of **29** was subsequently also confirmed by X-ray analysis [95]. The mixed 3:1 aggregate **29** shows indeed a significantly lower reactivity than the 2:2 aggregate **28**. Attempts to perform an X-ray analysis of **28** failed, instead a hexameric 4:2 aggregate was obtained [95]. The lower reactivity of **29** can be explained by the remote positions of acetylide from the free (i.e., THF and not pyrrolidine coordinated) lithium ion, to which the ketone should initially bind prior to the acetylide transfer [88]. In contrast, the more reactive 2:2 aggregate **28** has free (THF coordinated) lithium ions and acetylide groups in close proximity (Scheme 3). After the first 50% conversion, the lithiated product **27** and remaining acetylide **25** might form a 3:1 alkoxide-acetylide aggregate similar to **29**. Hence, both intriguing phenomena, the aging effect and the requirement for the 2:1 stoichiometric have as same origin, i.e., the mixed 2:2 aggregate **28**, which can be regarded as the genuine alkynylating reagent.

Semiempirical MNDO computations suggest an intramolecular H-bonding interaction between the NH function and carbonyl group, resulting in a pre-oriented CO function. The computations also support a stereochemical model which correctly reproduces the sense of selectivity for the acetylide transfer within the ketone-coordinated, mixed 2:2 aggregate (**30**) [88].

Apparently, the unreactive species formed after 50% conversion contains one equivalent of the product alcoholate of **27**, one equivalent of the remaining lithium acetylide **25** and two equivalents of the chiral alkoxide **26**, forming a mixed 3:1 alkoxide-acetylide tetramer, which shows a similar reactivity as **29**. The product alcoholate of **27** was suggested to exhibit chelating abilities similar to the amino alcoholate **26**. Then **26** and lithiated **27** could block via chelation the lithiums close to the acetylide moiety and cause the observed low reactivity, which is also apparent in **29**. However, lithiated [^{15}N]ketoaniline **27** did not show N-Li contacts according to ^{6}Li-NMR studies in the reaction mixture [95].

There is a close structural analogy between the mixed 2:2 (**28**) and 3:1 (**29**) aggregates in the efavirenz synthesis and those complexes which are obtained from anisyl fencholates and *n*-BuLi (**19** and **20**, Scheme 2) [76,77,78]. As suggested to explain the lower reactivity of **29** relative to **28**, the lithium ions neighbouring to the carbanionic moiety in **19** and in **29** are blocked by the chelating ligand, whereas the *trans*-situated lithium ion has a free coordination site in **19**, or is coordinated by (labile) THF in **29**. In contrast, the lithiums neighbored to the carbanionic moiety are free in **20**, which corresponds to the proposed structure of **28**.

The development of the efavirenz synthesis illustrates very nicely the fruitful interplay between synthetic demand on the one side and helpful elucidations by NMR and X-ray structural analyses as well as computational chemistry on the other.

As with aldehydes, catalytic additions to ketones are more promising with less polar organometallics and some enantioselective catalytic organozinc additions to ketones have been developed [96, 97]. For the synthesis of efavirenz (**23**), stoichiometric amounts of chiral zinc alkoxide additives proved to be highly useful, and high enantioselectivities could be achieved without protection of the amino function [98].

4 Additions to Epoxides and Oxetanes

Enantioselective additions of organolithiums to epoxides and oxetanes are synthetically highly valuable, as (at least in principle) alcohols with up to two, or up to three, new stereocenters respectively, are available in one addition step, e.g., from *meso*-substrates [Eqs. (4) and (5)].

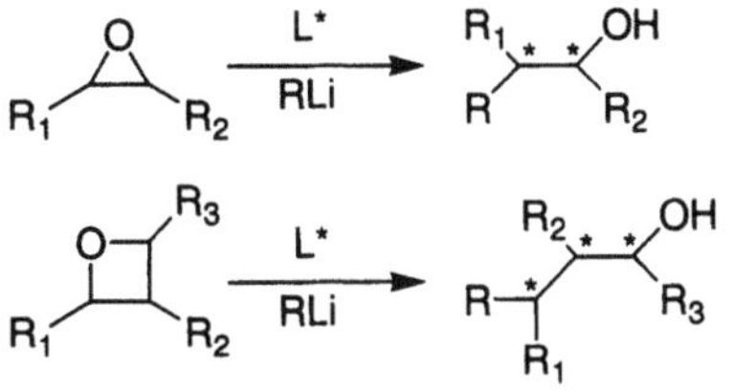

(4)

(5)

There is a frequently noted incompatibility of organolithiums with many chiral Lewis acids, which are often employed for asymmetric openings of epoxides [99]. Analogous Lewis base-catalyzed reactions are rare [100]. However, activation by Lewis acids, such as $BF_3 \cdot OEt_2$, is necessary for the opening of less reactive epoxides [101]. While organolithiums are rarely employed, applications of hetero-nucleophiles are well known in enantioselective desymmetrizations of *meso*-epoxides [102].

The group of Tomioka studied chiral ether ligands as mediators in phenyllithium additions to cyclohexene oxide [Eq. (6), R=Ph] [103, 104]. The best enantioselectivity (47% ee) was observed with the Lewis acid $BF_3 \cdot OBu_2$ in toluene at –78 °C with ligand **31** (Scheme 4). Analogous additions to 3-phenyloxetane gave 47% ee with $BF_3 \cdot OBu_2$ in Et_2O.

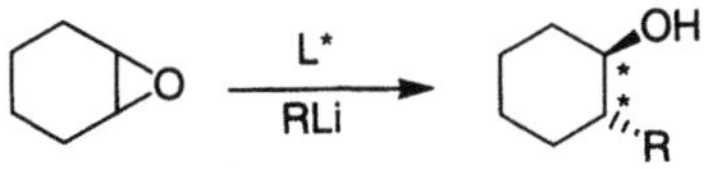

(6)

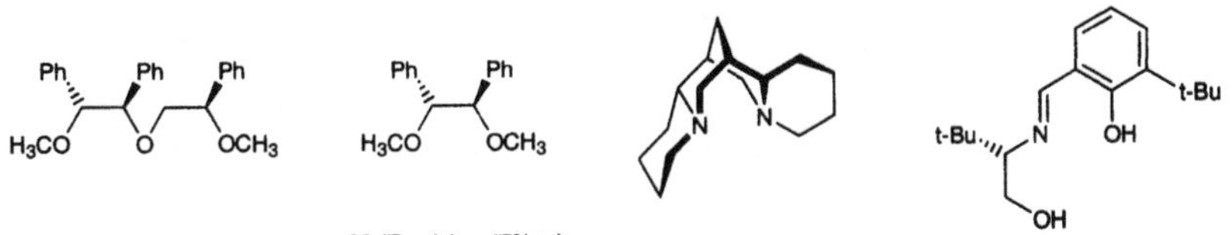

31 (Tomioka, 47%ee) **32** (Tomioka, 47%ee) **1** (Alexakis, 48%ee) **33** (Oguni, 90%ee)

Scheme 4. For enantioselectivities refer to Eq. (6), R=Ph

Alexakis et al. the used the combination of the diamine sparteine **1** (Scheme 4) with the Lewis acid $BF_3 \cdot OEt_2$ for phenyllithium additions to cyclohexene epoxide [Eq. (6)] [105]. They observed that addition of 1.5 equivalents $BF_3 \cdot OEt_2$ to a mixture of 2 equivalents PhLi, 2 equivalents of sparteine and 1 equivalent of cyclohexene oxide in Et_2O at –78 °C initiated a rapid reaction with quantitative formation of *trans*-phenylcyclohexanol in 48% ee. The 1:1 ratio of sparteine and organolithium reagent was found to be optimal. Only 5% ee or 6% ee were observed for *n*-BuLi or MeLi, respectively, while with cyclohexenyllithium no enantioselectivity was observed. Screening of a variety of other aryllithiums showed that 1-naphthyllithium yields 87% ee, while *o*-anisyllithium gives only 15% ee, apparently due to an unfavorable coordination effect of the methoxy group. PhLi additions to cyclooctene oxide gave the corresponding alcohol in 62% ee [106].

Oguni et al. employed even catalytic amounts (5 mol %) of the chiral Schiff base **33** without additional Lewis acid in a highly enantioselective addition (90% ee) of phenyllithium to cyclohexene oxide [107]. The *t*-Bu group and the imino hydrogen atom in **33** were found to be essential for high enantioselectivities.

5 Additions to Acetals

While the diastereoselective cleavage of cyclic chiral acetals is quite well established [108–110], enantioselective variants, e.g., desymmetrizations of *meso*-acetals [Eq. (7)], are much less reported. The group of Harada used chiral arylboron Lewis acids for the enantioselective ring cleavage of 1,3-dioxolanes by silyl enol ethers as carbon nucleophiles [111, 112].

i-Pr O O —(L*, RLi)→ i-Pr O OH * R (7)

Müller and Nury performed an enantioselective desymmetrization of 1,3-dioxolanes by organolithiums [113], using a protocol similar to the method which Alexakis et al. employed for the opening of epoxides [105]: Three equivalents of $BF_3 \cdot OEt_2$ were added to a mixture of the acetal (one equivalent), the organolithium reagent (four equivalents) and sparteine (**1**, four equivalents) in Et_2O at –78 °C [113]. The cleavage of benzaldehyde dimethyl acetal gave with MeLi 11% ee and with *n*-BuLi 25% ee. Higher selectivities than with alkyllithiums were achieved with aryllithiums: *ortho*-ethylphenyllithium cleaves *ortho*-*i*-Pr-2-phenyl-1,3-dixolane with 75% ee [Eq. (7), R=*ortho*-ethylphenyl]. With smaller amounts of sparteine, i.e., 3.0, 2.0 and 1.5 equivalents, the enantioselectivity was even increased to 80–81% ee, but 1.0 and 0.5 equivalents gave 68 and 23% ee. Changing the solvent from Et_2O to THF eliminated the enantioselectvity completely, but non-polar solvents like pentane or toluene gave 41 and 24% ee, respectively, i.e., lower selectivities than Et_2O. Müller et al. employed the asymmetric acetal cleavage for the synthesis of (*S*)-neobenodine using the dimethoxyethane ligand **32** (Scheme 4). Sparteine (**1**) was found to be less satisfactory for this purpose [114].

Acknowledgement. The author is grateful to the Deutsche Forschungsgemeinschaft (DFG) for support by a Heisenberg-Stipend and thanks especially Dr. Göran Hilmersson (Göteborg, Sweden) for fruitful suggestions.

Note added in proof. Complexes between MeLi and Chiral 3-aminopyrrolidine lithium amides bearing a second asymmetric center on their lateral amino group have been studied using multinuclear low-temperature NMR spectroscopy, and a relationship between the topology of these complexes and the sense of induction in the enantioselective alkylation of aaromatic aldehydes by alkyllithiums has been proposed [115]. 1,2-Amino sulfides have been used as chiral ligands in the enantioselective addition of BuLi and MeLi to various aldehydes (PhCHO, EtCHO) at low temperatures in up to 98.5% ee [16].

References

1. In this review, the optical purities are reported uncorrected as presented in the original publications
2. Corey EJ, Cheng XM (1995) The logic of chemical synthesis. Wiley, New York
3. Huryn DM (1991) In: Trost BM, Fleming I (eds), Comprehensive organic synthesis. Pergamon Press, Oxford, p 49
4. Sapse A-M, Schleyer, P v R (eds) (1995) Lithium chemistry. Wiley, New York
5. Lambert C, Schleyer P v R (1994) Angew Chem 106: 1187; Angew Chem Int Ed Engl 33:1129
6. Lambert C; Schleyer P v R (1993) In: Methoden org chem (Houben Weyl). 4th edn, vol E19d, Thieme, Stuttgart, 1
7. Bauer W, Schleyer P v R (1992) In: Snieckus V (ed), Advances in carbanion chemistry, vol 1. Jai Press, Greenwich, CT
8. Berrisford DJ, Bolm C, Sharpless KB (1995) Angew Chem 107:1159; Angew Chem Int Ed Engl 34:1059
9. Noyoi R, Kitamura M (1991) Angew Chem 103:34; Angew Chem Int Ed Engl 30:49
10. Cohen HL, Wright GF (1953) J Org Chem 18:432
11. Allentoff N, Wright GF (1957) J Org Chem 22:1
12. French W, Wright GF (1964) Can J Chem 42:2474
13. Iffland DC, Davis JE (1977) J Org Chem 42:4150
14. Goldfuss B (2001) Nachr Chem 49:1333
15. Hoppe D, Hense T (1997) Angew Chem 109:2376; Angew Chem Int Ed Engl 36:2282
16. Basu A, Thayumanavan S (2002) Angew Chem 114:740; Angew Chem Int Ed 43:716
17. Noyori R, Suga S, Kawai K, Okada S, Kitamura M (1988) Pure Appl Chem 60:1597
18. Tomioka K (1990) Synthesis 541
19. Noyori R (1994) Asymmetric catalysis in organic synthesis. Wiley, New York
20. Juaristi E, Beck AK, Hansen J, Matt T, Mukhopadhyay T, Simson M, Seebach D (1993) Synthesis 1271
21. For a review on the Mukaiyama aldol reaction, see: Carreira EM (1999) In: Jacobsen EN, Pfaltz A, Yamamoto H (eds), Comprehensive Asymmetric Catalysis. Vol. III, Springer, Berlin Heidelberg New York, pp 997–1065
22. Sawamura M, Ito Y (2000) In: Ojima I (ed), Catalytic asymmetric synthesis. 2nd edn, Wiley-VCH, Weinheim, p 493
23. Nozaki H, Aratani T, Toraya T (1968) Tetrahedron Lett 4097
24. Nozaki H, Aratani T, Toraya T, Noyori R (1971) Tetrahedron 27:905
25. Seebach D, Dörr H, Bastani B, Ehrig V (1969) Angew Chem 81:1002; Angew Chem Int Ed Engl 8:982
26. Seebach D, Kalinowski H-O, Bastani B, Crass G, Daum H, Dörr H, DuPreez NP, Ehrig V, Langer W, Nüssler C, Oei H-A, Schmidt M (1977) Helv Chim Acta 60:301
27. Seebach D, Oei H-A, Daum H (1977) Chem Ber 110:2316
28. Seebach D, Langer W (1979) Helv Chim Acta 62:1701 and 1710
29. Seebach D, Crass G, Wilka E-M, Hilvert D, Brunner E (1979) Helv Chim Acta 62:2695
30. Mukaiyama T, Soai K, Kobayashi S (1978) Chem Lett 219
31. Soai K, Mukaijama T (1978) Chem Lett 491
32. Mukaiyama T, Soai K, Sato T, Shimizu H, Suzuki K (1979) J Am Chem Soc 101:1455

33. Mukaiyama T, Suzuki K (1980) Chem Lett 255
34. Johnson WS, Frei B, Gopalan AS (1981) J Org Chem 46:1512
35. Mazaleyat J-P, Cram DJ (1981) J Am Chem Soc 103:4585
36. Eleveld MB, Hogeveen H (1984) Tetrahedron Lett 25:5187
37. Arvidsson PI, Hilmersson G, Davidsson Ö (1999) Chem Eur J, 5:2348
38. Whitesell JK, Jaw B-R (1981) J Org Chem 46:2798
39. Colombo L, Gennari C, Poli G, Scolastico C (1982) Tetrahedron 38:2725
40. Alberts AH, Wynberg H (1989) J Am Chem Soc 111:7265
41. For an early report on the role of mixed aggregates see: Seebach D, Amstutz R, Dunitz JD (1981) Helv Chim Acta 64:2622
42. Ye M, Logaraij S, Jackman LM, Hillegass K, Hirsh KA, Bollinger AM, Grosz AL (1994) Tetrahedron 50:6109
43. Kang J, Kim JI, Lee JH (1994) Bull Korean Chem Soc 15:865
44. Corruble A, Valnot J-Y, Maddaluno J, Duhamel P (1997) Tetrahedron: Asymmetry 8:1519
45. Corruble A, Valnot J-Y, Maddaluno J, Duhamel P (1998) J Org Chem 63, 8266
46. Schön M, Naef R (1999) Tetrahedron: Asymmetry 10:169
47. Knollmüller M, Ferencic M, Gärtner P (1999) Tetrahedron: Asymmetry 10:3969
48. Aspinall HC, Dwyer JLM, Greeves N, Steiner A (1999) Organometallics 18:1366
49. Review on RLi/ROM compounds: Lochmann L (2000) Eur J Inorg Chem 1115
50. McGarrity JF, Ogle C A, Brich Z, Loosli H-R (1985) J Am Chem Soc 107:1810
51. Snieckus V (1990) Chem Rev 90:879
52. Hommes NJR v E, Schleyer P v R (1994) Tetrahedron 50:5903
53. Hommes NJR v E, Schleyer P v R (1992) Angew Chem 104:768; Angew Chem In Ed Engl 31:755
54. Goldfuss B, Schleyer P v R, Handschuh S, Hampel F (1998) J Organomet Chem 552:285
55. Activation via deaggregation of organolithiums by coordinating solvents (THF) or ligands (TMEDA) is frequently used in lithiations of hydrocarbons: Brandsma L, Verkruijsse H (1987) Preparative polar organometallic chemistry. Springer, Berlin, Heidelberg, New York
56. Streitwieser A, Wang DZ-R (1999) J Am Chem Soc 121:6213
57. Wang DZ, Kim Y-J, Streitwieser A (2000) J Am Chem Soc 122:10754
58. Novak DP, Brown TL (1972) J Am Chem Soc 94:3793
59. Kieft RL, Novak DP, Brown TL (1974) J Organomet Chem 77:299
60. Eppers O, Günther H (1990) Helv Chim Acta 73:207
61. Seebach D (1988) Angew Chem 100:1685; Angew Chem Int Ed Engl 27:1624
62. Weidemann B, Seebach D (1983) Angew Chem 95:12, Angew Chem Int Ed Engl 22:40
63. For a recent review on alkylations of carbonyl compounds by organozinc and titanium species see: Pu L, Hong-Bin Y (2001) Chem Rev 101:757
64. For titanium TADDOLATES: Seebach D, Beck AK, Imwinkeiried R, Roggo S, Wonnacott A (1987) Helv Chim Acta 70:954
65. For a review on catalytic asymmetric aryl additions see: Bolm C, Hildebrand JP, Muniz K, Hermanns N (2001) Angew Chem 113:3383; Angew Chem Int Ed 40:3284
66. Weber B, Seebach D (1992) Angew Chem 104:96; Angew Chem Int Ed Engl 31:84
67. Weber B, Seebach D (1994) Tetrahedron 50:6117
68. Hilmersson G, Davidsson O (1995) J Organomet Chem 489:175
69. Arvidsson PI, Ahlberg P, Hilmersson G (1999) Chem Eur J 5:1348
70. Arvidsson PI, Davidsson Ö, Hilmersson G (1999) Tetrahedron: Asymmetry 10:527
71. Granander J, Scott R, Hilmersson G (2002) Tetrahedron 58:4717–4725
72. Williard PG, Sun C (1997) J Am Chem Soc 119:11693
73. Hilmersson G, Malmros B (2001) Chem Eur J 7:331
74. Goldfuss B, Steigelmann M, Rominger R (2000) Eur J Org Chem 1785
75. Goldfuss B, Steigelmann M, Khan SI, Houk KN (2000) J Org Chem 65:77
76. Goldfuss B, Khan SI, Houk KN (1999) Organometallics 18:2927
77. Goldfuss B, Steigelmann M, Rominger F (2000) Angew Chem 112:4299, Angew Chem Int Ed 39:4133
78. Goldfuss B, Steigelmann M, Rominger F, Urtel H (2001) Chem Eur J 7:4456

79. Kottke T, Stalke D (1993) Angew Chem 105:619; Angew Chem Int Ed Engl 32:580
80. Briggs TF, Winemiller MD, Xiang B, Collum DB (2001) J Org Chem 66:6291
81. Sun X, Winemiller MD, Xiang B, Collum DB (2001) J Am Chem Soc 123:8039
82. Jiang B, Feng Y (2002) Tetrahedron Lett 43: 2975
83. Tzalis D, Knochel P (1999) Angew Chem 111:1547; Angew Chem Int Ed 38:1463
84. Frantz DE, Fässler R, Carreira E (2000) J Am Chem Soc 122:1806
85. Corey EJ, Guzman-Perez A (1998) Angew Chem 110:402, Angew Chem Int Ed 37:388
86. Wakefield BJ (1974) The chemistry of organolithium compounds. Pergamon Press, Oxford, p 129
87. Smith MB, March J (2001) Advanced organic chemistry. 5th edn, Wiley, New York, p 1208
88. Thompson A, Corley EG, Huntington MF, Grabowski EJJ, Remenar JF, Collum DB (1998) J Am Chem Soc 120:2028
89. Pierce ME, Parsons RL Jr, Radesca LA, Lo YS, Silverman S, Moore JR, Islam Q, Choudhury A, Fortunak JMD, Nguyen D, Luo C, Morgan SJ, Davis WP, Confalone PN, Chen C-y, Tillyer RD, Frey L, Tan L, Xu F, Zhao D, Thompson AS, Corley EG, Grabowski EJJ, Reamer R, Reider PJ (1998) J Org Chem 63: 8536
90. Thompson AS, Corley EG, Huntington MF, Grabowski EJJ (1995) Tetrahedron Lett 36: 8937
91. Huffman MA, Yasuda N, DeCamp AE, Grabowski EJJ (1995) J Org Chem 60:1590
92. Parsons RL Jr, Fortunak JM, Dorow RL, Harris GD, Kauffman GS, Nugent WA, Winemiller MD, Briggs TF, Xiang B, Collum DB (2001) J Am Chem Soc 123:9135
93. Rutherford JL, Hoffmann D, Collum DB (2002) J Am Chem Soc 124:264
94. Lucht BL, Collum DB (1994) J Am Chem Soc 116:7949
95. Xu F, Reamer RA, Tillyer R, Cummins JM, Grabowski EJJ, Reider PJ, Collum DB, Huffman JC (2000) J Am Chem Soc 122:11212
96. Ramon JD, Yus M (1998) Tetrahedron Lett 39:1239
97. Dosa PI, Fu GC (1998) J Am Chem Soc 120:445
98. Tan L, Chen C-Y, Tillyer RD, Grabowski EJJ, Reider PJ (1999) Angew Chem 111:724; Angew Chem Int Ed 38: 711
99. Yamamoto H (ed) (2000) Lewis acids in organic synthesis. Wiley-VCH, Weinheim
100. Schneider C, Brauner J (2001) Eur J Org Chem 23:4445
101. Yamaguchi M; Hirao I (1983) Tetrahedron Lett 24:391
102. Hodgson DM, Gibbs AR, Lee GP (1996) Tetrahedron 52:14361
103. Mizuno M, Kanai M, Iida A, Tomioka K (1996) Tetrahedron: Asymmetry 7:2483
104. Mizuno M, Kanai M, Iida A, Tomioka K (1997) Tetrahedron 53:19699
105. Alexakis A, Vrancken E, Mangeney P (1998) Synlett 1165
106. Alexakis A, Vrancken E, Mangeney P (2000) J Chem Soc Perkin Trans 1 3354
107. Oguni N, Miyagi, Y, Itoh K (1998) Tetrahedron Lett 39:9023
108. Seebach D, Imwinkelried R, Weber T (1986) In: Scheffold R (ed) Modern synthetic methods, vol 4. Springer, Berlin, Heidelberg, New York, p 125
109. Alexakis A, Mangeney P (1990) Tetrahedron: Asymmetry 1:477
110. Alexakis A, Mhamdi F, Lagasse F, Mangeney P (1996) Tetrahedron: Asymmetry 7:3343
111. Kinugasa M, Harada T, Oku A (1996) J Org Chem 61:6772
112. Kinugasa M, Harada T, Oku A (1998) Tetrahedron Lett 39:4529
113. Müller P, Nury P (2000) Org Lett 2:2845
114. Müller P, Nury P, Bernardinelli G (2001) Eur J Org Chem 4137
115. Corruble A, Davoust D, Desjardins S, Fressigné C, Giessner-Prettre C, Harrison-Marchand A, Houte H, Lasne M, Maddaluno J, Oulyadi H, Valnot J (2002) J Am Chem Soc 124:15267
116. Granander J, Sott R, Hilmersson G (2003) Tetrahedron: Asymmetry 14:439

Topics Organomet Chem (2003) 5: 37–59
DOI 10.1007/b10342

Enantioselective Conjugate Addition and 1,2-Addition to C=N of Organolithium Reagents

Mayu Iguchi, Ken-ichi Yamada, Kiyoshi Tomioka

Graduate School of Pharmaceutical Sciences, Kyoto University, Yoshida, Sakyo-ku, Kyoto 606-8501, Japan. *E-mail: tomioka@pharm.kyoto-u.ac.jp*

Organolithium reagents are highly reactive species and are used in a variety of organic transformations. Development of new methodology for asymmetric reactions of organolithium reagents constitutes a remarkable area of fundamental progresses in recent synthetic organic chemistry. The conjugate addition (or Michael reaction, 1,4-addition) and 1,2-addition to C=N double bonds are powerful methods for forming a carbon-carbon bond. This chapter reviews the recent progress in enantioselective conjugate addition and 1,2-addition to C=N double bonds by organolithium reagents. Sect. 1 considers conjugate addition reactions with activated olefins and α,β-unsaturated imines, focusing on reactions controlled by external chiral ligands. Chiral oxazoline chemistry is also briefly presented. Sect. 2 summarizes the 1,2-additions to imines and imine congeners (hydrazones, oxime ethers, nitrones) using chiral auxiliaries and chiral ligands. We do not describe additions which use chiral auxiliary groups in carbonyl compounds.

Keywords. Asymmetric reaction, Organolithium, Conjugate addition, Imine, Chiral ligand

1
Enantioselective Conjugate Addition of Organolithium Reagents

Conjugate addition of organometallic reagents to electrophilically activated olefins constitutes versatile methodology for forming a carbon-carbon bond [1]. The products are the corresponding β-substituted carbonyl compounds. Because of the usefulness of the reaction as well as the product, a considerable number of approaches to asymmetric conjugate addition have been reported and several excellent reviews have been published [2–10]. Recently, prominent results for asymmetric conjugate addition have been achieved in copper-catalyzed reactions of diethylzinc [11] and rhodium-catalyzed reactions of arylboron reagents [12] with cycloalkenones (Scheme 1). However, the use of organometallics for enantioselective conjugate addition is an underdeveloped area.

Organolithium reagents are highly reactive species and are used in a variety of organic transformations, thus selective organolithium-based asymmetric conjugate addition reactions are challenging. In this chapter we focus on recent progress in enantioselective conjugate addition of organolithium reagents with achiral activated olefins under control of an external chiral ligand, or a chiral catalyst. Reactions using a chiral auxiliary are also briefly presented.

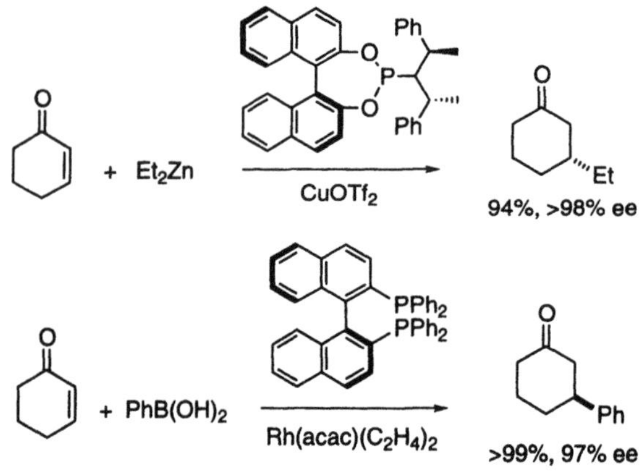

+ Et_2Zn → ($CuOTf_2$) 94%, >98% ee

+ $PhB(OH)_2$ → ($Rh(acac)(C_2H_4)_2$) >99%, 97% ee

Scheme 1.

1.1
Addition to Acceptors Having a Chiral Auxiliary

The chelation control methodology of Meyers provided a highly selective conjugate addition of organolithium reagents to a chiral oxazoline. The diastereoselective asymmetric addition of various organolithiums to α,β-unsaturated chi-

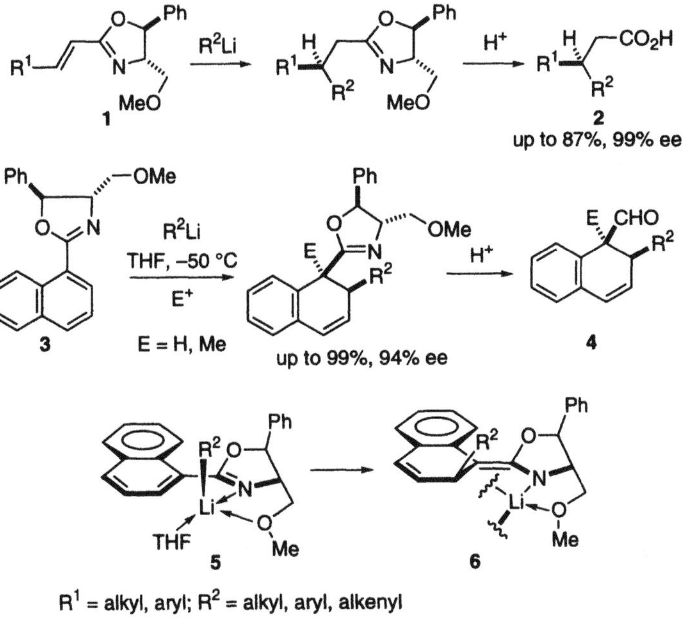

R^1 = alkyl, aryl; R^2 = alkyl, aryl, alkenyl

Scheme 2.

ral oxazolines **1**, chiral naphthyloxazolines **3** or chiral naphthyl imines leads to substituted alkanoic acids **2** and substituted dihydronaphthalenes **4** with high enantiomeric purity [13–16] (Scheme 2). The selectivity may be rationalized by assuming that the complexed organolithium R^2Li aligns itself as shown in **5**, where the C-Li bond is parallel to the π-orbitals of the naphthalene ring and undergoes a 1,5-sigmatropic rearrangement, and R^2 enters the aromatic π-system to give **6**.

A β-aryl-α,β-unsaturated *tert*-butyl ester **7** bearing a chiral imidazolidine (or oxazolidine) auxiliary at an *ortho* position undergoes stereoselective conjugate addition of an aryllithium reagent **8**, furnishing a β,β-diarylpropionic acid derivative **9** in up to 97% ee (Scheme 3) [17].

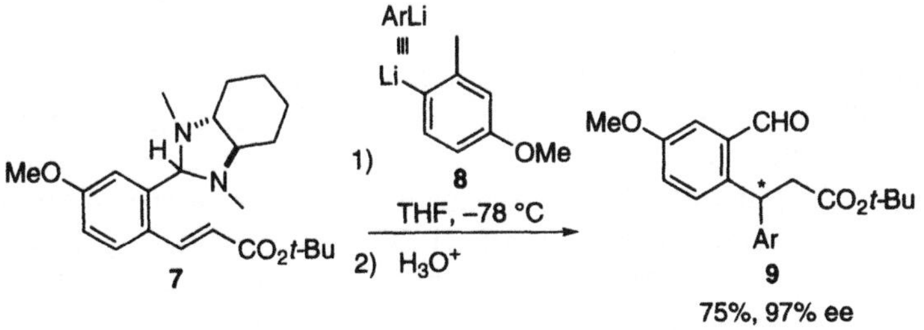

Scheme 3.

1.2 Reaction in the Presence of an External Chiral Ligand

1.2.1 *Addition to Activated Olefins*

Seebach studied the conjugate addition reaction of nitroolefin **10** with butyllithium in the presence of the multidentate ligand **11** (accessible from tartaric acid), which yielded the adduct **12** in up to 58% ee (Scheme 4) [18].

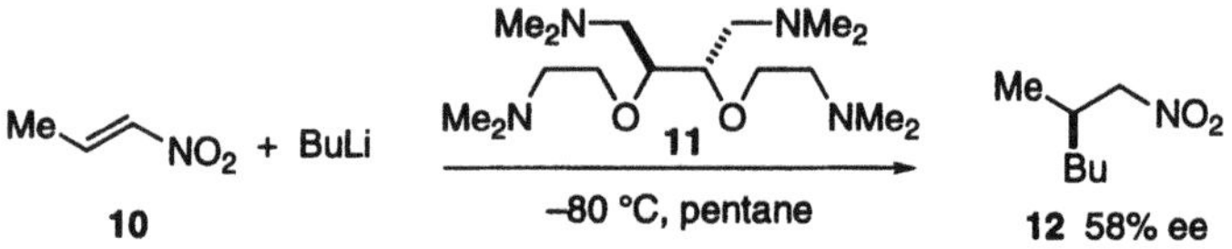

Me$_2$N NMe$_2$ Me$_2$N O 11 O NMe$_2$
Me NO$_2$ + BuLi → −80 °C, pentane → Me NO$_2$ Bu
10 12 58% ee

Scheme 4.

α,β-Unsaturated carbonyl compounds (typically esters) undergo conjugate addition with alkyl- and aryllithium reagents. Chemoselectivity (1,2- vs. 1,4-addition) in these reactions can be controlled by increasing the steric bulk in the nucleophile or the ester substituent. Because of the high nucleophilicity of organolithium reagents, a bulky dithiane reagent could be used as a Michael donor. The enantioselective conjugate addition reaction of lithiated dithioacetal derivative **14** with α,β-unsaturated ester **13** in the presence of a chiral ligand **15** provided adduct **16** in up to 67% ee (Scheme 5) [19].

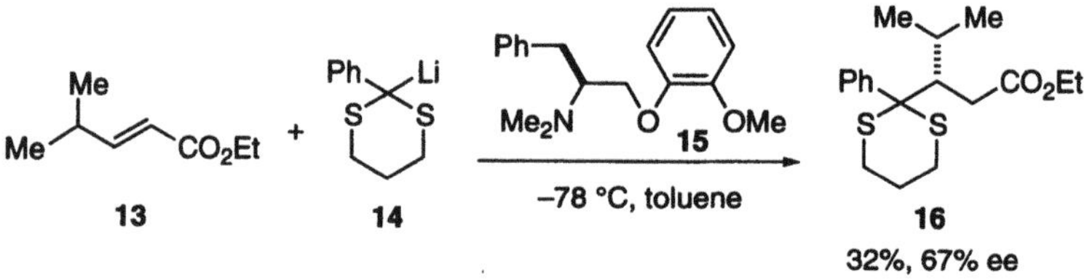

Me Me CO$_2$Et + Ph Li S S
13 14
Ph Me$_2$N O 15 OMe
−78 °C, toluene
Me Me Ph CO$_2$Et S S
16
32%, 67% ee

Scheme 5.

Recently, progress has been made based on the use of a bulky carbonyl masking group, so as to avoid 1,2-addition: 2,6-di-*tert*-butyl-4-methoxyphenyl (BHA) esters have been successfully employed in conjugate additions [20]. External chiral ligands, especially the C_2 symmetric diether **18** and the naturally occurring diamine, (–)-sparteine (**19**), have been used in a stoichiometric or catalytic manner with organolithiums in asymmetric conjugate addition reactions with BHA α,β-unsaturated ester **17** [21, 22]. The products are 3-substituted alkanoates **20**. The diether **18** is readily available by asymmetric dihydroxylation of stilbene followed by methylation [23]. Diether **18** and (–)-sparteine (**19**) are complementary to each other: **18** shows high efficiency with aryllithiums and **19** is effective with alkyllithiums (Scheme 6). The catalytic turnover of **19** is superior to that of **18**. The use of a poorly coordinating solvent, such as toluene or

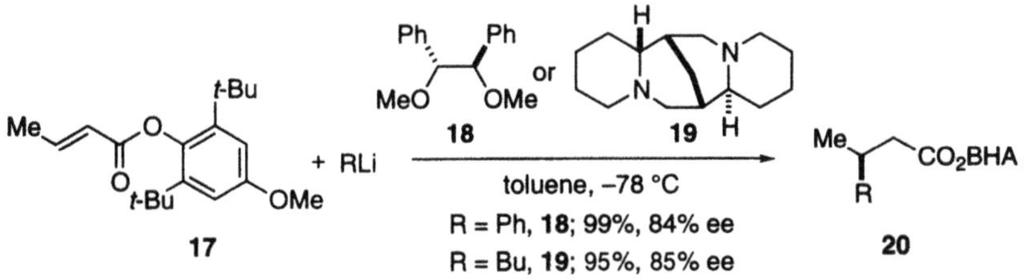

Scheme 6.

ether, is essential for high enantioselectivity because of the desirable formation of tight lithium-ligand chelated complexes as chiral carbonucleophiles.

The BHA group exerts remarkable activating and directing effects on the reaction of naphthalenecarboxylates with organolithium reagents [24, 25]. In the presence of either stoichiometric or catalytic amounts of the chiral diether **18**, aryllithiums reacted with BHA esters of 1- and 2-naphthalenecarboxylic acid **21/22** to afford 1,2-di-, 1,1,2- and 1,2,2-trisubstituted dihydronaphthalenes **23/24** in high ee, following a subsequent ketene-reduction in a one-pot process (Scheme 7).

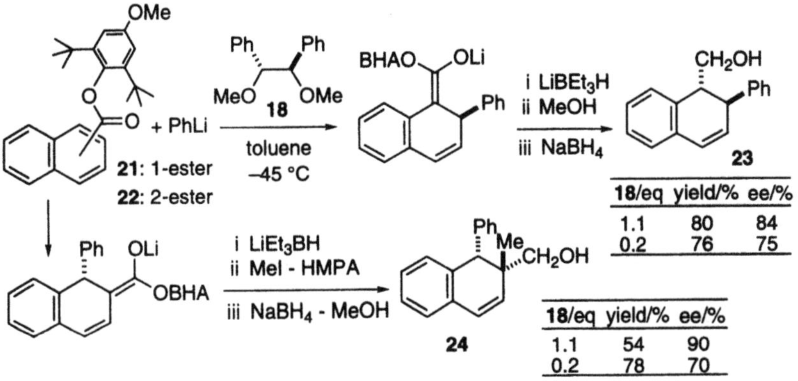

18/eq	yield/%	ee/%
1.1	80	84
0.2	76	75

18/eq	yield/%	ee/%
1.1	54	90
0.2	78	70

Scheme 7.

The external chiral ligand-controlled asymmetric conjugate addition technology has been proven to be applicable to an asymmetric synthesis of a dopamine D1 agonist, dihydrexidine (**28**) (Scheme 8) [26]. The synthetic pathway to **28** relies on three key processes: enantioselective conjugate addition reaction of phenyllithium with α,β-unsaturated ester **25**, Curtius rearrangement-type conversion of **26** into amine **27**, and finally Pictet-Spengler-type cyclization to complete the skeleton.

A similar transformation to that shown in Scheme 3, but enantioselective, was carried out with α,β-unsaturated *tert*-butyl ester **29** and aryllithium reagent **8**, providing the 1,4-addition product **30** in up to 88% ee (Scheme 9) [27]. The *ortho* (relative to the unsaturated ester portion) substituent is important for improving the enantioselectivity. A survey of various ligands showed that **18** was more effective than **19**, or chiral amino ether ligands.

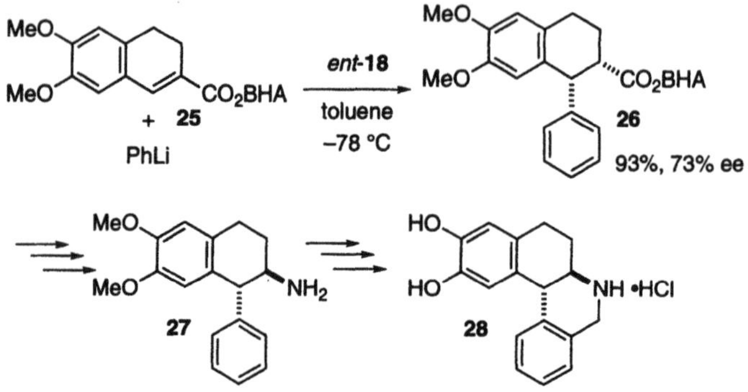

25 + PhLi → (ent-18, toluene, –78 °C) → 26 (93%, 73% ee) → → 27 → → 28

Scheme 8.

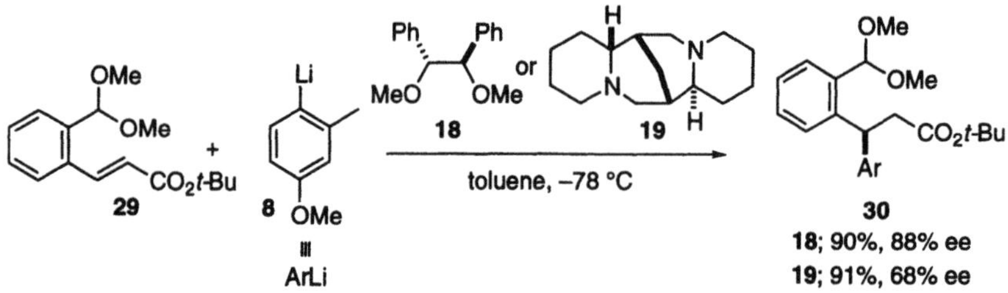

29 + 8 (≡ ArLi) → (18 or 19, toluene, –78 °C) → 30

18; 90%, 88% ee

19; 91%, 68% ee

Scheme 9.

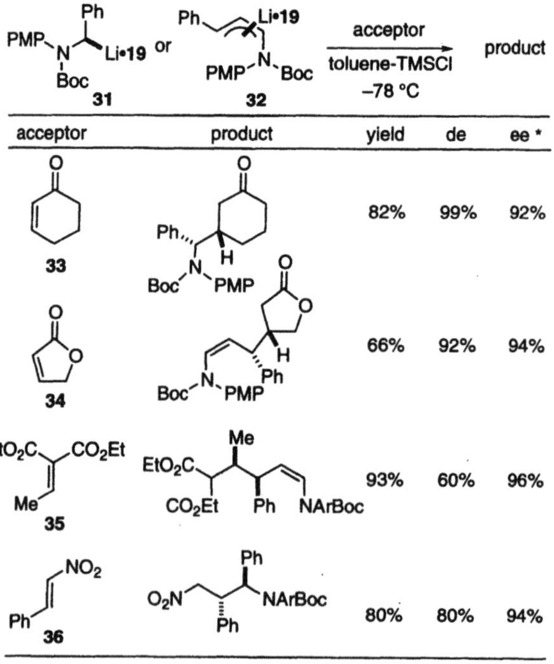

31 or 32 → (acceptor, toluene-TMSCl, –78 °C) → product

acceptor	product	yield	de	ee *
33		82%	99%	92%
34		66%	92%	94%
35		93%	60%	96%
36		80%	80%	94%

* for major diastereomer

Scheme 10.

(–)-Sparteine (**19**) is an excellent chiral ligand for the asymmetric Michael addition of enantioenriched anilinobenzylic and -allylic organolithiums **31**, **32** [PMP=*p*-methoxyphenyl] (Scheme 10, see also Chapter 5) [28, 29]. Complexation of the organolithium with **19** prevents the inversion at the carbanionic center. The choice of ligand for the lithium cation can provide control of 1,2- vs. 1,4-addition of organolithium species to α,β-unsaturated carbonyl compounds. Furthermore, in these addition reactions two contiguous stereocenters were constructed with high diastereo- and enantioselectivities. The availability of both organolithium epimers through a stannylation/lithiation sequence provides a choice in the configuration of the donor organolithium species. Cyclic enone **33**, unsaturated lactone **34**, acyclic doubly activated olefin **35** and nitroolefin **36** are acceptors compatible in this reaction. Further transformations of the addition products provide [3.3.0]-, [4.3.0]-, [5.3.0]-, and [5.4.0]-carbocycles and heterocycles with high stereoselectivities [30]. The presence of TMSCl in the addition reactions prevents side reactions. The allyllithium-(–)-sparteine (**19**) complex **32** was isolated and crystallographically characterized, providing a model for predicting the stereochemical outcome of the addition reaction [31].

1.2.2
Addition to α,β-Unsaturated Imines

The C_2 symmetric chiral diether **18** has been most efficient in the asymmetric conjugate addition of organolithium reagents to naphthaldehyde imine **37** and cyclic and acyclic α,β-unsaturated aldimines **38–40** (Scheme 11) [32, 33]. Following hydrolysis β-substituted aldehydes **41** were obtained, which were then reduced with $NaBH_4$ to afford the corresponding alcohols **42** with excellent enantioselectivity. The enantiofacial selection has been explained through a lithium-coordinated complex between organolithium, imine and chiral diether **18**, where the migrating C-Li bond is parallel to the reacting p-orbital of the imine

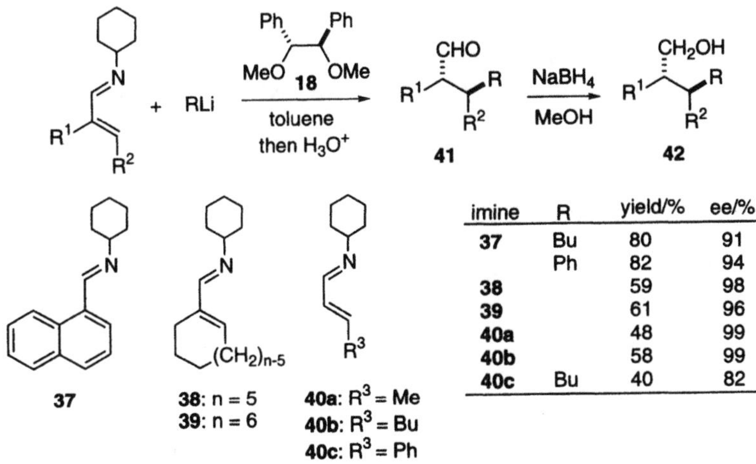

imine	R	yield/%	ee/%
37	Bu	80	91
	Ph	82	94
38		59	98
39		61	96
40a		48	99
40b		58	99
40c	Bu	40	82

Scheme 11.

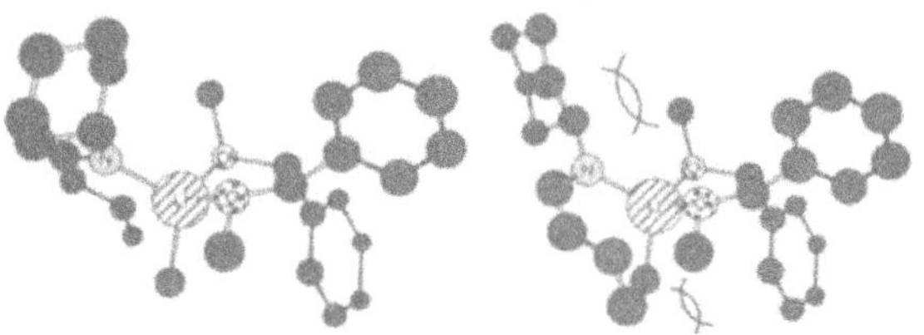

Fig. 1.

(Fig. 1, shown for the imine of crotonaldehyde). From the favored complex (shown on the left), the R group of the organolithium reagent is then transferred to the less-hindered face of the double bond of the unsaturated imine. These additions are the enantioselective evolution of Meyers's chiral oxazoline and chiral imine chemistry (Sect. 1.1).

1,4- vs. 1,2-addition regioselectivity in the reaction of imines derived from naphthaldehydes and α,β-unsaturated aldehydes is rationalized mainly by the relative magnitude of the LUMO coefficient on the 2 possible reaction sites [34, 35]. Replacement of the cyclohexyl group of the imine moiety by an aryl group leads to a larger coefficient at the imine carbon. Reaction of an imine bearing an aryl group on the imine nitrogen atom results in selective 1,2-addition, not Michael-type conjugate addition. Catalytic asymmetric 1,2-addition reaction of such an imine with an organolithium reagent was catalyzed by a chiral aminoether ligand to provide the corresponding chiral amine in high ee (Sect. 2.4).

The first prominent catalytic asymmetric conjugate addition reaction of an organolithium reagent was the reaction of 1-naphthyllithium **44** with naphthaldehyde imine **43** having a fluoro substituent as a leaving group under the control of the diether **18** [36]. Only a catalytic amount of **18** (5 mol%) was required to give binaphthyl **45** in 82% ee, in which an enantioselective conjugate addition-elimination mechanism is operative (Scheme 12). This nucleophilic aromatic substitution consists of two successive processes: The first is the conjugate addition of the naphthyllithium-diether complex to **43**, and the second involves elimination of the LiF-diether complex in which transfer of central chirality to axial chirality occurs to give **45**. Regeneration of the naphthyllithium-diether complex from the LiF-diether complex through ligand exchange is essential for the prop-

43 + 44 → (18, toluene, –45 °C) → 45

18/eq	time/h	ee/%	yield/%
2.3	1	90	>99
0.05	3.5	82	97

Scheme 12.

agation of the catalytic asymmetric process. High selectivity was realized by using a (2,6-diisopropylphenyl)imino group. Dramatic improvements in ee (those shown in Scheme 12) were achieved by using lithium bromide-free naphthyllithium, prepared from naphthylpropyltellurium and butyllithium [37].

The chromium complex **46** of benzaldehyde imine is also a good substrate for asymmetric conjugate addition of organolithium reagents, where the reaction was mediated by a stoichiometric amount of chiral diether **18** in toluene to give, following propargylic electrophile incorporation, the cyclohexadienal **48** in up to 93% ee (Scheme 13) [38]. SAMP [(*S*)-1-amino-2-(methoxymethyl)pyrrolidine] hydrazones were used effectively in diastereoselective nucleophilic additions to (arene)$Cr(CO)_3$ complexes [39].

Lithiated 2-trimethylsilylbenzenethiol is a good Michael donor nucleophile under the control of a chiral aminoether; however, a discussion of this chemistry lies outside the scope of this chapter [40].

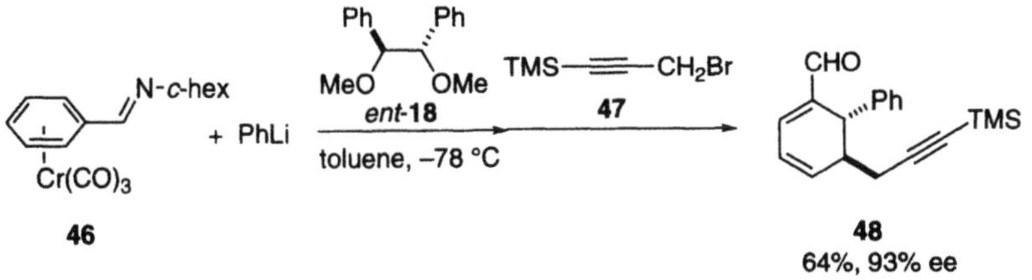

N-*c*-hex
$Cr(CO)_3$
46
+ PhLi
Ph Ph
MeO OMe
ent-**18**
toluene, –78 °C
TMS—≡—CH_2Br
47
CHO
Ph
TMS
48
64%, 93% ee

Scheme 13.

2
Enantioselective 1,2-Addition of Organolithium Reagents to C=N

2.1
Addition to Imines and Imine Congeners Having a Chiral Auxiliary

Asymmetric nucleophilic 1,2-addition of organolithium reagents to C=N double bonds provides a versatile method for preparation of chiral amines. Optically active amines are abundantly present in biologically active compounds and are also important chiral building blocks. Asymmetric addition to C=N double bonds has been achieved with the use of a chiral auxiliary or chiral ligands. The chiral auxiliary strategy involves a diastereoselective reaction, separation of the diastereomers and subsequent removal of an auxiliary giving enantiomerically pure products. The other strategy employs chiral ligands for the direct introduction of chirality to an imine and/or an organolithium reagent. This section presents an overview of these achievements.

2.1.1
Addition to Imines

Diastereoselective addition of organolithium reagents to imines which contain chiral auxiliaries gives chiral amines after the removal of the auxiliaries. Takahashi and coworkers reported pioneering work concerning this strategy. Organolithium reagents undergo addition to imines **49** derived from aryl aldehydes and valinol or phenylglycinol, giving amines **51** in good yields and in high diastereoselectivity (Scheme 14) [41, 42]. The observed diastereoselection was rationalized by considering formation of the five-membered ring **50**, where chelation of the alkoxide and the imino group to the lithium metal is operative. The organolithium reagent then approaches the imine from the sterically less-hindered *si*-face.

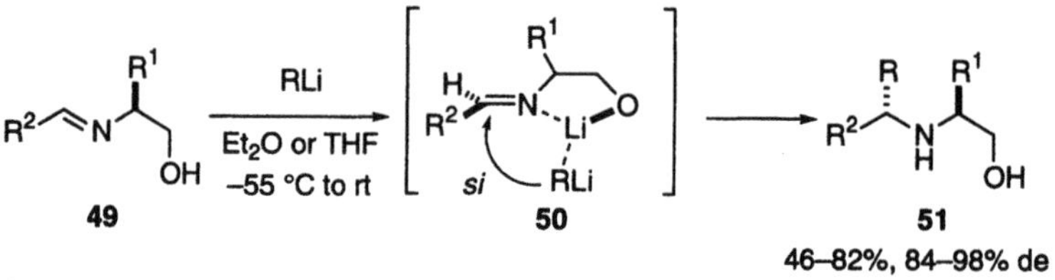

RLi, Et_2O or THF, –55 °C to rt

49 → [**50**] → **51**

46–82%, 84–98% de

R^1 = *i*-Pr, Ph
R^2 = Et, Bn, Ph, *p*-tol, 4-$MeOC_6H_4$, 4-ClC_6H_4, 2-furyl, 2-thienyl, 3-thienyl, ferrocenyl
R = Me, Bu, Ph, *p*-tol, 4-$MeOC_6H_4$, 4-ClC_6H_4

Scheme 14.

Amines having an ether moiety instead of a hydroxy group are also good chiral auxiliaries. Imines such as **52**, derived from chiral amino ethers, undergo addition reactions with organolithium reagents to give the corresponding amines **53** with good diastereoselectivity (Scheme 15). Methyl ethers of valinol [41c] and phenylglycinol [43], (1*R*,2*S*)-2-methoxy-1,2-diphenylethylamine [44], and 1-(2-methoxyphenyl)ethylamine [45] have been used as chiral amino ethers. In these reactions, the sense of diastereoselectivity is predictable using chelation models similar to **50**.

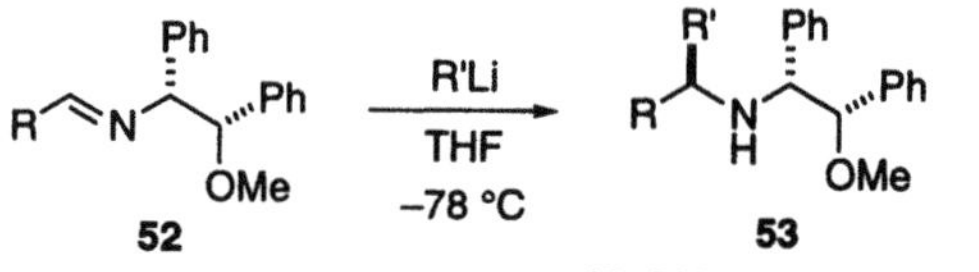

R'Li, THF, –78 °C

52 → **53**

43–89%, 88–100% de

R = Ph, 4-ClC_6H_4, 4-$MeOC_6H_4$, PhCH=CH, *i*-Pr
R' = Me, Bu, Ph, CH_2=$CHCH_2$, MeCH=$CHCH_2$, Me_2C=$CHCH_2$, TMS–≡–CH_2

Scheme 15.

1-Arylethylamines are widely used as chiral auxiliaries because of the availability of both enantiomers and the potential for auxiliary removal by hydrogenolysis. Reactions using 1-phenylethylimine [46] and 1-(1-naphthyl)ethylamine [47] have been reported. With 1-(1-naphthyl)ethylimines **54**, the use of organolithium-boron trifluoride reagents gave good diastereoselectivity (Scheme 16).

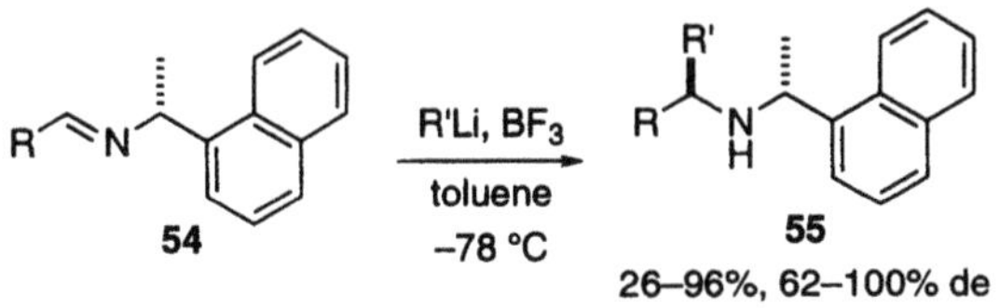

R'Li, BF_3, toluene, –78 °C

54 → 55: 26–96%, 62–100% de

R = *t*-Bu, $PhCH_2CH_2$, Ph, 1-naphthyl, 2-$MeOC_6H_4$, 2-pyridyl, 2-quinolyl, 3-indolyl
R' = Me, Bu

Scheme 16.

2.1.2
Addition to Hydrazones

The addition of organolithium reagents to hydrazones gives substituted hydrazines. Since amines can be obtained by the reductive cleavage of the N-N bond of hydrazines, hydrazones can be considered as useful, more stable, equivalents of imines. Thus, the asymmetric addition of organolithium reagents to chiral hydrazones is a potentially attractive alternative method for the preparation of optically active amines. The first examples of the addition of organometallic reagents to chiral hydrazones, derived from chiral hydrazines, was reported by Takahashi and coworkers using Grignard reagents [48]. More general addition of organolithium reagents has been extensively examined using a chiral hydrazone, prepared from a chiral hydrazine, (*S*) and (*R*)-1-amino-2-(methoxymethyl)pyrrolidine (SAMP and RAMP). A variety of organolithium reagents added diastereoselectively to SAMP hydrazones **56**, affording highly optically pure amines **58**, after reductive N-N bond cleavage (Scheme 17) [49–51]. The cleavage, however, generally requires rather harsh conditions and sometimes suffers from partial racemization and/or saturation of an aromatic moiety. (*S*)-1-Amino-2-(methoxymethyl)indoline (SAMI) hydrazones were reported advantageous in terms of mild conditions for selective cleavage of the N-N bond of the corresponding hydrazines, as well as for the excellent levels of asymmetric induction [52]. The diastereoselectivity was rationalized using the model **59** (shown using a SAMP hydrazone), where chelation of the lithium atom by the

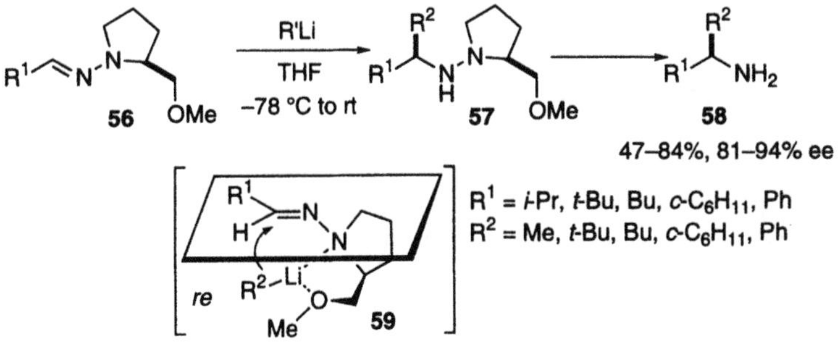

R'Li, THF, –78 °C to rt

56 → 57 → 58: 47–84%, 81–94% ee

R^1 = *i*-Pr, *t*-Bu, Bu, *c*-C_6H_{11}, Ph
R^2 = Me, *t*-Bu, Bu, *c*-C_6H_{11}, Ph

Scheme 17.

pyrrolidine nitrogen and the methoxymethyl side-chain restricts rotation around the N-N bond and increases the conformational rigidity [49c]. The organolithium reagent is then aligned below the C=N double bond plane and attacks from the *re*-face.

2.1.3
Addition to Oxime Ethers

Although oxime ethers are generally less electrophilic than the corresponding imines, the lability of the N-O bonds is appealing in terms of their transformation into amines. Asymmetric addition of organolithium reagents to oxime ethers having a chiral auxiliary gives chiral amines after the N-O bond cleavage of the resulting hydroxylamine ethers, and the latter can be achieved much more easily than cleavage of the amine C-N bond or the hydrazine N-N bond. The first oxime ether example of this type was reported by Miller; moderate diastereoselectivity was observed [53]. More general reactions have been developed by using oxime ethers derived from ephedrine [54] or chiral 1-phenylalkoxyamines [55] which showed good asymmetric induction in the addition of organolithium reagents in the presence of boron trifluoride. In the latter reaction with *O*-(1-phenylalkyl)oximes **60** (Scheme 18), the diastereoselectivity was dependent on the substituents on the chiral auxiliary [55c]. Replacement of the phenyl group by the bulkier naphthyl group decreased the selectivity from 71% de to 55% de (R=Me). In contrast, the larger R, isopropyl group gave the product **61** in >95% de, while the smaller R, methyl group gave **61** in 71% de. These results suggest that the addition proceeds via conformer **62**.

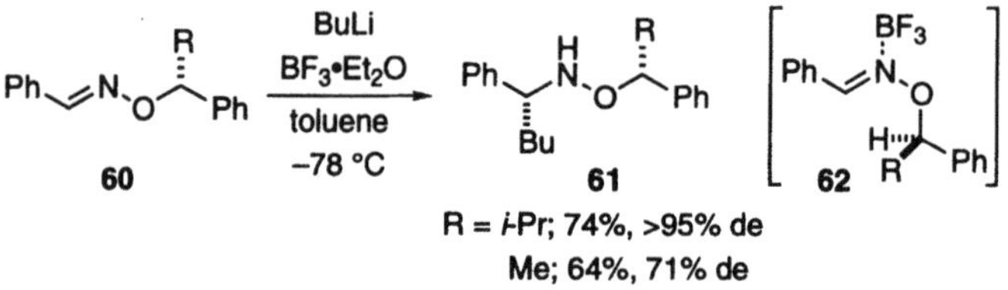

BuLi, $BF_3 \cdot Et_2O$, toluene, –78 °C

60 → **61**; **62**

R = *i*-Pr; 74%, >95% de
Me; 64%, 71% de

Scheme 18.

2.2
Addition to an Imine-Chiral Ligand Complex

Chiral *N*-borylimine **65** generated in situ by reduction of benzonitrile **63** with diisopinocamphenylborane **64** reacted with butyllithium to give the adduct **66** with 24% ee in 71% yield (Scheme 19) [56].

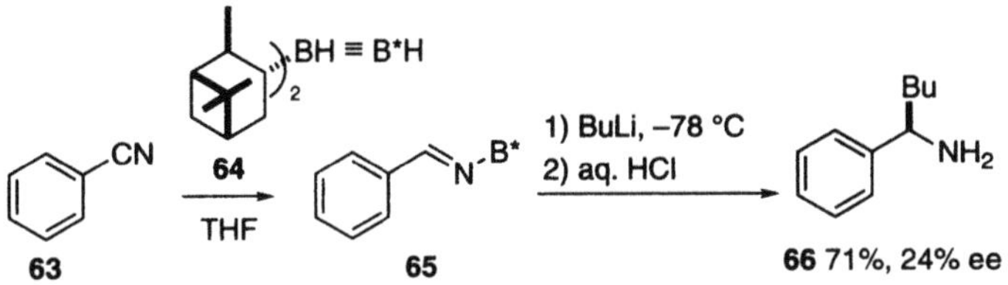

Scheme 19.

2.3
Addition of Organolithium Reagents Bearing a Chiral Sulfoxide Moiety

2.3.1
Addition to Imines

Another strategy for diastereoselective addition is the use of organolithium reagents bearing a chiral removable moiety. The addition of the lithium carbanion of (*R*)-methyl *p*-tolyl sulfoxide **67** to an imine **68** (R^1=R^2=Ph) afforded β-sulfinylamine **70** with high diastereoselectivity [57]. The addition of sulfoxide **67** to a variety of other imines proceeded in moderate to good diastereoselectivity (Scheme 20) [58]. The diastereoselectivity of the reaction was kinetically controlled, and the six-membered cyclic model **69** was proposed to rationalize the observed stereoselection. This methodology allowed the construction of (*R*)-tetrahydropalmatine in four efficient synthetic steps from 3,4-dihydroisoquinoline [58b].

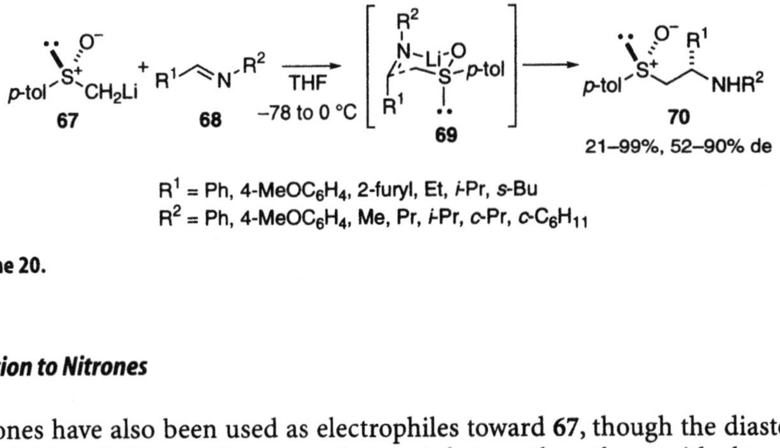

Scheme 20.

2.3.2
Addition to Nitrones

Nitrones have also been used as electrophiles toward **67**, though the diastereoselectivities of the reactions were less satisfactory than those with the corresponding imines [59, 60]. Addition of one equivalent of the lithium salt of quinidine (**72**) improves the diastereoselectivity dramatically for various 3,4-dihydroisoquinoline *N*-oxides **71** (Scheme 21) [60]. The formation of a facial discriminating reagent derived from quinidine (72) and α-sulfinylcarbanion **67** is likely responsible for the high selectivity. The sulfoxide and hydroxy groups are easily removable by hydrogenolysis, yielding the isoquinoline alkaloid **74** [R^1, R^2= OMe; (*R*)-(+)-salsolidine (**84**)].

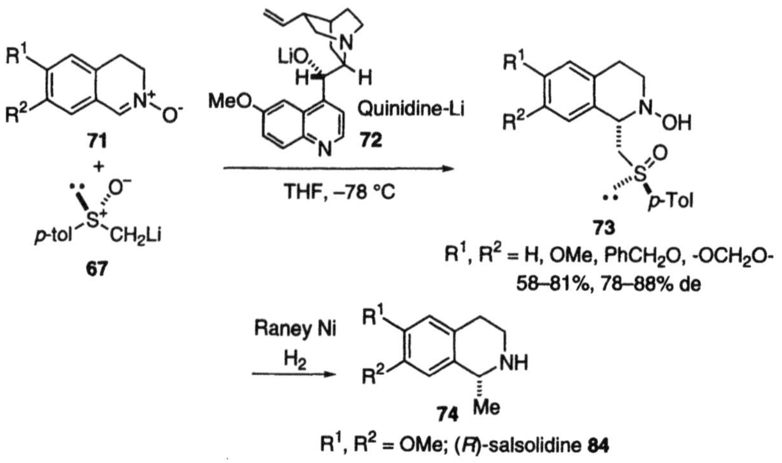

Scheme 21.

2.4
Reaction in the Presence of an External Chiral Ligand

2.4.1
Addition to Imines

In the last decade the asymmetric additions of organometallic reagents to the C=N bond of imines in the presence of a stoichiometric or catalytic amount of a chiral ligand have been developed as a new technology for the synthesis of optically active amines, including alkaloids. Such ligand-induced enantioselective synthesis avoids auxiliary attachment and removal steps. This strategy also has the potential for direct recovery and reuse of the unchanged chiral ligands. Addition to imines has been limited by the poor electrophilicity of the azomethine carbon atom, in comparison with that of a carbonyl group. However, the reaction can be considerably accelerated by the use of chiral ligands/catalysts. Several excellent articles reviewing the state-of-the-art of this reaction have been published [7, 9, 61–65]. In this section we describe the enantioselective addition to C=N of organolithium reagents in the presence of an external chiral ligand.

Tomioka and coworkers first reported the stoichiometric and catalytic asymmetric 1,2-addition of organolithium reagents to *N*-arylimines mediated by an external chiral ligand [66–69]. The chiral β-amino ether derivative **15** is an excellent asymmetric controller [70] which works as a catalyst even in substoichiometric amounts. Using organolithium reagents (R=Me, Bu, Ph, vinyl), aryl- or

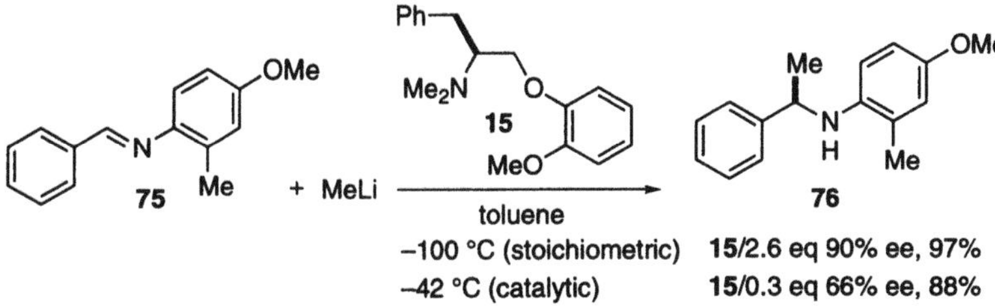

Scheme 22.

conjugated imines **75** were successfully converted into optically active amines **76** with high selectivities (Scheme 22).

The *N*-aryl substitution in the starting imines exerts a profound effect on enantioselectivities. Particularly in the case of addition to the imines bearing substituted *N*-4-methoxyphenyl or *N*-naphthyl groups **77**, excellent enantioselectivities (up to 97% ee) were obtained (Scheme 23) [71]. Reaction of heterocycle-involving imine and heterocyclic carbonucleophiles in the presence of a chiral ligand has been also reported [72, 73].

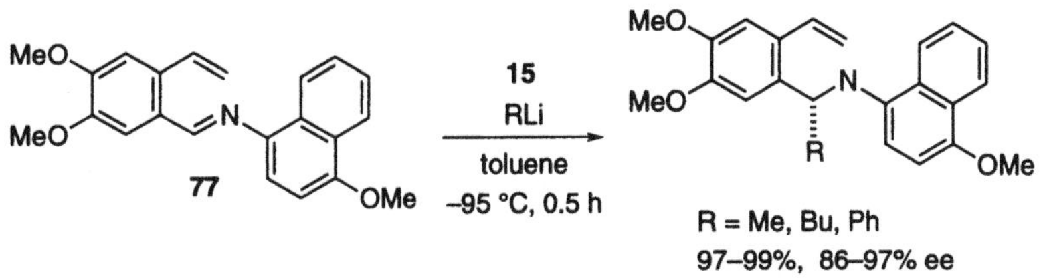

Scheme 23.

The synthetic utility of the enantioselective addition to an azomethine function relies on a practical method for the *N*-dearylation of the resulting chiral amines. The 4-methoxyphenyl groups of amines were removed by a two-step procedure, which begins with an *N*-protection step (BuLi/$ClCO_2CH_2Ph$) followed by oxidative cleavage of the *N*-aryl moiety (ceric ammonium nitrate: CAN). The *N*-naphthalene groups were removed by CAN oxidation followed by treatment with sodium borohydride and then acetic anhydride [71]. In both cases, the amines were isolated in good yield without significant loss of enantiomeric purity.

The highly selective asymmetric addition of organolithiums to acyclic imines and *N*-aryl group oxidative removal provided a facile and efficient synthetic route to the optically active 1-substituted tetrahydroisoquinoline (TIQ) **81** [74] (Scheme 24), (*R*)-(+)-salsolidine (**84**) [71] (Scheme 25), and optically pure α-amino acid derivatives **88** bearing a bulky α-substituent [75] (Scheme 26). The reaction of organolithium reagents with the acyclic imine **78** and subsequent cyclization of the secondary amine under Moffat oxidation conditions, after oxidative hydroboration of **79**, gave **80** (Scheme 24).

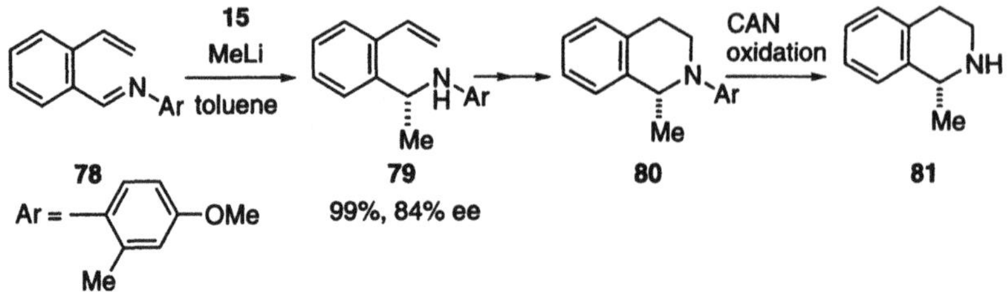

Scheme 24.

The synthesis of (*R*)-(+)-salsolidine (**84**), a representative isoquinoline alkaloid, was realized by employing the addition of methyllithium to *N*-naphthylimine **77**, cyclization of **83** and further elaboration for oxidative removal of the *N*-naphthyl group (Scheme 25).

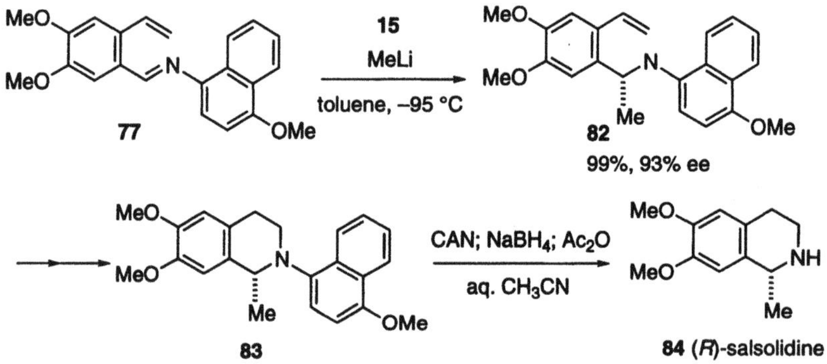

Scheme 25.

The 1,2-addition reaction of an anisidine imine **85** with phenyllithium was controlled by chiral diether **18** to give the corresponding secondary amine **86**. *N*-Dearylation followed by oxidative conversion of the phenyl group of **87** to a carboxyl group without racemization afforded the *N*-acetyl-α-amino acid **88** (Scheme 26).

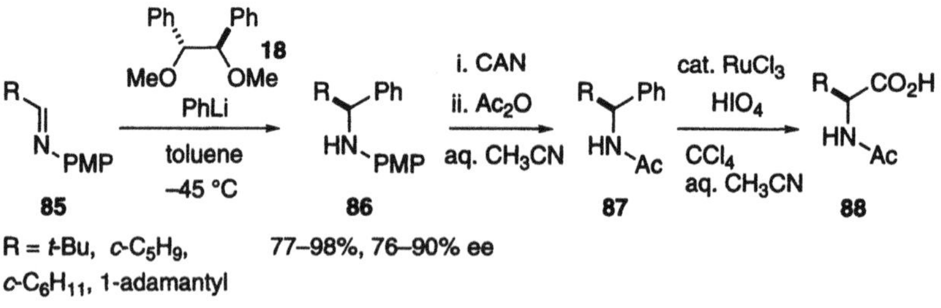

Scheme 26.

N-Silylimine **89** in ether at –78 °C was asymmetrically alkylated with butyllithium in the presence of the dilithium alkoxide of the chiral diol **93** (76%, 62% ee) (Scheme 27) [76]. Addition of the preformed (–)-sparteine (**19**)-BuLi complex to benzaldehyde *N*-diisobutylaluminoimine **90**, prepared in situ from partial reduction of benzonitrile with diisobutylaluminum hydride, in pentane at –78 °C gave the primary amine **92** in good ee (70% yield, 74% ee) [77]. The use of polymer-supported amino alcohol **94** in THF at –78 °C allows the asymmetric alkylation of an *N*-borylimine **91** to give the primary amine **92** with 44% ee [77].

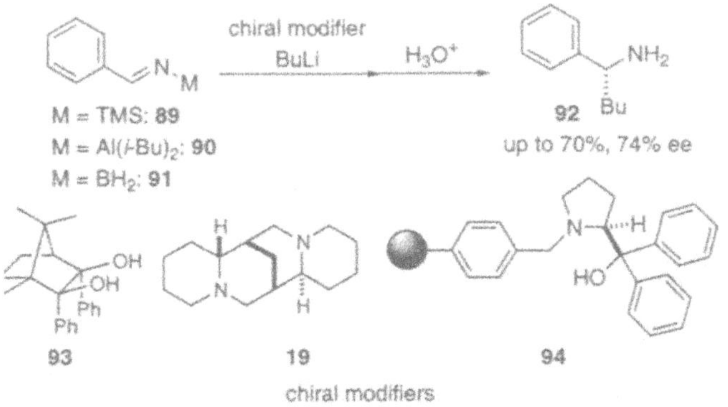

Scheme 27.

The bidentate C_2-symmetric bis-oxazolines **97** and (–)-sparteine (**19**) have been used as ligands with success in imine additions [78–80]. Excellent results have been obtained by addition of RLi (R=Me, Bu, Ph, vinyl) to aryl-, olefinic and aliphatic imines in the presence of (–)-sparteine (**19**) or chiral bis-oxazolines **97**. Notably, the enolizable aliphatic imines **95** also gave addition products **96** in high yields and high enantioselectivities (Scheme 28). The oxazoline **97** promotes both methyl- and vinyllithium additions with significant enantioselectivities (e.g., R=Me, 91% ee). On the other hand, **19** is beneficial for butyl- and phenyllithium additions (e.g., R=Bu, 91% ee). High enantioselectivities were obtained even when the chiral ligand was used in a catalytic amount. Investigations into the effect of the ligand architecture of the bis-oxazoline revealed that the size of the bridging substituents has a dramatic effect on the reaction selectivity, while ligand bite angle plays only a small role. The effect of the imine and ligand structure on the enantioselective addition mediated by (–)-sparteine (**19**) and bis-oxazolines has been studied [81].

An enantioselective addition of *i*-BuLi to methylimine **98** in the presence of the bis-oxazoline derivative **99** was used as the key step in the asymmetric synthesis of (*R*)-desmethylsibutramine (**100**), a single enantiomer version of a pharmacologically active metabolite of the anti-obesity drug sibutramine [82] (Scheme 29).

The asymmetric lithiation of *O*-benzyl carbamates **101**-imine addition sequence using the *s*-BuLi/(–)-sparteine (**19**) complex has been studied [83]. The reactions proceeded with high diastereoselectivity, giving the *threo*-β-amino alcohol derivatives **102** with modest to good ee (Scheme 30).

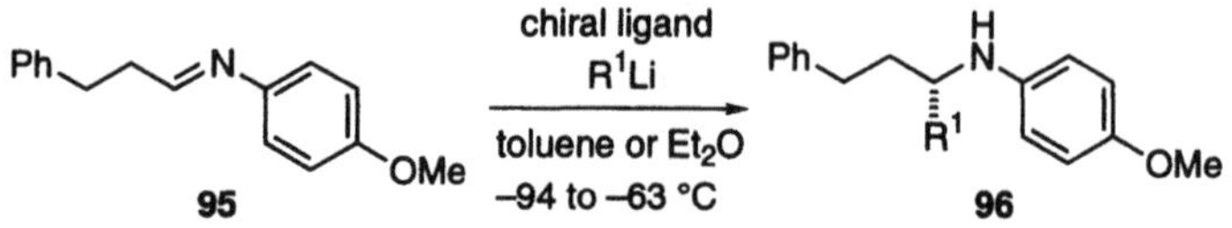

95 → 96: chiral ligand, R^1Li, toluene or Et_2O, −94 to −63 °C

R^1	chiral ligand/ eq	yield/%	ee/%
Me	**97**/1.0	96	91
Me	**97**/0.2	81	82
CH_2=CH	**97**/1.0	95	89
Bu	**19**/1.0	90	91
Ph	**19**/1.0	99	82

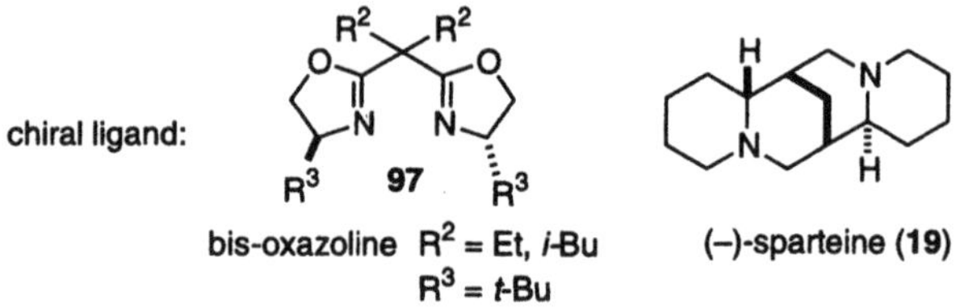

chiral ligand: **97** bis-oxazoline R^2 = Et, *i*-Bu; R^3 = *t*-Bu; (–)-sparteine (**19**)

Scheme 28.

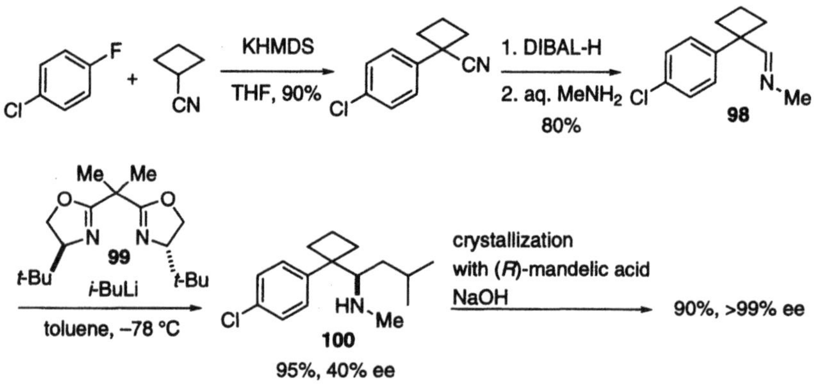

KHMDS, THF, 90%; 1. DIBAL-H, 2. aq. $MeNH_2$, 80% → **98**

99, *i*-BuLi, toluene, −78 °C → **100** 95%, 40% ee; crystallization with (*R*)-mandelic acid, NaOH → 90%, >99% ee

Scheme 29.

101: *s*-BuLi, (–)-sparteine **19**, Et_2O, −78 °C; then PhCH=N-PMP, −78 °C → **102**

84%, >90% de
56% ee (for major diastereomer)

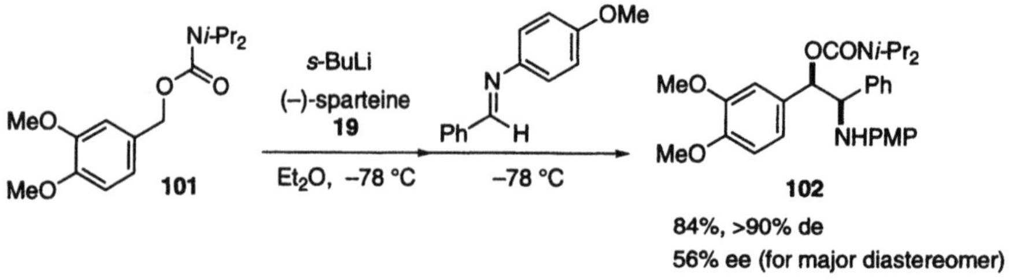

Scheme 30.

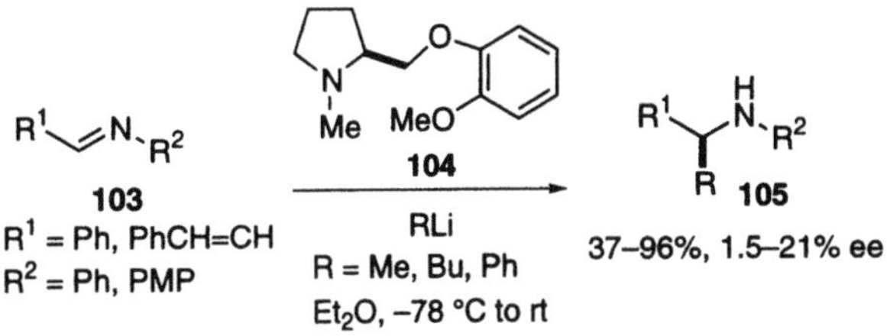

Scheme 31.

The (*S*)-proline-derived chiral ligand **104**, similar to the aminoether **15**, mediates the asymmetric addition of organolithium reagents to arylimines **103** producing (*S*)-amine **105** with relatively poor enantiomeric excess, up to 21% ee (Scheme 31) [84–86].

The ligand **108** has activity as an external controller of stereochemistry in the enantioselective addition of methyllithium to imines derived from veratraldehyde with up to 41% ee [87]. (–)-Sparteine (**19**)-mediated reaction of organolithium reagents afforded isoquinoline alkaloid **107** directly from 3,4-dihydroisoquinoline **106** with up to 47% ee (Scheme 32) [88].

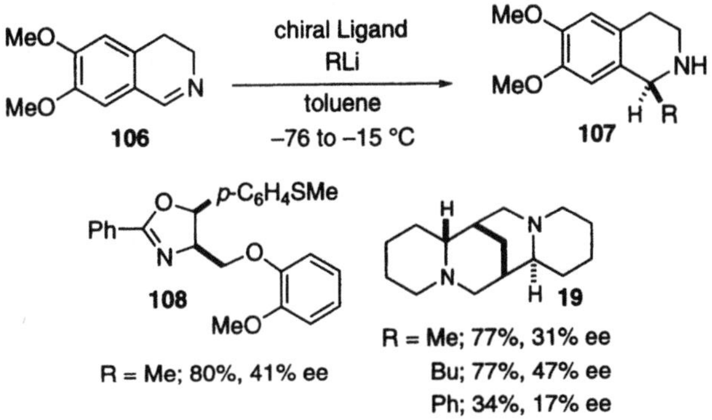

Scheme 32.

The lithium alkoxide of quinine (**111**) was used as a stoichiometric chiral additive to carry out the highly enantioselective addition of lithium acetylide **110** to cyclic *N*-acylketimines **109** (Scheme 33) [89]. Quinidine (**72**) (Scheme 21) was employed to give the opposite enantiomer. Using the bulky 9-anthrylmethyl protecting group at a distal position on the imine, adduct **112** of 97% ee was obtained and applied to the asymmetric synthesis of the HIV reverse transcriptase inhibitor **113**, through enantioenrichment of the (+)-CSA salt and deprotection of the 9-anthrylmethyl group.

Chiral bisaziridines such as **114** in the addition of organolithiums to an arylimine exerted asymmetric induction in up to 89% ee. The best result was obtained with vinyllithium (Scheme 34) [90].

Although outside the scope of this chapter, catalytic asymmetric reactions of lithium ester enolates with imines have been developed using a ternary complex

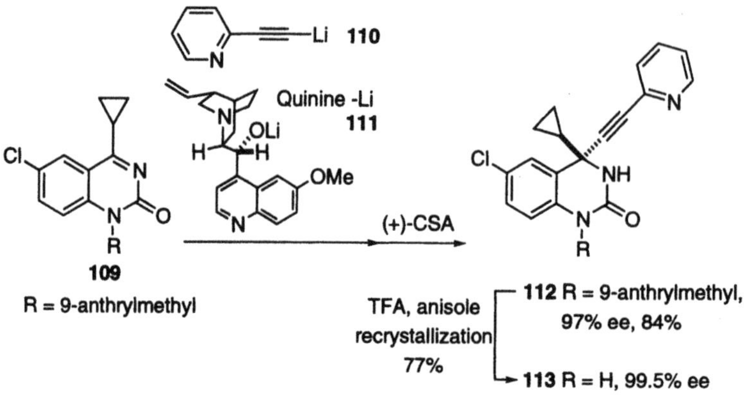

Scheme 33.

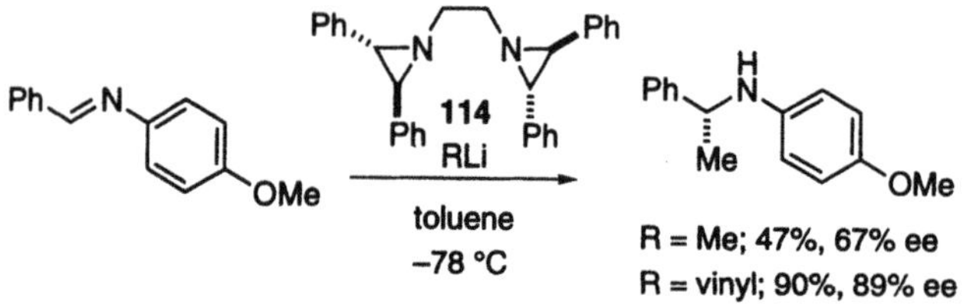

Scheme 34.

reagent, which comprises a chiral ether ligand, an achiral lithium amide, and a lithium ester enolate; these reactions leading to the corresponding β-lactams in high enantiomeric excesses [91].

3 Conclusions

Organolithium compounds are highly reactive and readily available organometallic reagents, thus the development of enantioselective reactions of organolithium reagents is ideal. Although conjugate additions of organolithiums are uncommon compared to organocopper-mediated conjugate additions, recent investigations of asymmetric conjugate additions of organolithiums have opened up a new, convenient and general methodology. Auxiliary-controlled conjugate addition has reached a high degree of diastereoselectivity based on chiral oxazoline chemistry. However, the approach towards enantioselective conjugate addition reactions of organolithiums with achiral activated olefins under control of an external chiral ligand or catalyst is a current focus of study. In the field of nucleophilic additions of organolithium reagents to the C=N bond of imine and imine congeners, considerable progress has been made during the last 30 years. Several natural products and biologically active compounds containing amino groups have been synthesized. Particularly, the past decade has witnessed impressive developments in external chiral ligand-mediated reactions. However, the level of enantioselectivity in the reactions with use of a catalytic amount of chiral ligand remains moderate. The search for a practical catalytic enantioselective addition to azomethine functionality remains a challenge in synthetic chemistry.

Note added in proof. The reactions of various (1-sulfinyl-2-naphthyl)methanimines with alkyllithium reagents have been examined: naphthylmethanimines bearing a 2,4,6-triisopropylphenylsulfinyl group gave the addition products as single diastereomers, possibly derived from the predominant rotamer around the C-S bond axis [92]. The reaction of chiral [1-(2,4,6-triisopropylphenylsulfiniyl)-2-naphthyl]methanimine with MeLi and subsequent elimination of the sulfinyl group afforded optically active 1-(2-naphtyl)ethylamine. The *R*- and *S*-enantiomers of salsolidine **107** (R=Me, Scheme 32) were prepared in good yield and moderate enantioselectivity (33 and 27% ee, respectively) by the addition of MeLi to 6,7-dimethoxy-3,4-dihydroisoquinoline **106** in the presence of chiral oxazoline ligands similar to **108** [93]. 1-Methyl-1,2-dihydroisoquinoline and 1-butyl-1,2-dihydroisoquinoline were obtained by enantioselective addition of organolithium reagents to isoquinoline in the presence of (−)-sparteine (up to 57% ee) [94]. Various organolithiums together with different chiral ligands were studied in additions to an azirine [3-(2-naphthyl)-2*H*-azirine]; however, only low ees were obtained [up to 17% ee using (−)-sparteine] [95]. Diastereoselective addition of organolithiums to chiral imidoylphenols to prepare phenolic Mannichtype bases has beeen reported [96].

References

1. Perlmutter P (1992) Conjugate addition reactions in organic synthesis, Tetrahedron Organic Chemistry Series, Vol. 9. Pergamon Press, Oxford
2. Tomioka K (1990) Synthesis 541
3. Rossiter BE, Swingle NM (1992) Chem Rev 92:771
4. Leonald J, Díez-Barra E, Merino S (1998) Eur J Org Chem 2051
5. Sibi MP, Manyem S (2000) Tetrahedron 56:8033
6. Krause N, Hoffmann-Röder A (2001) Synthesis 171
7. Jacobsen EN, Pfaltz A, Yamamoto H (1999) Comprehensive asymmetric catalysis. Springer, Berlin, Heidelberg, New York
8. Otera J (2000) Modern Carbonyl Chemistry. Wiley-VCH, Weinheim
9. Tomioka K, Hasegawa M (2000) J Syn Org Chem Jpn 58:848
10. Tomooka K (2001) J Syn Org Chem Jpn 59:322
11. Feringa BL, Pineschi M, Arnold LA, Imbos R, de Vries AHM (1997) Angew Chem Int Ed Engl 36:2620
12. Hayashi T, Sakai M, Miyaura N (1998) J Am Chem Soc 120:5579; Kuriyama M, Tomioka K (2001) Tetrahedron Lett 42:921; Kuriyama M, Nagai K, Yamada K, Miwa Y, Taga T, Tomioka K (2002) J Am Chem Soc, 124:8932
13. Morrison JD (1984) Asymmetric Synthesis, Vol 3 Part B. Academic Press, Orlando, Florida
14. Meyers AI, Whitten CE (1975) J Am Chem Soc 97:6266; Meyers AI, Smith RK, Whitten CE (1979) J Org Chem 44:2250
15. Barner BA, Meyers AI (1984) J Am Chem Soc 106:1865; Meyers AI, Roth GP, Hoyer D, Barner BA, Laucher D (1988) J Am Chem Soc 110:4611; Robichaud AJ, Meyers AI (1991) J Org Chem 56:2607
16. Meyers AI, Brown JD, Laucher D (1987) Tetrahedron Lett 28:5283
17. Frey LF, Tillyer RD, Caille A-S, Tschaen DM, Dolling U-H, Grabowski EJJ, Reider PJ (1998) J Org Chem 63:3120
18. Seebach D, Crass G, Wilka EM, Hilvert D, Brunner E (1979) Helv Chim Acta 62:2695
19. Tomioka K, Sudani M, Shinmi Y, Koga K (1985) Chem Lett 329
20. Cooke Jr MP (1986) J Org Chem 51:1637
21. Asano Y, Iida A, Tomioka K (1997) Tetrahedron Lett 38:8973

22. Asano Y, Iida A, Tomioka K (1998) Chem Pharm Bull 46:184
23. Nakajima M, Tomioka K, Iitaka Y, Koga K (1993) Tetrahedron 49:10793; Mckee BH, Gilheany DG, Sharpless KB (1992) Organic Synthesis 70:47
24. Tomioka K, Shindo M, Koga K (1993) Tetrahedron Lett 34:681
25. Shindo M, Koga K, Asano Y, Tomioka K (1999) Tetrahedron 55:4955; Shindo M, Koga K, Tomioka K (1999) Chem Pharm Bull 47:1318
26. Asano Y, Yamashita M, Nagai K, Kuriyama M, Yamada K, Tomioka K (2001) Tetrahedron Lett 42:8493
27. Xu F, Tillyer RD, Tschaen DM, Grabowski EJJ, Reider PJ (1998) Tetrahedron: Asymmetry 9:1651
28. Curtis MD, Beak P (1999) J Org Chem 64:2996
29. Park YS, Weisenburger GA, Beak P (1997) J Am Chem Soc 119:10537
30. Lim SH, Curtis MD, Beak P (2001) Organic Lett 3:711
31. Pippel DJ, Weisenburger GA, Wilson SR, Beak P (1998) Angew Chem Int Ed 37:2522
32. Tomioka K, Shindo M, Koga K (1989) J Am Chem Soc 111:8266
33. Shindo M, Koga K, Tomioka K (1998) J Org Chem 63:9351
34. Tomioka K, Shioya Y, Nagaoka Y, Yamada K (2001) J Org Chem 66:7051
35. Tomioka K, Okamoto T, Kanai M, Yamataka H (1994) Tetrahedron Lett 35:1891
36. Shindo M, Koga K, Tomioka K (1992) J Am Chem Soc 114:8732
37. Hiiro T, Kambe N, Ogawa A, Miyoshi N, Murai S, Sonoda N (1987) Angew Chem Int Ed Engl 26:1187
38. Amurrio D, Khan K, Kündig EP (1996) J Org Chem 61:2258
39. Kündig EP, Liu R, Ripa A (1992) Helv Chim Acta 75:2657
40. Nishimura K, Ono M, Nagaoka Y, Tomioka K (1997) J Am Chem Soc 119:12974; Tomioka K, Okuda M, Nishimura K, Manabe S, Kanai M, Nagaoka Y, Koga K (1998) Tetrahedron Lett 39:2141; Nishimura K, Ono M, Nagaoka Y, Tomioka K (2001) Angew Chem Int Ed 40:440; Nishimura K, Tomioka K (2002) J Org Chem. 67:431
41. (a) Takahashi H, Suzuki Y, Inagaki H (1982) Chem Pharm Bull 30:3160; (b) Takahashi H, Suzuki Y, Hori T (1983) Chem Pharm Bull 31:2183; (c) Suzuki Y, Takahashi H (1983) Chem Pharm Bull 31:2895
42. (a) Wu M-J, Pridgen LN (1991) J Org Chem 56:1340; (b) Higashiyama K, Inoue H, Takahashi H (1992) Tetrahedron Lett 33:235; (c) Glorian G, Maciejewski L, Brocard J, Agbossou F (1997) Tetrahedron: Asymmetry 8:355
43. Ukaji Y, Watai T, Sumi T, Fujisawa T (1991) Chem Lett 1555
44. Hashimoto Y, Takaoki K, Sudo A, Ogasawara T, Saigo K (1995) Chem Lett 235
45. Hashimoto Y, Kobayashi N, Kai A, Saigo K (1995) Synlett 961
46. Alvaro G, Savoia D, Valentinetti MR (1996) Tetrahedron 52:12571
47. (a) Kawate T, Yamada H, Yamaguchi K, Nishida A, Nakagawa M (1996) Chem Pharm Bull 44:1776; (b) Yamada H, Kawate T, Nishida A, Nakagawa M (1999) J Org Chem 64:8821
48. Takahashi H, Tomita K, Otomasu H (1979) J Chem Soc Chem Commun 668
49. (a) Enders D, Reinhold U (1986) Angew Chem Int Ed Engl 25:1109; (b) Enders D, Bartzen D (1991) Liebigs Ann Chem 569; (c) Enders D, Nübling C, Schubert H (1997) Liebigs Ann/Recueil 1089
50. (a) Enders D, Reinhold U (1995) Angew Chem Int Ed Engl 34:1219; (b) Enders D, Reinhold U (1996) Liebigs Ann 11
51. (a) Enders D, Lochtman R, Raabe G (1995) Synlett 126; (b) Enders D, Lochtman R (1997) Synlett 355
52. Kim YH, Choi Y (1996) Tetrahedron Lett 37:5543
53. Kolasa T, Sharma SK, Miller MJ (1988) Tetrahedron 44:5431
54. Dieter RK, Datar R (1993) Can J Chem 71:814
55. (a) Gallagher PT, Lightfoot AP, Moody CJ, Slawin AMZ (1995) Synlett 445; (b) Brown DS, Gallagher PT, Lightfoot AP, Moody CJ, Slawin AMZ, Swann E (1995) Tetrahedron 51:11473; (c) Gallagher PT, Hunt JCA, Lightfoot AP, Moody CJ (1997) J Chem Soc Perkin Trans 1 2633
56. Itsuno S, Hachisuka C, Kitano K, Ito K (1992) Tetrahedron Lett 33:627

57. Tsuchihashi G, Iriuchijima S, Maniwa K (1973) Tetrahedron Lett 36:3389
58. (a) Roman B, Marchalin S, Samuel O, Kagan HB (1988) Tetrahedron Lett 29:6101; (b) Pyne SG, Dikic B (1990) J Org Chem 55:1932; (c) Bravo P, Capelli S, Crucianelli M, Guidetti M, Markovsky AL, Meille SV, Soloshonok VA, Sorochinsky AE, Viani F, Zanda M (1999) Tetrahedron 55:3025
59. Pyne SG, Hajipour AR (1992) Tetrahedron 48:9385
60. Murahashi S, Sun J, Tsuda T (1993) Tetrahedron Lett 34:2645
61. Denmark SE, Nicaise OJC (1996) Chem Commun 999
62. Enders D, Reinhold U (1997) Tetrahedron: Asymmetry 8:1895
63. Bloch R (1998) Chem Rev 98:1407
64. Kobayashi S, Ishitani H (1999) Chem Rev 99:1069
65. Berrisford DJ (1995) Angew Chem Int Ed Engl 34:178
66. Tomioka K, Inoue I, Shindo M, Koga K (1990) Tetrahedron Lett 31:6681
67. Inoue I, Shindo M, Koga K, Tomioka K (1994) Tetrahedron 50:4429
68. Tomioka K, Inoue I, Shindo M, Koga K (1991) Tetrahedron Lett 32:3095
69. Inoue I, Shindo M, Koga K, Tomioka K (1993) Tetrahedron: Asymmetry 4:1603
70. Inoue I, Shindo M, Koga K, Kanai M, Tomioka K (1995) Tetrahedron: Asymmetry 6:2527
71. Taniyama D, Hasegawa M, Tomioka K (2000) Tetrahedron Lett 41:5533
72. Taniyama D, Kanai M, Iida A, Tomioka K (1997) Heterocycles 46:165
73. Tomioka K, Satoh M, Taniyama D, Kanai M, Iida A (1998) Heterocycles 47:77
74. Taniyama D, Hasegawa M, Tomioka K (1999) Tetrahedron: Asymmetry 10:221
75. Hasegawa M, Taniyama D, Tomioka K (2000) Tetrahedron 56:10153
76. Itsuno S, Yanaka H, Hachisuka C, Ito K (1991) J Chem Soc Perkin Trans 1 1341
77. Itsuno S, Sasaki M, Kuroda S, Ito K (1995) Tetrahedron: Asymmetry 6:1507
78. Denmark SE, Nakajima N, Nicaise OJC (1994) J Am Chem Soc 116:8797
79. Denmark SE, Nakajima N, Nicaise OJC, Faucher AM, Edwards JP (1995) J Org Chem 60:4884
80. Denmark SE, Sttif CM (2000) J Org Chem 65:5875
81. Arrasate S, Lete E, Sotomayor N (2001) Tetrahedron: Asymmetry 12:2077
82. Krishnamurthy D, Han ZX, Wald SA, Senanayake CH (2002) Tetrahedron Lett 43:2331
83. Arrasate S, Lete E, Sotomayor N (2002) Tetrahedron: Asymmetry 13:311
84. Jones CA, Jones IG, North M, Pool CR (1995) Tetrahedron Lett 36:7885
85. Jones CA, Jones IG, Mulla M, North M, Sartori L (1997) J Chem Soc Perkin Trans 1 2891
86. Jones CA, North M (1997) Tetrahedron: Asymmetry 8:3789
87. Brózda D, Chrzanowska M, Gluszynska A, Rozwadowska MD (1999) Tetrahedron: Asymmetry 10:4791
88. Chrzanowska M, Sokolowska J (2001) Tetrahedron: Asymmetry 12:1435
89. Huffman MA, Yasuda N, DeCamp AE, Grabowski EJJ (1995) J Org Chem 60:1590
90. Tanner D, Harden A, Johansson F, Wyatt P, Andersson PG (1996) Acta Chem Scand 50:361; Andersson PG, Johansson F, Tanner D (1998) Tetrahedron 54:11549
91. Fujieda H, Kanai M, Kambara T, Iida A, Tomioka K (1997) J Am Chem Soc 119: 2060; Kambara T, Hussein MA, Fujieda H, Iida A, Tomioka K (1998) Tetrahedron Lett 39:9055; Kambara T, Tomioka K (1999) Chem Pharm Bull 47:720; Tomioka K, Fujieda H, Hayashi S, Hussein MA, Kambara T, Nomura Y, Kanai M, Koga K (1999) Chem Commun 715; Kambara T, Tomioka K (1999) J Org Chem 64:9282; Hussein MA, Iida A, Tomioka K (1999) Tetrahedron 55:11219
92. Nakamura S, Yasuda H, Toru T (2002) Terahedron: Asymmetry 13:1509
93. Chrzanowska M (2002) Terahedron: Asymmetry 13:2497
94. Alexakis A, Amiot F (2002) Terahedron: Asymmetry 13:2117
95. Risberg E, Somfai P (2002) Terahedron: Asymmetry 13:1957
96. Cimarelli C, Palmieri G, Volpini E (2003) J Org Chem 68:1200

Topics Organomet Chem (2003) 5: 61–138
DOI 10.1007/b10337

Enantioselective Synthesis by Lithiation Adjacent to Oxygen and Electrophile Incorporation

Dieter Hoppe, Felix Marr, Markus Brüggemann

Westfälische Wilhelms-Universität, Organisch-Chemisches Institut, Corrensstraße 40, 48149 Münster, Germany. *E-mail: dhoppe@uni-muenster.de*

Enantioenriched sp^3-hybridized 1-oxy-alkyllithium compounds are accessible by lithiodestannylation of the appropriate chiral stannanes or by deprotonation of 1-alkyl carbamates by means of *sec*-butyllithium/(–)-sparteine. These are usually configurationally stable at temperatures below –40 °C and are substituted by a wide array of electrophiles with strict stereoretention. When applying chiral substrates, bearing an adjacent stereogenic center, often a high internal chiral induction occurs, being the basis for an efficient kinetic resolution in the deprotonation step. α-Oxybenzyllithium derivatives are usually more easily accessible due to mesomeric stabilization, but most of these compounds undergo facile racemization or epimerization at temperatures around –70 to –78 °C. The sense of stereospecificity of the electrophilic substitution is less predictable: both – retention or inversion – are common, depending on the individual situation. 1-Oxy-2-alkenyllithium reagents have similar stereochemical properties. In a number of cases configurational stability is recorded. Procedures for efficient dynamic kinetic resolution, involving a crystallization step, have been developed. In particular, 1-lithio-2-alkenyl carbamates – after titanation – are valuable homoenolate reagents for achieving highly stereoselective homoaldol reactions. Chiral 1-oxy-2-alkynyllithium derivatives allow for a facile entry to enantioenriched allenes.

Key words. Organolithiums, Carbamates, Chiral Ligands, Enantioselectivity, Deprotonation

Abbreviations

Ac	acetyl
Boc	*tert*-butyloxycarbonyl
BOM	benzyloxymethyl
Box	bis[oxazolinyl](-ligand)
Cb	*N,N*-diisopropylcarbamoyl
Cbse	*N*-[2-(*tert*-butyldiphenylsilyloxy)ethyl]-*N*-isopropylcarbamoyl
Cbx	spiro(4,4-dimethyl-1,3-oxazolidine-2,1'-cyclohexane)-3-carbonyl
Cby	(2,2,4,4-tetramethyl-1,3-oxazolidin)-3-carbonyl
DME	1,2-dimethoxyethane
DMF	*N,N*-dimethylformamide
DMPU	1,3-dimethyl-3,4,5,6-tetrahydro-2(1*H*)-pyrimidinone
inv.	inversion
LDA	lithium diisopropylamide
LDBB	lithium 4,4'-di(*tert*-butyl)biphenylenide
LDMAN	lithium 8-(*N,N*-dimethylamino)-naphthalenide
LN	lithium naphthalenide
m-CPBA	*meta*-chloroperbenzoic acid
MOM	methyloxymethyl
n.d.	not determined

Ox	oxazolidinyl
PDC	pyridinium dichromate
ret.	retention
r.s.	regioselectivity
SET	single electron transfer
TBS	*tert*-butyldimethylsilyl
TIPS	triisopropylsilyl
TMEDA	*N,N,N',N'*-tetramethylethylenediamine

1
Introduction and Scope

1-Oxygen-substituted alkyllithium derivatives of type **2** were the first "chiral carbanions" which were accessible in enantiopure form and proved to be configurationally stable, at least at temperatures below –40°C [Eq. (1)]: Still and Sreekumar cleaved (*R*)-1-(benzyloxymethoxy)-1-tributylstannylpropane (**1**) with *n*-butyllithium and trapped the intermediate lithium compound **2** [1] by dimethyl sulfate to give the BOM ether of (*S*)-2-butanol (**3**) [2, 3].

H_3C H SnBu$_3$ O O Bn (**1**) —[*n*-BuLi, THF, –78 °C, – Bu_4Sn]→ H_3C H Li O O Bn (**2**) —[Me_2SO_4]→ H_3C H CH_3 O O Bn (**3**) (1)

Overall, the two-step sequence proceeds with strict retention of configuration, and it was assumed that both the tin-lithium exchange and the methylation occur with retention. The starting material **1** had to be prepared by a four-step sequence including a chromatographic racemate resolution via diastereomers.

The substitution of chiral reagents of the more general type **4** by electrophiles leads to enantioenriched products **5**; thus reagents **4** serve as synthetic equivalents for chiral 1-hydroxyalkanides **6** [Eq. (2)].

R^1 R^2 OR^3 Li (**4**) + ElX ⟶ R^1 R^2 OR^3 El (**5**) R^1 R^2 OH ⊖ (**6**) (2)

In the following years, chiral carbanionic reagents of type **4** [4] became valuable reagents in enantioselective synthesis, mainly due to improved access to chiral stannanes of type **1** (Sect. 2.1) and – more importantly – when simple deprotonation procedures became available (Sect. 2.4).

Usually, the selectivities of both the individual steps – formation of the lithium intermediate and electrophilic substitution – cannot be monitored separately, and the product of the two selectivities is measured. Fortunately, it turned out, for most carbanion pairs and electrophilic reagents, that the substitution step is completely stereospecific (mostly retention, in some cases inversion), within the limits of detection. Larger errors only may occur when the rates of (diastereomer) interconversion are similar to the rates of electrophilic substitution (see Beak et al in this volume).

The utility of a chiral carbanionic reagent is mainly:

1. Access to stereochemically homogeneous reagents, either by stereospecific or highly stereoselective transformations of an already enantiopure precursor or a high chiral induction in the preparation from an achiral precursor to form an organometallic intermediate, which is configurationally stable under the reaction conditions. Generally high configurational stability is found below −40 °C for the sp^3-hybridized 1-oxyalkyllithium derivatives of type **4** (R^1, R^2=alkyl; R^1=alkyl, R^2=H).
   Configurational stability is decreased by adjacent mesomeric groups R^1 or R^2, such as phenyl, 1-alkenyl, or 1-alkynyl.
   For configurationally unstable "carbanions", a (thermodynamically driven) equilibration between diastereomeric isomers is the method of choice, if the difference in free energy is large enough (≥2 kcal·mol^{-1}). In rare cases, a dynamic kinetic resolution is possible by preferential crystallization of one diastereomer.
2. Sufficient reactivity towards the added electrophiles below the temperatures of carbanion decomposition or racemization. Usually, aldehydes, methyl iodide, allylic and benzylic bromides, and trialkylsilyl chlorides are the least problematic electrophiles in this context.
3. Convenient access to a broad array of related reagents in a predictable manner. Certainly, the deprotonation of a simple precursor by strong commercially available bases, such as alkyllithium reagents, is the first choice (see Sects. 2.4–2.5). Reductive lithiations are also useful, particularly in the diastereoselective generation of 2-lithio-tetrahydropyrans and 2-lithio-1,3-dioxanes (Sect. 2.1).

Following the first example published by Evans et al. [5], racemic 1-oxy-2-alkenyllithium reagents have subsequently found wide application as homoenolate reagents [Eq. (3)] [6, 7]. Electrophiles can attack the γ-position in the allylic lithium reagent **7** leading to the γ-adduct **9** in addition to the "normal" α-product **8**. Hydrolysis of the enol derivative **9** leads to aldehydes or ketones **10**. Thus, reagents **7** are equivalents for homoenolate anions **11**. The first configurationally stable enantioenriched allyllithium derivatives were discovered by Hoppe and Krämer [8], opening the door for enantioselective homoaldol reactions (Sect. 4).

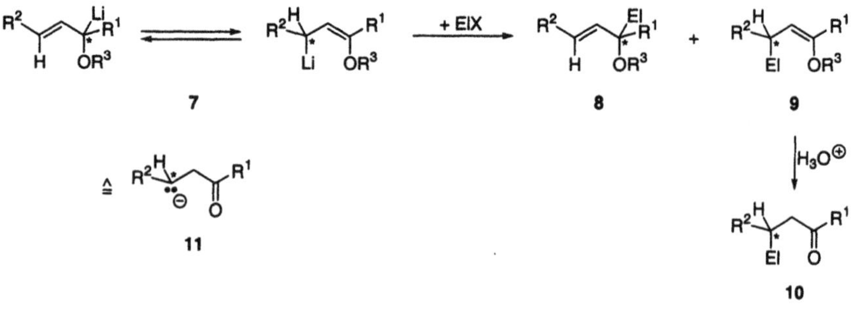

7 $\xrightarrow{+\,ElX}$ **8** + **9** $\xrightarrow{H_3O^{\oplus}}$ **10** ≙ **11** (3)

Similar issues of regioselectivity arise from the electrophilic substitution of 1-oxy-2-alkynyllithiums **12**, which can lead to mixtures of the alkyne **13** and allene **14** (Sect. 5). The first enantioenriched lithium compound of this type was described in 1991 by Hoppe and coworkers [9].

(4)

12 **13** **14**

2
Non-Mesomerically Stabilized, *sp*^{3}-Hybridized Lithium 1-Oxyalkanides

2.1
Preparation by Metal-Lithium Exchange

The classical access, developed by Still, requires the resolution of a racemic 1-trialkylstannyl-1-alkanol [2, 3, 10]. A great improvement was made by asymmetric reduction of acylstannanes **15**, which provides the stannyl alcohols **16** with approx. 80 to 96% ee [Eq. (5)] [11]. An alternative access, developed by Nakai, uses the nucleophilic ring opening of chiral α-stannyl acetals [12]. Another highly enantioselective approach starts from chiral (1-chloroalkyl)boronates [13].

(5)

15 (*S*)-BINAL-H **16** **17**

a R' = CH_2Ph (≙ BOM)
b R' = CH_3 (≙ MOM)

(1-Oxy-2-alkenyl)stannanes can be prepared similarly [11, 14, 15]; these are usually applied directly in carbonyl addition reactions.

The lithiodestannylation by butyllithium in THF proceeds with strict stereoretention, since the reaction starts with an attack at tin in compound **17** to form the unstable stannate intermediate **18** and tetraalkyltin is split off [Eq. (6)]. It is most likely that the equilibrium of the reaction is shifted towards the products due to favorable chelate formation, such as **19a** [16].

(6)

17 **18** **19**

L = THF

19a

Although carried out with achiral or racemic substrates, a comprehensive study by Macdonald and McGarvey uncovered several important features of the tin-lithium exchange reaction [16]. The order of thermodynamic stabilities is [Eq. (7)]:

MOMOCH$_2$Li > MOMOCH(alkyl)Li > H$_3$CLi > 1-(OMOM)cyclohexyllithium > *n*-BuLi > cyclohexyllithium (7)

The alkoxymethoxy substituent greatly contributes to the kinetic and thermodynamic stability of the lithium 1-alkoxyalkanides. However, the *O*-benzyllithium compound **21** undergoes [1,2]-Wittig rearrangement with a high rate, comparable to its formation [Eq. (8)].

20 →(*n*-BuLi, THF, –78 °C)→ [**21**] →(90%)→ **22** (8)

Further evidence for stereoretention in the lithiodestannylation step and in a subsequent methylation results from experiments with diastereomeric cyclohexyl derivatives. Transmetallation of the axial and equatorial stannanes *ax*-**23** and *eq*-**23** followed by methylation with dimethyl sulfate yielded ethers of the respective 1-methylcyclohexanols *ax*-**25** and *eq*-**25** without mutual contamination [Eq. (9)] [16].

ax-**23** →(*n*-BuLi, DME, –30 °C)→ *ax*-**24** →(Me$_2$SO$_4$, 81%)→ *ax*-**25**

eq-**23** →(*n*-BuLi, DME, –30 °C)→ *eq*-**24** →(Me$_2$SO$_4$, 90%)→ *eq*-**25** (9)

The reaction is less diastereoselective when using alkyl bromides and iodides, pointing to radicals being involved [via single-electron transfer (SET) processes] [16].

The addition of lithium 1-oxyalkanides **26a** to the carbonyl group of aldehydes and ketones usually takes place with very low enantiofacial differentiation, giving rise to mixtures of (half-protected) *syn*- and *anti*-diols **27** [Eq. (10a)] [16, 17]. A slight improvement is achieved in the presence of magnesium bromide [17]. Even the strained lithium compound **29**, bearing a lactone moiety, could be prepared via lithiodestannylation of **28** and was added onto the enal **30** to give the allylic alcohol **31** during the synthesis of the taxol ABC system [Eq. (10b)] [18].

Lithium 1-methoxyalkanides were found to add to aldehydes with high *syn*-diastereoselectivity after conversion to the corresponding tributylplumbanes **32** and a titanium tetrachloride-mediated addition step [Eq. (10c)] [19, 20]. Unfortunately, acetal-type *O*-protecting groups are not suitable in these Lewis acid-mediated reactions.

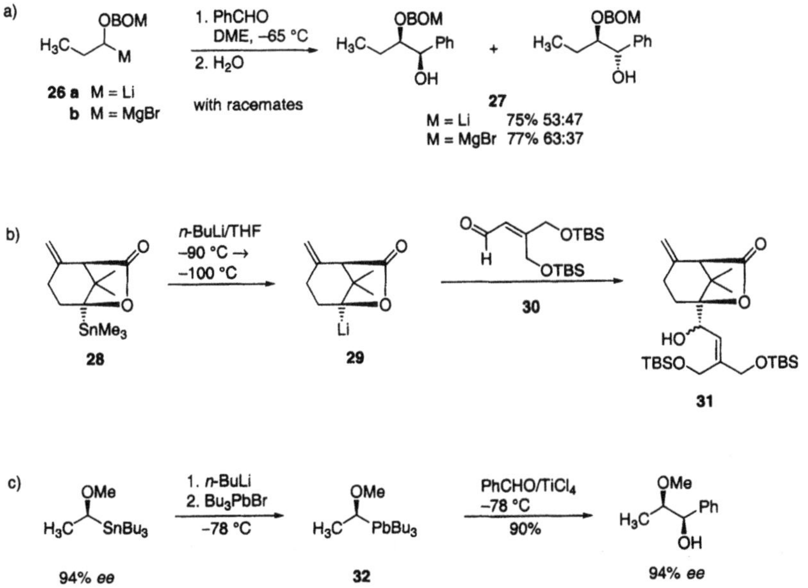

(10)

An alternative consists of acylation with *N,N*-dimethylamides and a subsequent diastereoselective reduction [17]. The sequence was applied in the enantioselective total synthesis of (+)-*endo*-brevicomin **33** [Eq. (11)] [21].

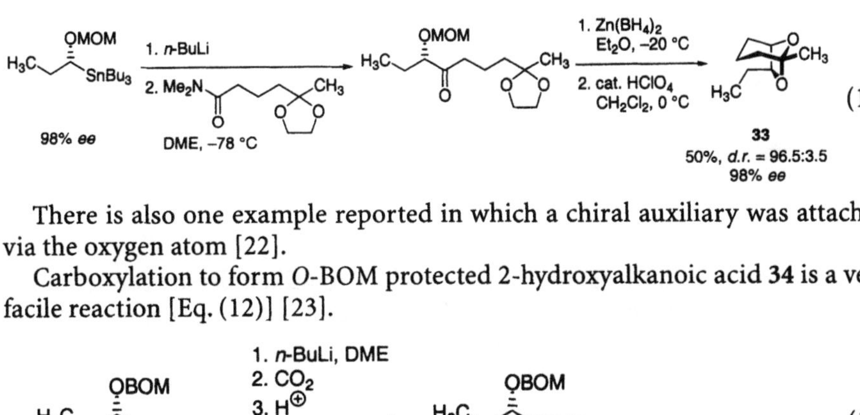

(11)

There is also one example reported in which a chiral auxiliary was attached via the oxygen atom [22].

Carboxylation to form *O*-BOM protected 2-hydroxyalkanoic acid **34** is a very facile reaction [Eq. (12)] [23].

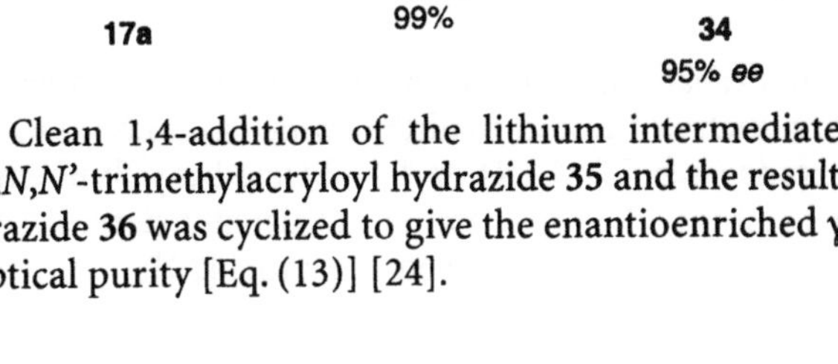

(12)

Clean 1,4-addition of the lithium intermediate was also observed onto *N,N,N'*-trimethylacryloyl hydrazide **35** and the resulting 4-MOMO-alkanoyl hydrazide **36** was cyclized to give the enantioenriched γ-lactone **37** without loss of optical purity [Eq. (13)] [24].

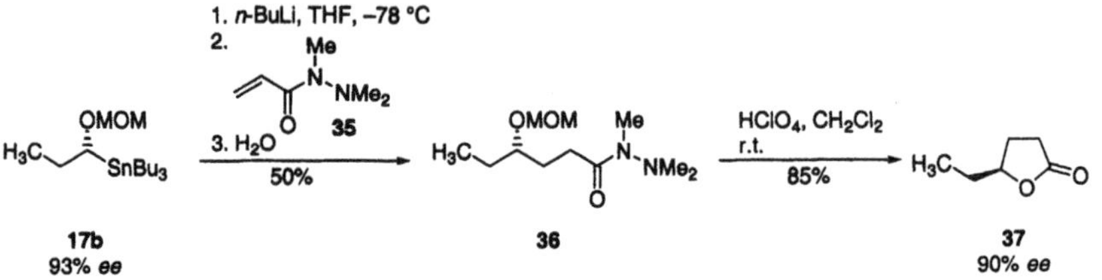

(13)

Attempts have been made to transform the enantioenriched lithium 1-alkoxyalkanides into cuprates to enable 1,4-addition onto 2-alkenones. The yields are good, however in most examples racemization is observed [25, 26].

In contrast, with cuprates derived from 2-lithiotetrahydropyrans and 4-lithio-1,3-dioxanes, bearing at least one further stereogenic center, the stereochemical integrity is maintained.

The lithium cuprates **39**, prepared from α- and β-2-deoxy-D-glucopyranosylstannanes α- and β-**38** are configurationally stable and provide the corresponding Michael addition products **40** on reaction with methyl vinyl ketone [Eq. (14)] [27]. The cuprates α-**39** [28] and **41** [29] have been used by Kocienski et al. for allylic substitution at η^3-molybdenum complexes.

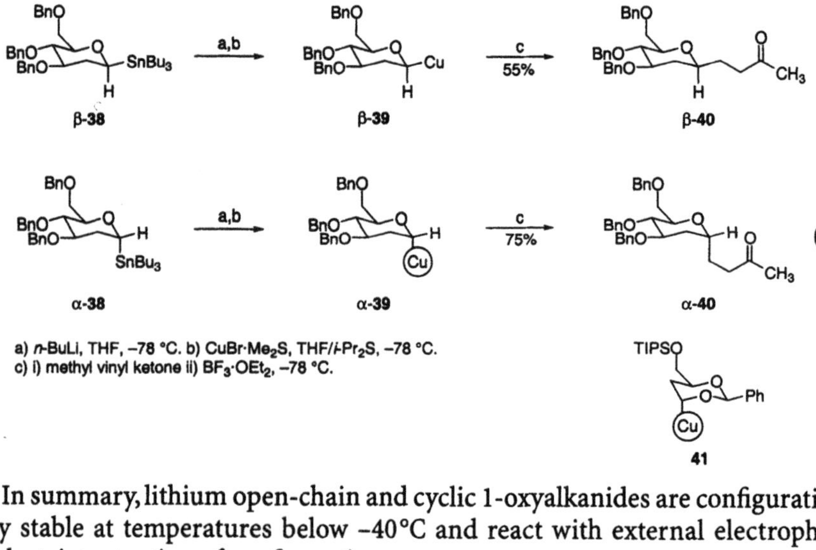

(14)

a) n-BuLi, THF, –78 °C. b) $CuBr{\cdot}Me_2S$, THF/i-Pr_2S, –78 °C.
c) i) methyl vinyl ketone ii) $BF_3{\cdot}OEt_2$, –78 °C.

In summary, lithium open-chain and cyclic 1-oxyalkanides are configurationally stable at temperatures below –40 °C and react with external electrophiles with strict retention of configuration.

2.2 Preparation by Reductive Lithiation

A powerful access to lithium carbanions is from phenylthio ethers and acetals using lithium radical anions such as lithium naphthalenide (**42a**, LN), 8-(*N,N*-dimethylamino)-naphthalenide (**42b**, LDMAN) or 4,4'-di(*tert*-butyl)biphenylenide (**42c**, LDBB) [Eq. (15)] [30–32]. The reaction is not usually stereospecific since a configurationally labile radical **43** is involved [Eq. (15)] [33].

(15)

When a pre-existing stereocenter is present, such as in a substituted tetrahydropyran **45**, very valuable stereochemical features are observed [Eq. (16)] [33].

(16)

Starting from each diastereomer of the racemic *trans*-fused 2-phenylthio-1-oxadecalin **45** on treatment with 2 equivalents of LDMAN (**42b**) in THF at –78°C, and subsequent trapping of the lithium intermediates **47** with benzaldehyde, the same ratio of alcohols *ax*-**48** and *eq*-**48** (95:5) is produced [33]. If the lithium compounds **47** are kept at −30 °C in the presence of TMEDA before aldehyde addition, the ratio *ax*-**48**:*eq*-**48** shifts to 13:87. It is generally found for the reductive lithiation of tetrahydropyrans and 1,3-dioxanes, that reduction of the radical **46** by a further radical anion and scavenging of the lithium carbanion pair preferentially proceeds from the axial face to produce the thermodynamically less stable axial lithiated heterocycle. On equilibration at −30 °C, the more stable equatorial epimer is formed; the addition step occurs again with retention of the configuration.

Similar results were recorded by Rychnovsky when reducing the β-*S*-phenyl-2-deoxythiopyranoside **49** [Eq. (17)] [34]. This approach had been reported previously by Sinaÿ and coworkers [35] and was utilized for efficient carbohydrate substitution [36, 37].

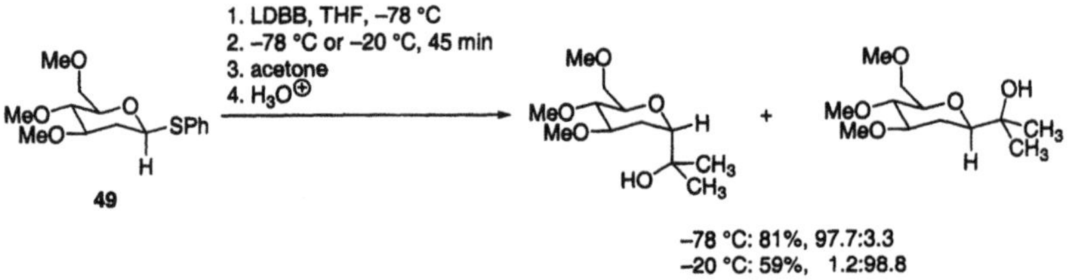

1. LDBB, THF, −78 °C
2. −78 °C or −20 °C, 45 min
3. acetone
4. $H_3O^{\oplus}$

49

−78 °C: 81%, 97.7:3.3
−20 °C: 59%, 1.2:98.8

(17)

In Sinaÿ's research group, the use of 2-deoxy-D-glucopyranosyl phenyl sulfones as carbanion precursors has been extensively investigated [38–40]. Deprotonation adjacent to the phenylsulfonyl moiety allows for the introduction of carbon electrophiles at the anomeric center. Reductive desulfonylation of **50** (with LN, **42a**) leads via the axial lithium compound **51** to the 1-β-substituted deoxyglucopyranose **52** following stereospecific protonation using methanol [Eq. (18)] [39].

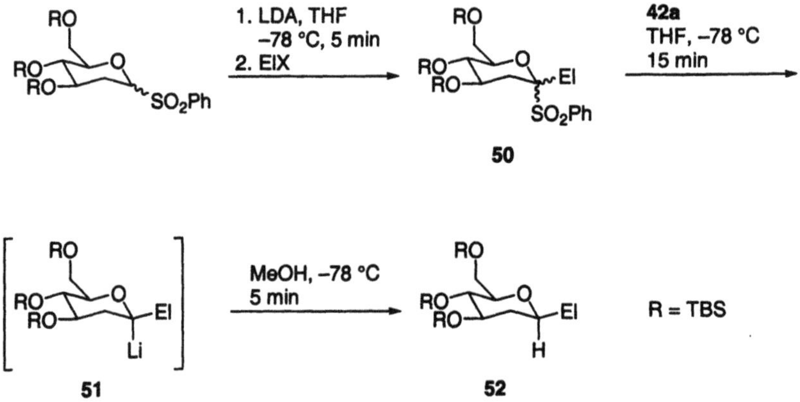

1. LDA, THF, −78 °C, 5 min
2. ElX

50

42a, THF, −78 °C, 15 min

51

MeOH, −78 °C, 5 min

52

R = TBS

(18)

The discussed methods are not applicable to pyranosides, bearing a protected 2-oxy group, because 1,2-elimination of alkoxide to form the glycal becomes the main reaction [37].

Glycosyl chlorides can also serve as substrates for reductive lithiations [28, 35]. Kessler et al. managed to "protect" a free 2-hydroxy group against elimination by deprotonation of the α-3,4,6-*O*-tribenzyl-D-glucopyranosyl chloride (**53**) before reductive lithiation [Eq. (19)] [41]; carboxylation of the α-lithio compound **54** and subsequent *O*-acylation provided the heptonic acid α-**55**. For the synthesis of the β-carboxylic acid β-**55**, lithium-tin exchange in the corresponding β-glucopyranosyltriphenylstannane **56** was the key step [41]. The methods could be extended to 2-amino-2-deoxy derivatives [42].

1. *n*-BuLi, THF, –78 °C; 2. **42a**
1. CO_2; 2. H_2O; 3. Ac_2O, pyridine; 57%
53 **54** α-**55**

(19)

1. *n*-BuLi, THF, –78 °C; 2. CO_2; 3. H_2O; 4. Ac_2O, pyridine
56 β-**55**

Rychnovsky et al. initially prepared 4-lithio-1,3-dioxanes by reductive lithiation of the 4-phenylthio derivatives **57** [Eq. (20)] [43, 44]. The stereochemical features resemble those of the 2-lithiotetrahydropyrans (vide supra). Again, the axial lithium compound *ax*-**58** is formed under kinetic conditions, but the equatorial diastereomer *eq*-**58** is favored by >4 kcal·mol^{-1}; this has been verified by quantum-chemical calculations at a high level [44]. Organocuprates derived from reagents such as *ax*-**58** behave well [44].

LDBB, THF, –78 °C
epimerization, –20 °C, 30 min
57 *ax*-**58** *eq*-**58**
H_3C–CO–CH_3, –78 °C

(20)

ax-**59** 78%, 98:2
eq-**59** 52%, 95:5

The reductive lithiation has been extended to 4-cyano-1,3-dioxanes such as **60** [Eq. (21)] [45]. Obviously, protonation of the axial lithium intermediate **63** is the most useful substitution reaction. Carbon-carbon coupling is accomplished at the stage of the nitrile anion such as **61**. Under carefully worked out conditions (lithium diethylamide, DMPU), equatorial double alkylation is observed. Decyanation by means of lithium in ammonia (**62**→**63**→**64**) proceeds with retention of configuration. The method offers also a powerful tool for the synthesis of 1,3,5,7-polyols by sequential 1,3-induction. These elegant steps have been applied successfully several times in the total syntheses of natural macrocylic polyhydroxy-containing antibiotics [46–52]; commentaries [53, 54].

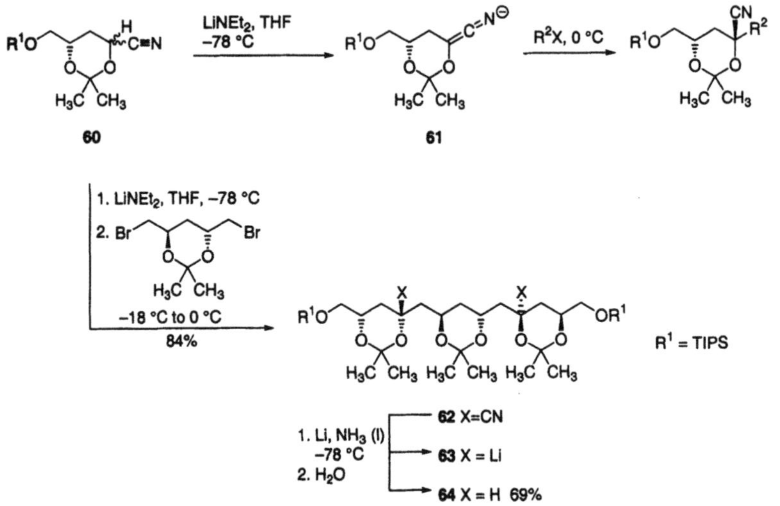

(21)

2.3 Preparation by Stereospecific Deprotonation

The combination of the enhanced *s*-character of the CH-bond by ring strain together with a strong inductively electron-withdrawing substituent leads to a sufficient acidification for the deprotonation of (*S*)-2-(trifluoromethyl)oxirane **65** which leads to the chemically and configurationally stable lithium compound **66** [Eq. (22)] [55]. The latter adds to carbonyl compounds in high yields and also reacts with triphenylsilyl and -stannyl chloride, and as well with methyl iodide with complete retention of stereochemistry in the product **67**.

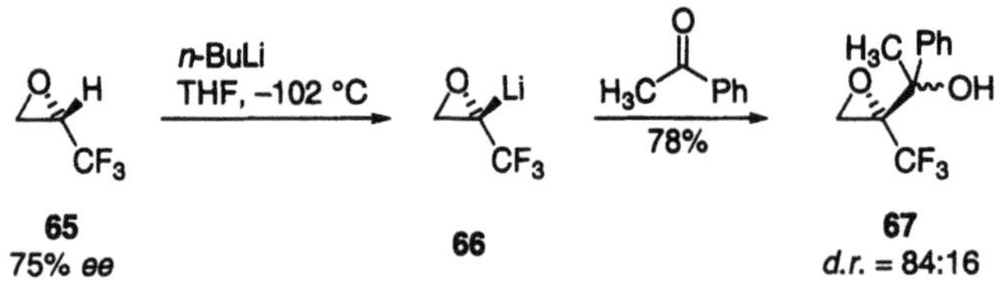

(22)

Cyclopropyl carbamates are sufficiently acidic to be deprotonated by *n*- or *sec*-butyllithium, leading to configurationally stable lithium compounds (Sect. 2.5.3) [56, 57].

2.4
Preparation by Diastereoselective Deprotonation

Dialkyl ethers and alkyl carboxylates, lacking mesomerically stabilizing substituents, cannot be deprotonated (due to insufficient acidity) under mild enough conditions which guarantee the persistence of the resulting "carbanions"; this is also true if the powerful Schlosser-Lochmann bases are used [58–61]. We were able to solve this problem by using alkyl *N,N*-dialkylcarbamates **68** [Eq. (23)] [62–64] in non-polar solvents (diethyl ether, pentane, toluene), with *sec*-butyllithium, in the presence of TMEDA. Under these conditions smooth removal of an α-proton proceeds with formation of the chelate complex **69**, which is trapped by electrophiles to give the racemic adducts **70**. We preferentially used 2,2,4,4-tetraalkyl-1,3-oxazolines as the amide component, because the aminoacetal is more easily cleaved to the alcohol **72** than in the corresponding *N,N*-diisopropylcarbamates.

s-BuLi, TMEDA, –78 °C, 4-6 h; ElX

68 R^1+R^1 = $(CH_2)_5$ (= *Cbx*); R^1 = CH_3 (= *Cby*) **69** **70**

CH_3SO_3H, MeOH; NaOH

71 **72** (23)

Acidic hydrolysis affords the hydroxyalkylurethanes **71**, which are cleaved under basic conditions with neighboring group participation by the hydroxy group. The use of less basic barium hydroxide, as proposed in ref. [62], is superior when chiral substrates, prone to racemization, are handled.

Secondary alkyl carbamates, such as the *O*-isopropyl or the *O*-cyclohexyl derivatives could not be deprotonated. However, the optically active 1-(trimethylsilyl)ethyl carbamate **73** reacts smoothly and is substituted with overall retention of configuration [Eq. (24)] [56].

s-BuLi, TMEDA, Et_2O, –78 °C; AcOD, 94%

73 > 95% *ee* **74** > 95% *ee* (24)

In the presence of (–)-sparteine, efficient enantiotopic differentiation is caused by this chiral external ligand (Sect. 2.5). In the current section we deal with internal chiral induction in chiral alkyl carbamates.

Two effects are operative in the kinetic diastereotopic selection during the deprotonation reaction of a chiral carbamate: Minimizing steric bulk in the transition state and, eventually, achieving favorable complexation of the lithium cation by further heteroatom substituents in the substrate. The first effect was studied with 2-arylpropyl-type carbamates **75** and **80** [Eq. (25)] [65]. Surprisingly, the reactions of the 2-phenylpropyl and the 1,2,3,4-tetrahydronaphth-1-yl derivatives (**75** and **80**) take opposite directions. After kinetically controlled deprotonation of *rac*-**75** with *sec*-butyllithium, followed by carboxylation and ester formation, the diastereomeric carboxylic esters *rac*-**78** and *rac*-**79** were formed in a ratio of 95:5. Since all reactions proceed with retention of the configuration and the lithium carbanions are configurationally stable, it is concluded that the removal of the *pro-S* proton from (*R*)-**75** to form **76**·TMEDA is about 20 times more rapid than formation of **77**·TMEDA. In (*R*)-**80**, however, preference for the abstraction of the *pro-R* proton is approx. 7 times greater than for the *pro-S* proton [Eq. (25)]. The origin of the difference results from the conformational restriction of substrate **80** by inclusion of the carbanionic center into the ring system. Even simple PM3 calculations on the transition state energies reflect this trend semi-quantitatively [66].

s-BuLi, TMEDA, Et_2O, –78 °C
rac-**75** → *rac*-**76** / *rac*-**77** (Li·TMEDA)
1. CO_2 2. CH_2N_2
→ *rac*-**78** : *rac*-**79** = 95 : 5

s-BuLi, TMEDA, Et_2O, –78 °C
rac-**80** → Li·TMEDA intermediates
1. CO_2 2. CH_2N_2
→ *rac*-**81** : *rac*-**82** = 13 : 87

(25)

(*S*)-2-(*N*,*N*-Dibenzylamino)alkyl carbamates **83**, prepared by a few synthetic steps from L-amino acids, are easily deprotonated by means of *sec*-butyllithium/TMEDA [Eq. (26)] [63, 67, 68]. A selection between the diastereotopic protons *pro-R*-H and *pro-S*-H takes place to give the lithium carbanions **85** and **87** which are formed in unequal amounts, and these are trapped with strict stereoretention by several electrophiles (MeOD, CO_2, $ClCO_2Et$, Me_3SiCl, R_3SnCl, MeI, aldehydes, ketones and acid chlorides). The substrate-induced stereoselection is best (approx. 9:1 in favor of the *pro-R* proton) when R is α -unbranched residue such as CH_3CH_2 or $C_6H_5CH_2$ of medium size leading to product **86** (Table 1) [68]. The opposite diastereomer **88** predominates slightly when R is CH_3, C_6H_5 or $(CH_3)_2CH$.

(26)

El = a D; b CO_2Me; c $SiMe_3$; d $SnBu_3$; e $SnMe_3$; f CH_3; g $PhCH_2$; h $(CH_3)_2COH$; i PhCO; j *i*-PrCO; k EtCO

The deprotonation step is kinetically controlled. The experimental evidence [68] leads to the conclusion that the dibenzylamino group does not act as a complexing Lewis base but by exerting its steric bulk. The relative rate of competing diastereomeric transition states is determined by the relative free energies of **TS84a‡** and **TS84b‡** (Fig. 1); calculations are in progress.

The deprotection of the hydroxy group proceeds smoothly by methanolysis in the presence of methanesulfonic acid, followed by KOH or by 5 N aqueous hydrochloric acid, followed by KOH, or in a one-step reductive deprotection procedure with $LiAlH_4$ in THF [68]. The removal of the *N*-benzyl groups by Pd-catalyzed hydrogenolysis is possible.

A more convenient *N*-protecting group for aminoalkyl carbamates, diphenylmethylidene, [69] was explored by Boie and Hoppe [Eq. (27)] [70]. Alkyl carbamate **89**, derived from (*R*)-2-amino-1-butanol was smoothly deprotonated by

Table 1. Substrate-induced diastereoselectivity in the deprotonation of carbamates **83** by *s*-BuLi/TMEDA

83	R	85:87[a]
b	CH_3	38:62
c	CH_3CH_2	88:12
d	$(CH_3)_2CHCH_2$	82:18
e	$(CH_3)_2CH$	40:60
f	$C_6H_5CH_2$	90:10
g	C_6H_5	40:60
h	$Bn_2N(CH_2)_3$	83:17

[a]Concluded from the average ratio of products **86** and **88** from several experiments.

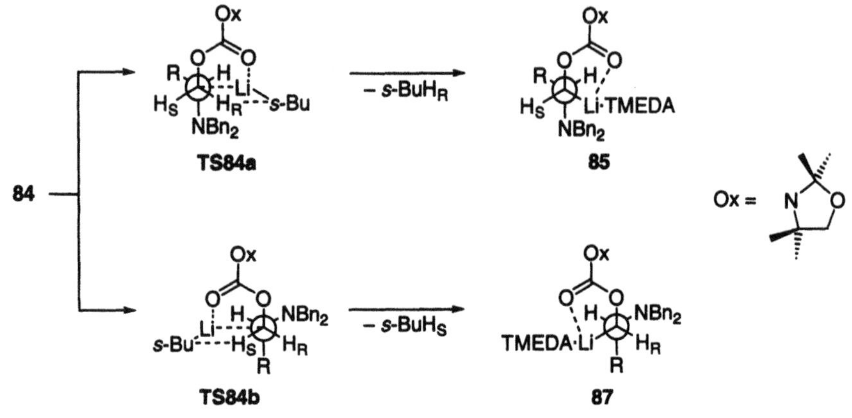

Fig. 1.

sec-BuLi/TMEDA in diethyl ether and trapped with electrophiles (such as trimethylsilyl chloride, trimethyltin chloride, *n*-alkyl iodides and pivaloyl chloride) to form the diastereomers **90** and **91** in yields between 34 and 91%. The ratios **90**/**91** (~3:1) are decreased in comparison to the corresponding *N*,*N*-dibenzylamines (9:1). Removing both protecting groups in **90**/**91** turned out to be possible by refluxing with 6 N hydrochloric acid in THF/water. The crude hydrochlorides were converted using benzoyl chloride to the benzamides **92**/**93** which could be separated by chromatography. Further examples demonstrate that the method is also applicable to the analogous phenylalaninol derivative [70]. Unfortunately, if the slower-reacting base (–)-sparteine/*s*-BuLi is applied, only low yields are obtained since addition of *sec*-butyllithium onto the phenyl rings in the imino moiety competes [70].

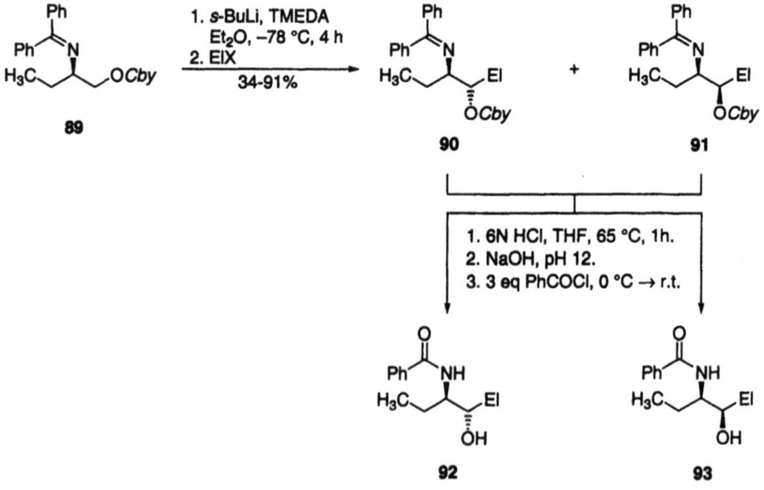

(27)

The *N*,*N*-dibenzylphenylalaninol derivative **94** offers a facile entry to stereochemically homogeneous α,δ-diamino-β,γ-alkanediols such as the bis-Boc-protected compounds **97** and **100**, which represent partial structures of some efficient HIV protease inhibitors [Eq. (28)] [71].

(28)

Deprotonation of **94** and subsequent addition of **95** to (*S*)-*N*,*N*-dibenzylphenylalaninal afforded a diastereomerically homogeneous adduct **96** (yield: 61%). Surprisingly, no diastereomeric impurities, arising from the epimer of type **91** could be detected. Most probably, double stereoselection [72] leads to a further enrichment via the matched pair. Acylation of **95** using (*S*)-*N*,*N*-dibenzylalanine benzyl ester yielded the ketone **98**; its reduction by $LiAlH_4$ gave rise to the epimeric alcohol **99**. Each compound was converted by standard procedures to the bis(Boc)diamino derivatives **97** and **100**, respectively. A number of constitutional analogues and stereoisomers could be prepared by combining substrates of type **95** with optically active aldehydes or esters in a highly stereoselective brick-box system [71].

Compared to open-chain 2-(*N*,*N*-dibenzylamino)alkyl carbamates, the (*S*)-*N*-benzylprolinol carbamate **101** exhibits atypical reactivity [Eq. (29)] [67, 73]. Deprotonation by *sec*-butyllithium in ether at −78 °C proceeds rapidly even if no complexing diamine [TMEDA or (−)-sparteine] is added, and scavenging of the lithium intermediate **102** by electrophiles forms diastereomerically pure products **103a–e** (Table 2). These are derived by substitution of the *pro-R* proton. When TMEDA or (−)-sparteine is added no change of the diastereoselectivity is observed. We conclude from these facts that the prolinyl nitrogen atom takes part in the deprotonation by complexing the lithium base, eventually giving rise to the bicyclic chelate complex **102a**. Obviously, the intramolecular chelation is supported by the more rigid conformation of the cyclic carbamate and the less shielded lone electron pair at the prolinyl nitrogen atom. Allylation leads to a diastereomeric mixture of **103g** and **104g**, presumably caused by single-electron transfer processes (SET) during alkylation.

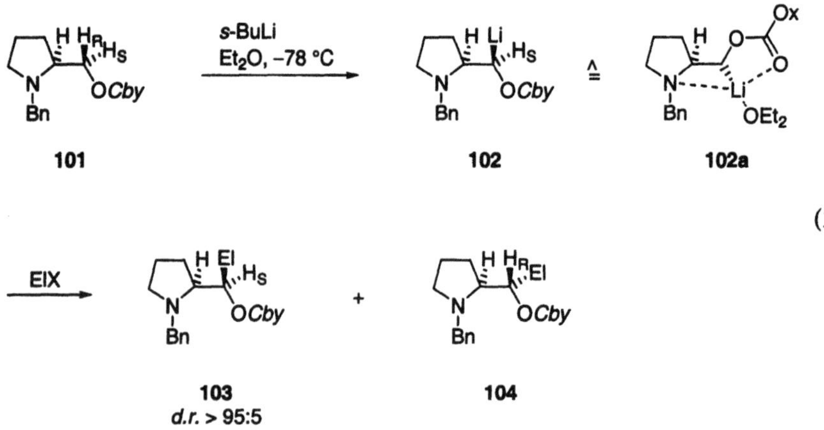

(29)

Table 2. [Table to Eq. (29)]

103	El(X)	Yield [%]
a	Me(I)	73
b	MeOCO(Cl)	84
c	Me_3Si(Cl)	33[a]
d	Bu_3Sn(Cl)	71[a]
e	Me_2C-OH	33[a]
f	*i*-PrCH-OH	85[a,b]
g	CH_2=CH-CH_2(Br)	82[a,c]

[a] TMEDA was added in these experiments.
[b] Diastereomeric mixture (*d.r.*=73:27) in respect to the carbinol moiety.
[c] *d.r.*=60:40

Similar features are observed when the homologous *N*-benzyl-2-piperidinemethanol carbamate *rac*-**105** was treated with achiral base [Eq. (30)] [73]. Again, one diastereomer *rac*-**108**, formed by an *ul*-process in the deprotonation step, was highly favored. In the presence of TMEDA, the yield is increased but the *d.r.* decreased. The intermediate *rac*-**106** arises from the abstraction of the *pro-S* proton in (*R*)-**105**, and the *pro-R* proton in (*S*)-**105**. Deprotonation under the influence of (–)-sparteine with concomitant kinetic resolution is a convenient access to enantioenriched material; **108** was produced with 94% *ee* (Sect. 2.5.4) [73, 74].

s-BuLi, TMEDA
Et_2O, –78 °C

rac-**105** → *rac*-**106** + *rac*-**107**

$ClCO_2Me$ → *rac*-**108** + *rac*-**109** (30)

without TMEDA 49%, *d.r.* = 92:8
with TMEDA 80%, *d.r.* = 85:15

In the lithiated dicarbamate **111** of (*S*)-2-(dibenzylamino)-1,4-butanediol (derived from L-aspartic acid) the 4-carbamoyloxy group also possesses a high tendency for intramolecular complexation [Eq. (31)] [75, 76]. The favorable equatorial positions of the dibenzylamino and 1-carbamate groups are displayed in the transition state of the deprotonation. When treated with *sec*-butyllithium in ether or THF, the bicyclic chelate complex **111** is formed exclusively by removal of the *pro-S*-1H atom. Trapping of **111** by many types of electrophiles gives stereohomogeneous substitution products **112** [Eq. (31), Table 3]. Since deprotection proceeds easily by the usual means, anion **111** constitutes a synthetic equivalent of the synthon **114**. No deprotonation in the 4-position was detected, however this can be achieved by "protecting" the *pro-S*-1H by conversion to deuterium (see below and Sect. 2.5).

s-BuLi, Et_2O or THF
–78 °C, 5 h

110 → **111** (+ Li·solvent species)

ElX → **112** + **113** *d.r.* > 98:2 **114** (31)

Table 3. Compounds **112** prepared from dicarbamate **110** via bis-chelate complex **111** [Table to Eq. (31)]

112[a]	El	ElX	Yield [%]
a	D	MeOD	96
b	Me	MeI	93
c	CO_2Me	CO_2[b]	92
d	MeCO	MeCOOEt	47
e	PhCO	PhCOCl	78
f	EtCO	EtCOCl	67
g	(*E*)-$CH_3CH=CHCO$	(*E*)-$CH_3CH=CHCOCl$	80
h	$MeCH(NBn_2)CO$	$MeCH(NBn_2)COOBn$	45
i	Me_2COH	Me_2CO	85
j	*i*-Pr_2COH	*i*-Pr_2CO	74
k	Ph_2COH	Ph_2CO	74
l	$C_5H_{10}COH$	$C_5H_{10}CO$	52
m	Me_3Si	Me_3SiCl	76
n	Me_3Sn	Me_3SnCl	75
o	PhS	PhSSPh	73
p	MeS	MeSSMe	53
q	PhSe	PhSeCl	61
r	Ph_2P	Ph_2PCl	56
s	Hex_2B	Hex_2BCl	52

[a] All products with *d.s.* and *r.s.* ≥98%.
[b] The crude acid was converted into the methyl ester by treatment with diazomethane.

Diastereoselectivity, surprisingly high for the reaction type (83:17 to 93:7), at the C=O group is observed in the addition of the chelate complex **111** onto achiral aldehydes [Eq. (32), Table 4] [76]. The major product **115** was deprotected to form the *N*,*N*-dibenzylaminotriol **117**.

1. *s*-BuLi Et$_2$O, −78 °C, 5 h 2. RCHO

CbyO OCby Bn$_2$N **110**

R OCby OCby HO NBn$_2$ **115** + R OCby OCby HO NBn$_2$ **116**

LiAlH$_4$, THF reflux, 16 h

R OH OH HO NBn$_2$ **117** R OH OH HO NBn$_2$

(32)

The importance of the donor group in the 4-position is evident from the following experiments: the 4-*O*-TBS 1-monocarbamate **118** could not be deprotonated at all [77], but the 4-*O*-methyl monocarbamate **119** is lithiated and substi-

Table 4. Addition products **115** and **116** from aldehydes [Table to Eq. (32)]

Aldehyde	R	Products	Combined yield [%]	Ratio 115:116
MeCHO	Me	**115a, 116a**	68	93:7[a]
EtCHO	Et	**115b, 116b**	75	88:12[a]
i-PrCHO	*i*-Pr	**115c, 116c**	74	83:17
t-BuCHO	*t*-Bu	**115d, 116d**	77	83:17
PhCHO	Ph	**115e, 116e**	78	84:16
c-PrCHO	*c*-Pr	**115f, 116f**	76	83:17

[a]Not separated.

tuted in the 1-position to form the product **120** with the usual high diastereoselectivity [Eq. (33), Table 5] [77].

CbyO OR NBn2
118 R = TBDMS
119 R = Me
s-BuLi, Et_2O, –78 °C, 5 h, R = Me
Bn_2N O-Me Li--L O O Ox
s-BuLi, Et_2O, –78 °C, 5 h, R = TBDMS
no deprotonation
ElX, –78 °C, 3 h → r.t.
El CbyO OMe NBn2
120

(33)

Table 5. Substitution products **120** via lithiation of functionalized 4-*O*-methyl monocarbamate **119** [Table to Eq. (33)]

120[a]	El	ElX	Yield [%]
a	CO_2Me	CO_2[b]	86
b	$SiMe_3$	Me_3SiCl	65
c	Me_2COH	Me_2CO	64
d	C_2H_5CO	C_2H_5COCl	64
e	(*E*)-$CH_3HC=CHCO$	(*E*)-$CH_3HC=CHCOCl$	78

[a]All products with *d.r.* ≥98:2.
[b] The crude acid was converted into the methyl ester by treatment with diazomethane.

In the 1-monosubstituted dicarbamates of type **112**, 4-deprotonation and 4-substitution are possible, but the induced diastereoselectivity is low and therefore the application of (–)-sparteine as an external source of stereodirecting power is required [76].

Highly diastereoselective, substrate-induced lithiation and substitution in the 4-position is possible when the 1-*Cby* group is exchanged for a non-complexing ligand and the amino group [Eq. (34)] is turned into a *N,N*-dimethylamino group [77]. Carboxylic ester **122** was obtained from the 4-carbamate **121**, albeit in modest yield [Eq. (34)].

TBSO, OCby, Me_2N H_S H_R — s-BuLi, Et_2O, −78 °C, 5 h → TBSO, O, Ox, Me_2N---Li---O, L — $ClCO_2Me$, 52% → TBSO, OCby, Me_2N CO_2Me (34)

121 **122** *d.r.* > 98:2

Similar trends are observed for lithiations of the (*S*)-2-(*N,N*-dibenzylamino)pentane-1,5-diol derivative [75]; the development of reliable procedures is in progress.

Oxygen groups with high chelating ability, when positioned at a stereogenic center in the 3- or 4-position of the alkyl carbamate, are able to direct efficiently the selection between diastereotopic α-protons. Use of (*S*)-1,3-butanediol dicarbamate **123** in the substrate-directed deprotonation leads preferentially to the *exo*-methyl-substituted bicyclic chelate complex **124** which is trapped by methyl iodide to form the *meso*-2,4-pentanediol dicarbamate (*R,S*)-**125** [Eq. (35)] [78].

H_3C, OCbx, CbxO H_S H_R — s-BuLi, TMEDA, Et_2O, −78 °C → H_3C, O, Ox, Li---O, Ox, O, OEt_2 — MeI, 44% →

123 **124**

H_3C, OCbx, CbxO, CH_3 ≙ H_3C, CH_3, CbxO, OCbx + H_3C, CH_3, CbxO, OCbx (35)

(*R,S*)-**125** (*S,S*)-**125**

d. r. = 98:2

Cbx =

Similarly, the homologous dicarbamate **126** gives, by the same reaction sequence, the chiral dicarbamate (*S,S*)-**127** in excess; ratio (*S,S*):(*R,S*)-**127**=88:12 [Eq. (36)] [78].

H_R H_S, H_3C, OCby, OCby — 1. s-BuLi 2. MeI, 60% → CH_3, H_3C, OCby, OCby + CH_3, H_3C, OCby, OCby (36)

126 (*S,S*)-**127** (*S,R*)-**127** *d.r.* = 88:12

In **123**, the internal chiral induction can be overridden by the external ligand (–)-sparteine (Sect. 2.5.4) [78].

An interesting carbanionic building block **129** serves as an enantiopure synthetic equivalent for the 1,3,4-trihydroxybutanide ion **133** and was prepared by substrate-directed deprotonation of (*S*)-3,4-*O*-isopropylidene-1,3,4-butanetriol carbamate **128** [Eq. (37)] [79], which was prepared in a few steps from (*S*)-malic acid. When *s*-BuLi in ether is used for deprotonation, subsequent electrophilic trapping provides almost diastereomerically pure (*d.r.*>95:5) products **130** (Table 6). The presence of TMEDA decreases the stereoselectivity to 75:35. Presumably, this bidentate ligand competes with the acetonide oxygen atom for lithium complexation. The substitution products **130** arise from the formal exchange of the *pro-S* proton by the electrophile. From this, the tricyclic chelate complex **129** is concluded to be the decisive intermediate. On the other hand, addition of (–)-sparteine, which supports the *pro-S*-selective deprotonation (Sect. 2.5) increases the diastereomeric ratio to greater than 99:1. The conformational rigidity of the dioxolane ring is important for achieving high diastereoselectivity, since the 3,4-dimethoxy derivative **134** reacts quite unselectively (*d.r.*=65:35) [79].

128
s-BuLi, Et_2O, –78 °C
129 131
ElX ElX
130 132
133 134

(37)

The (1-deuterated) diastereomers **135** are accessible via deuteration and TMEDA-assisted removal of the remaining proton [Eq. (38)] [79]. In other words, the extraordinarily high kinetic H/D isotope effect permits the utilization of deuterium as a "protecting group" against deprotonation.

Table 6. Deprotonation of the acetonide **128** and reaction with various electrophiles [Table to Eq. (37)]

Additive (L)	Product	El(X)	Yield [%]	130:132
Et_2O	**130a**	$Me_3Sn(Cl)$	63	98:2
TMEDA	**130a**	$Me_3Sn(Cl)$	71	75:35
(−)-sparteine	**130a**	$Me_3Sn(Cl)$	61	>99:1
(+)-sparteine	**130a**	$Me_3Sn(Cl)$	75	28:72
Et_2O	**130b**	Me(I)	70	>95:5
Et_2O	**130c**	MeOCO(OMe)	35	98:2
Et_2O	**130d**	HCO(OEt)	69	96:4
Et_2O	**130e**	*i*-PrCO(Cl)	57	>95:5
Et_2O	**130f**	$ElX{=}Ph_2CO$	76	>95:5
Et_2O	**130g**	(*E*)-$CH_3HC{=}CHCO(Cl)$	39	>95:5
Et_2O	**130h**	$MeCH(NBn)_2CO(OBn)$	51	>95:5
Et_2O	**130i**	ElX=-valerolactone	62	>95:5

128 —(1. *s*-BuLi, Et_2O; 2. MeOD)→ **(*S*)-[1D]-128** —(*s*-BuLi, TMEDA, −78 °C)→ —(Me_3SiCl, 45%)→ **135** (*d.r.* = 98:2) (38)

Although these deprotonations are kinetically controlled we were surprised to learn that simple semiempirical (PM3) calculations on the diastereomeric lithium intermediates are consistent with the observed selectivities [80]. Apparently, similar structural features determine the relative energies of the diastereomeric transition states and diastereomeric ground states in these internal chelate-directed lithiations. A further conclusion can be drawn (with more uncertainty): due to the favorable complexation of the lithium cation by four donor ligands, most intermediates (if not all) are monomeric and have a very low tendency for oligomerization.

2.5
Preparation by (–)-Sparteine-Assisted Deprotonation of Alkyl Carbamates

2.5.1
Scope of the Reaction with Achiral Alkyl Carbamates

As Hoppe, Hintze and Tebben published in 1990 [62], the alkyl esters of 2,2,4,4-tetraalkyl-1,3-oxazolidine-3-carboxylic acids are smoothly deprotonated by *sec*-butyllithium/(–)-sparteine [64, 81, 82] to form the lithium intermediate **140** [Eq. (39)]. The *pro-S* proton is removed with high selectivity over the *pro-R* proton (approx. $k_{rel} \approx 50:1$). (–)-Sparteine (as the free diamine base) is commercially available [83]. The enantiomer (+)-sparteine, which supports the removal of the *pro-R* proton, is less conveniently accessible: in gram-scale from bitter lupine seed, including a racemate resolution [84]. The chelate complexes **140** are configurationally stable below –40°C and do not interconvert with the epimers **141**; the reaction with many electrophiles yields the products **142** with strict stereoretention. It is likely that **138**, *s*-BuLi and (–)-sparteine associate in a complex **139** in which base and C-H bond are brought tightly together and, in addition, the deprotonation proceeds in the vicinity of the chiral ligand (–)-sparteine. The space left at the lithium cation by the *cis*-annulated six-membered ring of (–)-sparteine [right-hand peripheral piperidine ring in Eq. (39)] seems to be important for forming the reactive conformation of the deprotonation: it was observed that the C_2-symmetric diastereomer (–)-α-isosparteine **144** (Fig. 2), having both piperidine rings in *trans*-arrangements, is not able to support the deprotonation of alkyl carbamates [85].

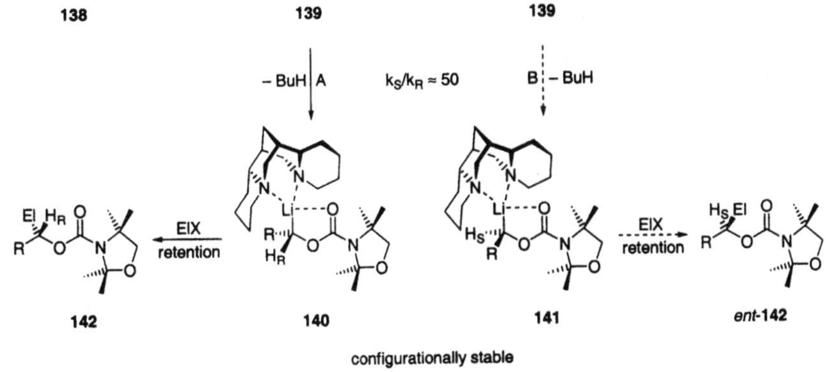

(39)

An unusually high kinetic acidity results for the protons at the methylene group, although the reaction is close to its thermodynamic limits. We were sur-

Fig. 2.

prised to learn that the slightly less basic *n*-butyllithium is not strong enough for achieving this deprotonation. It should be noted that *tert*-butyllithium also does not support the deprotonation, presumably for steric reasons (see below). Highly branched *N*-alkyl groups in the carbamate residue are required in order to prevent carbanionic attack at the carbonyl group. Unlike the *N,N*-diisopropylamino group, which we used in allyl carbamates to serve this latter purpose, the oxazolidine ring in **138/142** offers an aminoacetal moiety for acidic cleavage [86]. The 2,2,4,4-tetramethyl substitution in the *Cby* group gives rise to better resolved ^{1}H-NMR spectra than the initially applied (cyclohexane)spiro(oxazolidine) group (*Cbx*) group does [Eq. (35)].

Usually, stirring the alkyl carbamate with 1.5 equiv. of the chiral base in diethyl ether, toluene or pentane at –78 °C for 4 to 6 h gives the best results. THF is suitable only for non-enantioselective deprotonations, since it displaces (–)-sparteine at lithium [87]. The mechanistic features are discussed in Sect. 2.5.2. In order to demonstrate the scope of the reaction, a number of representative examples are collected in Eq. (40). A fair number of electrophiles have been introduced.

Unbranched and substituted alkyl carbamates, such as **146** or **147** [62, 88] do not cause any problems in the deprotonation step. Deuteration (with CH_3OD or dissolved CH_3CO_2D), methoxycarbonylation (gaseous CO_2, followed by diazomethane after work-up; or methyl chloroformate), alkylation with methyl iodide, substitution with trialkylsilyl chlorides, trialkyltin chlorides and even trimethyllead bromide, addition onto aldehydes and ketones, and acylation with acid chlorides or esters, all proceed without difficulties. Although a ketone is formed in the latter reactions, which is at least 15 orders of magnitude thermodynamically more acidic than the alkyl carbamate we never observed enolate formation, racemization or epimerization – with one exception: it occurred to some extent after formylation with formate esters [79].

Methylations with methyl iodide were observed to proceed with high yields and stereoselectivities. Longer-chain alkyl iodides failed in most attempts. Allyl bromide reacts smoothly – however, products of low enantioenrichment (see **146g**) result. We explain the fact by a single electron transfer (SET) during the alkylation. The intermediate formation of a mesomerically stabilized allyl radical supports the SET pathway [89]. A solution to this problem was most recently published by Taylor and Papillon who converted a lithio carbamate into the corresponding zinc cuprate prior to allylation [90]. Studies on the stereochemistry in a few metal-exchange reactions have been published by Nakai et al. [91].

H_3C OCby El

146a El = H,
b El = D, 86%, >96% *ee* [62]
c El = CO_2Me, 73% [62]
d El = $SiMe_3$, 86% [62]
e El = $SnMe_3$, 72% [88]
f El = $PbMe_3$, 61%, 93% *ee* [88]
g El = $CH_2CH{=}CH_2$, 60%, 42% *ee* [88]

$H_3C(CH_2)_6$ OCby El

147a El = CH_3, 61% [62]
b El = CO_2Me, 79% [62]

$H_3C(CH_2)_{11}$ OCby El

148 El = CH_3, 60%
98% *ee* [86]

OCby El

149a El = H
b El = CO_2Me, 52%
[85] [62]

OCby El

150a El = H
b El = CO_2Me, 0% [85]

OCby El

151 El = CH_3
64%, 92% *ee* [62]

Ph OCby El

152 El = CO_2Me
88%, > 95% *ee* [92]

OMe OCby El

153a El = H
b El = PhCO
85%, > 95% *ee*
[64] [94]

Fe OCby El

154 El = PPh_2
80%, > 95% *ee*
[64] [93]

(40)

Bn_2N OCby El

155a El = H
b El = CO_2Me, 56%, > 95% *ee*
c El = Bu_3Sn, 70%, > 95% *ee*
d El = Me_3Si, 70%, > 95% *ee*
e El = Me, 72%, > 95% *ee*
[67] [68]

Bn_2N OCby El

156a El = H
b El = CO_2Me, 94%, > 98% *ee*
c El = Me_3Sn, 93%, > 98% *ee*
d El = Me_3Si, 77%, 96% *ee*
e El = Me, 97%, > 98% *ee*
[95]

CbyO OCby El

157a El = H
b El = Me, 85%
> 95% *ee* [96]

CbyO OCby El

158a El = H
b El = CO_2Me, 80%
> 95% *ee* [96]

TBSO OCby El

159a El = H
b El = Bu_3Sn, 85%
> 95% *ee* [98]

NBn_2 CbyO OCby El

160aa n = 1, El = H
ab n = 1, El = CO_2Me, 85%, *d.r.* > 98:2
[75] [67]

The high directing effect of the O*Cby* group is best demonstrated with the 3-(2-methoxyphenyl)propyl carbamate **153a** [64, 94]. Although fairly acidic protons are available in the benzylic position and in the 3-position of the aryl residue, the deprotonation occurs exclusively in the α-position, leading after acylation by benzoyl chloride to the phenyl ketone **153b** in high yield and high enantiomeric purity.

The presence of 2- and 3-dibenzylamino [63, 67, 68, 95] and as well 3- or 4-O*Cby* [78, 96] or 4- or 5-TBSO groups [97, 98] (compounds **155a** – **159a**) does not interfere. The presence of a remote dibenzylamino group at a stereogenic center in combination with an ω-carbamoyloxy group in the dicarbamate **160a** does not decrease the *pro-S* selectivity in the (–)-sparteine-mediated deprotonation step [75, 77, 99].

2.5.2
Mechanism of Alkyl Carbamate Deprotonation and Substitution

From the experimental evidence is concluded that the ternary complex **139** is the decisive intermediate and the competition between the diastereomeric pathways A and B [Eq. (39)] determines the stereoselectivity by forming the diastereomeric complexes **140** and **141** with different rates in the magnitude $k_{HS}/k_{HR} \geq 50$.

The ethyl carbamate **146b**, obtained by (–)-sparteine-mediated lithiation and deuterolysis (performed twice) with a D-content of >99% and 98% ee, reacted extremely sluggishly with *sec*-butyllithium/(–)-sparteine and provided only traces of the expected silane **146d** [Eq. (41)] [56]. It is evident from this result that the electrophile enters the same topological position from which the proton was removed and, secondly, an unusually high kinetic isotope effect is operating. In order to estimate its magnitude **146b** was deprotonated by *sec*-butyllithium/TMEDA. Any discrimination must be caused by a kinetic isotope effect. After silylation of the intermediate (*R*)-**161**, the α-deuterated silane (*R*)-**162** was obtained with >96% ee and 98.2% of the original D-content. These data point to a value of ≥70 for k_H/k_D. We attribute the high isotope effect to efficient quantum chemical tunneling. A similar study was performed with the 2,2-dimethylpropane-1,3-diyl dicarbamate **158a** [56]. The high kinetic isotope effect has practical importance too: protons in the most acidic position have been protected by deuteration [75, 99]. Furthermore, it is – to our best knowledge – the first example for an efficient enantioselective synthesis which is based on kinetic isotope effects [100].

146a (H_R, H_S; CbyO, CH₃) —1. *s*-BuLi/sparteine, 2. MeOD→ 146b (H_R, D; CbyO, CH₃) —1. *s*-BuLi/sparteine, 2. Me₃SiCl ⇸ 146d (H_R, SiMe₃; CbyO, CH₃)

146b —*s*-BuLi/TMEDA→ route B: H_R, Li·TMEDA (CbyO, CH₃) —Me₃SiCl→ 146d (H_R, SiMe₃); route A: 161 (TMEDA·Li, D; CbyO, CH₃) —Me₃SiCl→ 162 (Me₃Si, D; CbyO, CH₃) > 96% *ee*, 98.2% D

(41)

Haller tried to model the diastereotopos-differentiation in the deprotonation step by MOPAC/PM3 calculations [64, 66]. The results were not really convincing, but give an idea why the *cis*-annulated ring in (–)-sparteine is so important. Close agreement between theoretical and experimental results was achieved for deprotonations, promoted by (*R*,*R*)-1,2-bis(dimethylamino)cyclohexane **145** (Fig. 2) [85]. The carbamates **146a, 149a** and **150a**, derived from ethanol, 2-

methylpropanol, or 2,2-dimethylpropanol were deprotonated with *sec*-butyllithium/(*R*,*R*)-**145** (1.45 equiv. each) in diethyl ether for 4h and the resulting carbanions quenched by gaseous CO_2. After work-up the crude acids were converted into the methyl esters **146c, 149b** or **150b** and the enantiomeric ratios determined [Eq. (42), Table 7]. The ees increase with the steric bulk of the alkyl residue and are raised up to 79% ee for **150b.** For **146a** and **150a,** PM3 calculations of $\Delta\Delta G^{\ddagger}$ of the transition states, leading to the ion-pairs, revealed energetic differences ΔH of 0.37 and 1.23 $kcal \cdot mol^{-1}$, respectively [85], which are in good agreement with the experimentally observed preference for *pro-S* proton removal.

146a R = Me
149a R = *i*-Pr
150a R = *t*-Bu
s-BuLi/(*R*,*R*)-**145**
Et_2O, –78 °C, 4 h
$-H_{pro\text{-}S}$
$-H_{pro\text{-}R}$
1. CO_2
2. $H_3O^{\oplus}$
3. CH_2N_2
146c R = Me
149b R = *i*-Pr
150b R = *t*-Bu
ent-**146c** R = Me
ent-**149b** R = *i*-Pr
ent-**150b** R = *t*-Bu

(42)

Table 7. [Table to Eq. (42)]

Starting material	Product	Yield [%]	*e.r.* (ee)	ΔH_{calc} [$kcal \cdot mol^{-1}$]
146a	146c/*ent*-146c	81	63:37 (26%)	0.37
149a	149b/*ent*-149b	96	77:23 (54%)	–
150a	150b/*ent*-150b	42	89.5:10.5 (79%)	1.23

Calculations on a higher level gave similar differences [85]. Quantum chemical simulations of the kinetically controlled formation of the appropriate (–)-sparteine complexes are in progress [101, 102].

2.5.3
Intramolecular Reactions of Lithiated Alkyl Carbamates

Lithiated alkyl carbamates rearrange on warming to form α-lithiooxycarboxamides. We noticed this and similar migrations as undesired side reactions [7, 103, 104]; the rearrangement has been investigated more closely by Nakai and coworkers [Eq. (43)] [105].

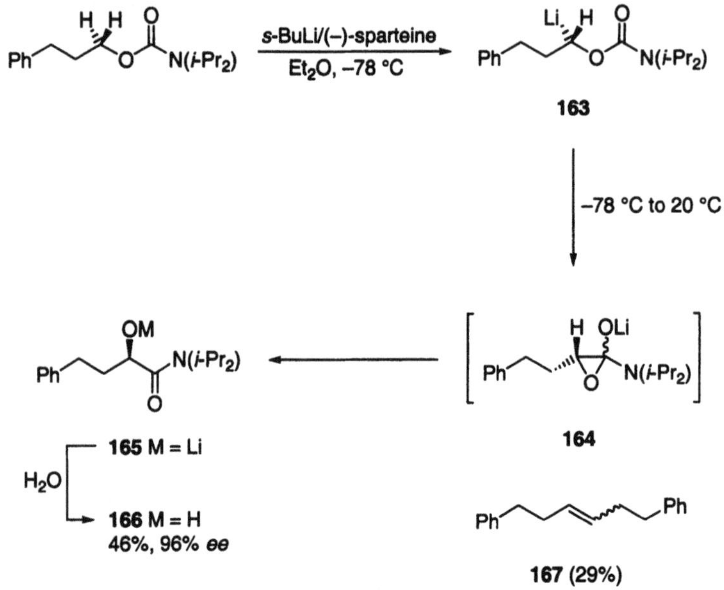

(43)

After warming the ethereal reaction mixture containing lithiocarbamate **163** to room temperature, the (*R*)-configured α-hydroxyamide **166** was isolated in 46% yield [105]. Overall retention in the rearrangement is best explained by assuming an intramolecular attack of the carbanionic center onto the carbonyl group via intermediate **164**. Interestingly, the alkene **167** (*E*/*Z*=86:14) was formed as a byproduct in 29% yield. In a formal sense it arises from a carbene dimerization [106, 107]. Lithiated 1,3-dicarbamates, such as **168**, were found to cyclize in an intramolecular substitution reaction forming cyclopropyl carbamates **169** [Eq. (44)] [56].

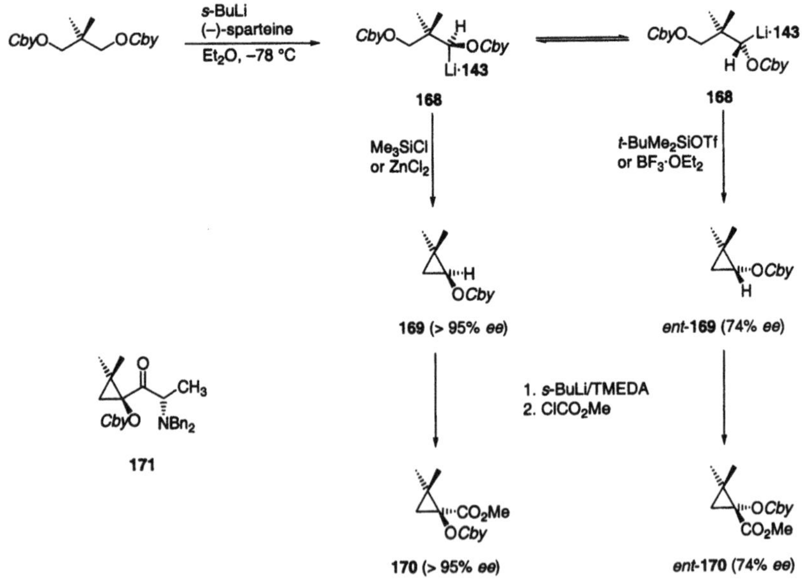

(44)

The cyclization step requires Lewis acid catalysis; lithium chloride, formed during an attempt at silylation, is sufficient. The stereochemical course under these conditions (retention at the carbanionic center, inversion at C-3) was rigorously proven by stereospecific deuteration, and an X-ray structure analysis of amino ketone **171**, as well [57]. Surprisingly, the use of the stronger Lewis acids *t*-$BuMe_2SiOTf$ or $BF_3{\cdot}OEt_2$ caused the formation of the opposite enantiomer *ent*-**169** [57]; the reasons are still unknown. The cyclopropanes **169** are easily deprotonated by *sec*-butyllithium/TMEDA and substituted by many electrophiles with complete retention [57]. The reactions are quite general; diastereomers are formed from 2-monosubstituted dicarbamates [108].

The strongly directing carbamate group allows for the presence of remote styryl or phenylethynyl groups during the enantioselective deprotonation step of the substrate. The phenyl group activates the double or triple bond for an intramolecular carbolithiation [109]. Since carbolithiation is the topic of Chapter 9, only few representative ideas and examples are discussed briefly below.

The (*Z*)- or (*E*)-phenylhexenyl carbamates **172** are smoothly deprotonated by *s*-BuLi/(−)-sparteine, and the lithium compound cyclizes during approximately 20 h at −78 °C to form the (cyclopentyl)benzyllithium **173** which is in equilibrium with its epimer **174** [Eq. (45)] [110]. Trapping this mixture yields the essentially enantiomerically and diastereomerically pure side-chain substituted *trans*-2-benzylcyclopentyl carbamates **175** in fair yields. Some of the intermediate **174** is lost due to 1,3-elimination resulting in formation of the achiral bicyclo[3.1.0]hexane derivative **176** [111, 112]. Related results have been reported by Nakai et al., when allowing the (*E*)-6-phenylhex-5-enyl *N*,*N*-diisopropylcarbamate to react under similar conditions [111].

s-BuLi/(−)-sparteine
Et_2O, −78 °C, 20-30 h

5-*exo-trig*

Li·(−)-sparteine

172a R = H
172b R = CH_3

ElX
inversion

173 **174** **175** **176**

(45)

A number of further suitable substituted δ-unsaturated carbamates underwent smooth enantioselective deprotonation/5-*exo-trig* and 5-*exo-dig* cyclization sequences, to form five-membered rings. In all examples, the 1-*pro-S* proton is removed by the chiral base with high selectivity, independent from the sense of a chiral center at C-4 [112–115]. These cyclization reaction can be extended to vinylogous phenylalkenes [115] and alkynes [115].

So far, we have not succeeded in a similar cyclization for six-membered ring formation. For the application of this method to the synthesis of enantioenriched indolizidines see Sect. 2.5.4 [74].

Due to the low acidity of alkyl carbamates it is often not advisable to carry out a deprotonation in the presence of sensitive groups. Here, the "stannyl trick" may be applied. The deprotonation-stannylation of a bifunctional carbamate is accomplished at an early stage and the stannyl group carried through the synthesis during creation of a sensitive acceptor group at the second functionality [98, 116].

For example, the 5-(*tert*-butyldimethylsilyloxy)pentyl carbamate **159a** was converted to the enantiomerically enriched (>95% ee) stannane **159b** via sparteine-mediated deprotonation. Then an allyl chloride unit was elaborated, finally, the (*S*)-lithium intermediate **179** was generated by lithiodestannylation. The (1*R*,2*S*)-2-vinylcyclopentyl carbamate **180** was produced with essentially complete enantio- and diastereoselectivity [Eq. (46)] [98]. Allyl chloride (**178**, H for Bu_3Sn) and epoxides do not survive direct lithiation [117].

TBSO–(CH2)5–OCby (**159a**) → 1. *s*-BuLi/(–)-sparteine, –78 °C, Et2O; 2. *n*-Bu3SnCl, –78 °C to 20 °C; 85% → **159b** (SnBu3) → 1. TBAF, Et2O, 20 °C; 2. Swern oxidation; 93% → aldehyde (SnBu3, OCby) → 1. (EtO)2P(O)CH2CO2Et, DBU, LiCl, CH3CN, 20 °C; 2. DIBALH, THF, –78 °C; 3. KHMDS, MsCl, LiCl, THF, –78 °C to 20 °C; 85% → **178** → *n*-BuLi, Et2O, –78 °C → [**179**] → – LiCl; 96%; *d.r.* > 97:3; > 95% *ee* → **180**

(46)

2.5.4
Kinetic Resolution and Desymmetrization

When a chiral compound, bearing diastereotopic protons, is subjected to a deprotonation reaction in the presence of a chiral inductor, one of two situations will occur:

-The internal and the external chiral induction each activate the same proton; they are "matched" [72] and enforce each other.

-The chiral inductions oppose; they are "mismatched" [72] and weaken each other. A lower, or an opposite diastereoselectivity is the result.

Deprotonation of the (*S*)-phenylalaninol derivative (*S*)-**181** in the presence of the achiral ligand TMEDA produces the diastereomers **182** and **183** in a ratio of 90:10 [Eq. (47, see Sect. 2.4) [68]. Here, due to internal substrate-inherent induction, the *pro-R*-H is removed preferentially (*ul* induction [118]). In the presence of (–)-sparteine, which has a high preference for the *pro-S* protons (*lk* induction), the diastereomer **183** is formed with an opposite 90:10 diastereoselectivity; obviously the substrate-inherent diastereoselectivity is overridden by the reagent-induced selectivity.

(*S*)-**181** —(s-BuLi/L_2)→ **182** + **183** (47)

L_2	182	:	183
TMEDA	90	:	10
(–)-sparteine	10	:	90

In several examples, such as the (*S*)-valinol derivative [68], (–)-sparteine prevents the mismatched reaction path completely; no deprotonation occurs. This is an ideal situation for the kinetic resolution of a racemic sample such as *rac*-**184** [Eq. (48)] [119].

(*R*)-**184** —(s-BuLi, (–)-sparteine)→ **185** —(ElX)→ **186**; (*R*)-**184** ⇢ **187** ⇢(ElX) **188**

\+ (*S*)-**184** ⇢(s-BuLi, (–)-sparteine) *epi*-**185** ⇢(ElX) *ent*-**188**; (*S*)-**184** ⇢ *epi*-**187** ⇢(ElX) *ent*-**186** (48)

R = cyclopropylmethyl; ElX = CO_2, *t*-BuCOCl, *i*-PrCOCl

The cyclopropylalaninol derivative *rac*-**184** was subjected to a deprotonation by *s*-BuLi/(–)-sparteine (ether, 7 to 10 h at –78°C) and the intermediate sparteine-lithium compounds **185/187** were quenched by CO_2 or acid chlorides to yield mixtures of diastereomers **186/188** in a ratio of approx. 90:10 with greater than 95% ee in an average yield of 37%. (*S*)-**184**, 42%, 80–85% ee, was reisolated. As expected, the *pro-S* proton in (*R*)-**184** was abstracted preferentially, since it leads to the matched pair.

Efficient kinetic resolution was also observed in reactions of carbamates *rac*-**75** and *rac*-**80** by means of *sec*-butyllithium/(–)-sparteine (see also Eq. 25) [65, 66]; Fig. 3 shows the more reactive enantiomer (*R*)-**75** and (*S*)-**80**, respectively; the substitution products, arising from the abstraction of the *pro-S* proton were formed with ≥95% ee. PM3 calculations correlate well with the experimentally recorded – opposite – selectivities [66].

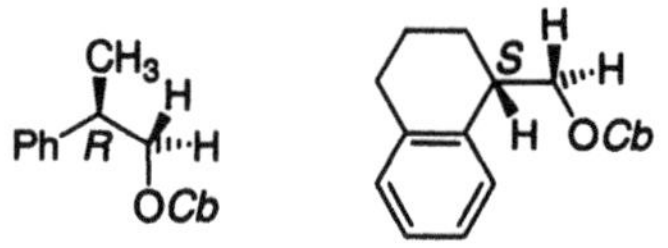

75 **80**

Fig. 3.

Dicarbamate **189** bears enantiotopic branches each having diastereotopic protons. Deprotonation by *s*-BuLi/(–)-sparteine, followed by treatment with $BF_3{\cdot}OEt_2$, results in the *trans*-cyclopropyl carbamate **191** in 76% ee [108]. From this result it is concluded that the preferential intermediate **190** stems from removal of the *pro-S*-H in the *pro-S* branch [Eq. (49)].

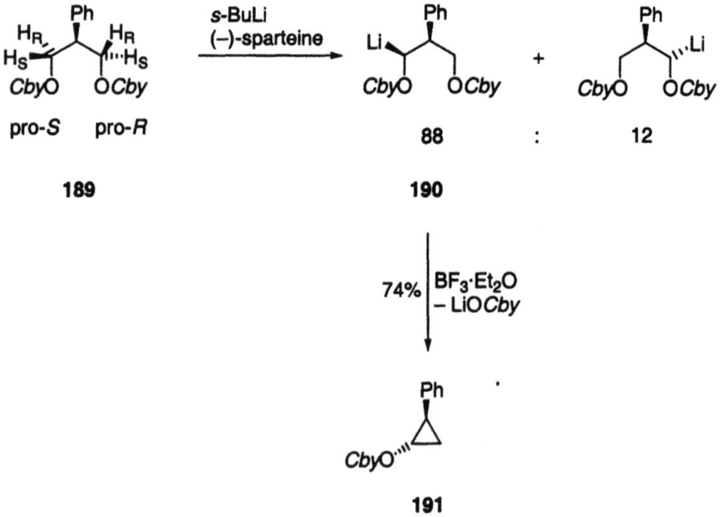

s-BuLi, (–)-sparteine; pro-*S*, pro-*R*; **189**; 88 : 12; **190**; 74%, $BF_3{\cdot}Et_2O$, – LiO*Cby*; **191**, 76% *ee*

(49)

A similar situation is given in the *meso*-dicarbamate **192** [see Eq. (61)] [120]. The *pro-S* proton at the *pro-R* branch exhibits the highest reactivity in the (–)-sparteine-mediated deprotonation to form the lithium compound **193** with a small amount of the diastereomer **195**. By applying prolonged reaction times (4–5 h), it is found that **195** is decomposed more rapidly than **193**, leading to a further enrichment. Trapping of the reaction mixture by different electrophiles leads to essentially enantiomerically and diastereomerically pure products **194a–c**. Allylation and benzylation result in lower diastereomeric ratios, probably due to SET mechanisms in the substitution step.

pro-R; pro-S; 192; s-BuLi, (–)-sparteine, toluene, -78 °C; 193; ElX; 194; 195; ElX; 196 (50)

Table 8. [Table to Eq. (50)]

194/196	El(X)	Yield [%]	194:196	ee [%][a]
a	$Me_3Sn(Cl)$	65	98:2	>95
b	$ElX=CO_2$[b]	63	96:4	>95
c	Me(I)	70	95:5	>95
d	$CH_2=CHCH_2(Br)$	59	68:32[c]	>95
e	$PhCH_2(Br)$	44	74:26[c]	>95

[a] *ee* of the major product.
[b] Isolated as its methyl ester after treatment of the free acid with diazomethane.
[c] Mainly *ent*-**196**.

Carboxylic ester **194b** was converted to the bicyclic γ-lactone **197** [Eq. (51)] [120].

194b; 1. 5 N HCl reflux, 2. CH_2N_2 r.t., 82%; $NaIO_4/RuCl_3$, 62%; 197 (51)

A kinetic resolution of lower efficiency could also be achieved when the racemic stannane *rac*–**198** was subjected to MeLi/(–)-sparteine [Eq. (52)] [120].

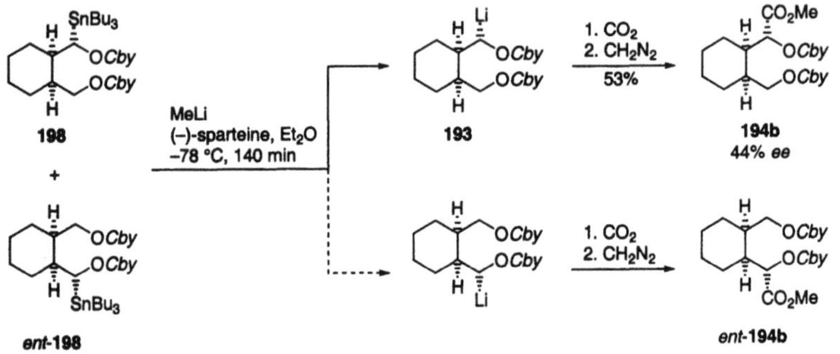

(52)

The ester **194b** was obtained with 53% yield and 44% *ee* (*e.r.* **194b**:*ent*-**194b**= 72:28) and the stannane *ent*-**198** was recovered (33%, 60% ee). The (1′*S*)-enantiomer **198** is cleaved more rapidly than (1′*R*)-**198** (*ent*-**198**). To our best knowledge, this reaction represents the first kinetic resolution by lithiodestannylation.

An efficient kinetic resolution was also observed during the (–)-sparteine-mediated deprotonation and cyclocarbolithiation [74] of the (2-piperidinyl)methyl carbamate *rac*-**199** [Eq. (53)]. The 1-*pro-S*-H in the enantiomer (*R*)-**199** is the most reactive proton, leading to the "matched" combination. The lithium compound **200** cyclizes with formation of the benzyllithium intermediate **201**; under optimized conditions [0.80 equiv. of (–)-sparteine, 0.75 equiv. of *s*-BuLi, 22 h at –78 °C], and after quenching with methanol the indolizidine **202** is isolated with high diastereomeric and enantiomeric purity (34%, *d.r.*=98:2, 95% ee). Optically active (–)-(*S*)-**199** (46%, 63% ee) is recovered. Scavenging of the benzyllithium **201** by electrophiles (CO_2, R_3SiCl, R_3SnCl) proceeds with medium to good diastereoselectivity [74]. Since some of the diastereomer **205** could be characterized in the reaction mixture, it is quite likely that the major stereodifferention occurs in the deprotonation step [Eq. (53)].

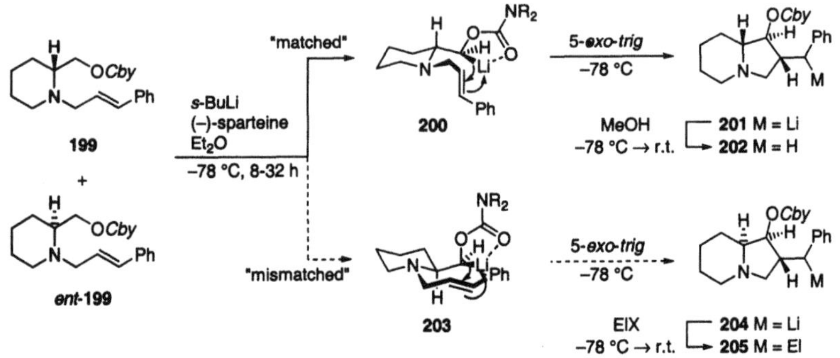

(53)

Kinetic resolution can also occur during the later steps of the reaction sequence [121, 122]. Almost no discrimination between the (*R*)- and the (*S*)-enantiomer is observed in the deprotonation of the 3-(indenyl)propyl carbamate *rac*-**206** [Eq. (65)] [121]. Of the two lithiated diastereomers (*S*,*S*)-**207** and (*R*,*S*)-**208**, only the (*R*,*S*)-stereoisomer is capable of intramolecular cyclization; after protonation of the reaction mixture, most of (*S*)-**206** (48%) is recovered, besides the enantiomerically pure products; for further discussion of carbolithiations see Normant, in this volume.

(S)-206 + (R)-206 —[s-BuLi, (–)-sparteine, Et_2O, 20 h]→ (S,S)-207 —[H_2O]→ (S)-206; + (R,S)-208 → → products (54)

An efficient desymmetrization in *meso*-epoxides of type **209** by *sec*-alkyllithium/(–)-sparteine was found by Hodgson et al. [see Eq. (66)] [123–125]. The intermediate lithiooxirane **210** usually undergoes carbenoid formation and intramolecular C-H insertion reactions (see Hodgson et al, in this volume); however, very recently it could be trapped by external electrophiles with stereoretention at low temperatures [126].

209 —[s-BuLi, (–)-sparteine]→ **210** → (El-substituted epoxide); **210** → [carbenoid, OLi] → further reactions (55)

3
Chiral Benzyl-Type α-Oxy-Organolithium Compounds

Benzyl-type organolithium compounds are stabilized by the mesomeric interaction of the negative charge of the carbanionic center with the aromatic π-system. The thermodynamic acidity of the benzylic compounds is 10 to 15 pK units higher than of the corresponding methyl derivatives [127]. Mesomeric stabilization requires a considerable flattening of the carbanionic center towards sp^2-hybridization [128]. As a consequence, the configurational stability is decreased, and – with few exceptions – most of the benzyllithium derivatives suffer from rapid racemization [Eq. (56)], see Sect. 3.2.

211 ⇌ **212** ($Li^{\oplus}$) ⇌ *ent*-**211** (56)

Notably, the close ion pairs **211** and *ent*-**211** are chiral species; this holds even if the carbanionic framework were completely planar. Enantiomerization results in migration of the lithium cation from one face of the carbanion to the other one, and is facilitated by the formation of a stabilized separated ion pair **212**. If no further chiral elements are interacting, such as chiral ligands at the lithium cation, the *e.r.* in the substitution products directly reflects the *e.r.* in the carbanionic intermediates **211**/*ent*-**211**. In most cases, the rate of racemization is much greater than the rate of the substitution step, and the formation of racemic products is the result.

The stereochemical course of the electrophilic substitution – retention, inversion, or partial racemization – of a given chiral benzylic organolithium by an electrophilic reagent still cannot be predicted on the basis of the present knowledge [Eq. (57)]; it depends mainly on the substituents in the carbanion, the electrophile and the complexing ligands at the lithium cation (Sect. 3.3) [64, 129, 130].

retention, inversion, ElX, 211, 213, *ent*-213 (57)

3.1 Stereospecific Deprotonation and Substitution of Enantioenriched Precursors

So far, the only examples of α-oxy-substituted benzylic precursors leading to lithium carbanions of considerable configurational stability are secondary carbamates **214** [131] and secondary 2,4,6-trisubstituted benzoates **215** [132, Eq. (58)].

s-BuLi, TMEDA Et_2O, –78 °C or *s*-BuLi, Et_2O/toluene, –78 °C; ElX retention or inversion

214 R^1 = N(*i*-Pr)$_2$ **216** **218**

215 R^1 = 2,4,6-triisopropylphenyl **217** **219** (58)

The deprotonation proceeds smoothly below −70 °C within a few minutes using *sec*-butyllithium in ether, hexane, or toluene in the presence of the complexing diamine TMEDA. The *N,N*-diisopropylcarbamates of type **216** are stable with respect to configuration and to decomposition under these conditions, whereas benzoates **217** decompose more rapidly.

The best solvent conditions for the deprotonation of the 2,4,6-triisopropylbenzoate **215a** were found to be toluene/20% diethyl ether [Eq. (59)] [132].

215a → 1. s-BuLi, Et_2O/toluene, −78 °C; 2. $ClSnMe_3$; 51% → **219a** (59)

We have studied the stereochemistry of lithiation-substitution of the benzylic carbamate (*R*)-**214a** [Eq. (60)] in detail [133, 134].

(*R*)-**214a** (97% *ee*) → (*R*)-**216a** → ElX: retention → **218**; inversion → *ent*-**218** (60)

Protonation with Brønsted acids proceeds with retention (Table 9). Alkylation, silylation, stannylation, carbon dioxide, and carbon disulfide addition (entries 3–10, 12, 25) are accompanied by inversion. This is also true for the reaction with acid chlorides (entry 13), but esters react with clean retention (entries 14–16). Aliphatic aldehydes and ketones add to the lithium compound **216a** with complete retention of configuration. A notable exception was found for benzaldehyde: it is added with inversion (entry 23).

Since the stannylation proceeds with inversion and the lithio-destannylation is a stereoretentative reaction, for the first time, the clean inversion of an enantioenriched benzyllithium compound could be accomplished with a chirality transmission of greater than 95% [Eq. (61)] [131, 133].

(*S*)-**216a** → $ClSnMe_3$ → (*S*)-**218h** → s-BuLi → (*R*)-**216a** (61)

Table 9. Stereospecific lithiation and substitution of carbamate (*R*)-**214a**

Entry	Product	El	ElX	Yield [%]	ee [%]	Course	Ref.
1	**218a**	D	DOMe	96	80	ret.	[135]
2	**214a**	H	HOAc	75	80	ret.[a]	[131]
3	*ent*-**218b**	CH_3CH_2	CH_3CH_2Br	85	>95	inv.	[135]
4	*ent*-**218c**	*n*-C_3H_8	*n*-C_3H_8Br	77	85	inv.[a]	[131]
5	*ent*-**218d**	CH_2=$CHCH_2$	CH_2=$CHCH_2Cl$	96	83	inv.[a]	[133]
6	*ent*-**218d**	Me_2C=$CHCH_2$	Me_2C=$CHCH_2Br$	52	>95	inv.	[136]
7	*ent*-**218e**	$PhCH_2$	$PhCH_2Cl$	96	>95	inv.	[136]
8	*ent*-**218f**	MeC≡CCH_2	MeC≡CCH_2Cl	61	>95	inv.	[136]
9	*ent*-**218g**	Me_3Si	Me_3SiCl	94	96	inv.	[131]
10	*ent*-**218h**	Me_3Sn	Me_3SnCl	92	≥95	inv.	[133]
11	*ent*-**218i**	Me_3Pb	Me_3PbCl	88	85	inv.	[133]
12	*ent*-**218j**	CO_2Me	i) CO_2; ii) CH_2N_2	84	84	inv.	[133]
13	*ent*-**218j**	CO_2Me	MeOCOCl	90	85	inv.	[133]
14	**218j**	CO_2Me	$(MeO)_2CO$	85	94	ret.	[133]
15	**218k**	CHO	HCOOMe	60	≥95	ret.	[136]
16	**218l**	COMe	$MeCO_2Me$	60	≥95	ret.	[133]
17	*ent*- **218l**	COMe	MeCOCN	43	92	inv.	[133]
18	**218m**	CO*i*-Pr	*i*-$PrCO_2Me$	94	≥95	ret.	[133]
19	**218n**	COPh	$PhCO_2Me$	95	≥95	ret.	[133]
20	*ent*-**218n**	COPh	PhCOCl	95	≥95	inv.	[133]
21	**218o**	Me_2COH	Me_2CO	71	54	ret.	[136]
22	**218p**	*i*-PrCHOH[b]	*i*PrCHO	92	>95	ret.	[136]
23	*ent*-**218q**	PhCHOH[c]	PhCHO	69	>95	inv.	[136]
24	*ent*-**218r**	CONH*i*-Pr	OCN*i*-Pr	90	85	inv.	[133]
25	*ent*-**218**	CS_2Me	i) CS_2; ii) MeI	67	≥95	inv.	[133]

[a] The originally assigned configuration is a mistake; for correction see [137].
[b] Mixture of diastereomers (*d.r.*=50:50).
[c] Mixture of diastereomers (*d.r.*=50:50); in addition 16% of regioisomers with rearranged *Cb* group.

Similar trends were recorded for further secondary benzyl carbamates. During these investigations it turned out that an extended mesomeric system in the aromatic substituent leads to a decreased configurational stability, and thus, to products with lower enantioenrichment. Some of the results are summarized in the following Eq. (62).

a) (*S*)-**220**, 91% *ee* — 1. *s*-BuLi, TMEDA, Et_2O; 2. Br–(CH$_2$)$_2$–(1,3-dioxolan-2-yl); inversion → (*S*)-**221**, 55%, 87% *ee* [136]

b) (*R*)-**222**, 91% *ee* — 1. *s*-BuLi, TMEDA, Et_2O; 2. $ClCO_2Me$; inversion → (*R*)-**223**, 91%, 80% *ee* [136]

H_2, Pd/C, MeOH; inversion → (*S*)-**224**, 95%, 60% *ee*

c) (*S*)-**225**, 95% *ee* — 1. *s*-BuLi, TMEDA, Et_2O; 2. CH_3CH_2Br; inversion → (*S*)-**226**, 80%, 88% *ee* [135]

d) (*R*)-**225** — 1. *s*-BuLi, TMEDA, Et_2O; 2. $ClSnMe_3$; inversion → (*R*)-**227**, 91%, 86% *ee* [135]

1. *s*-BuLi, Et_2O, 5 min; 2. CH_3CH_2Br; inversion → (*S*)-**226**, 72%, 45% *ee*

e) (*R*)-**228**, 92% *ee* — 1. *s*-BuLi, TMEDA, Et_2O; 2. $ClCO_2Me$; inversion → (*S*)-**229**, 75%, 60% *ee* [136]

H_2, Pd/C, MeOH; inversion → (*R*)-**230**, 93%, 47% *ee*

(62)

As it is seen from Eq. (62b) and (62e), the method allowed for the synthesis of moderately enantioenriched methyl esters of the commercially available drugs ibuprofen (**224**), 60% ee [136] or naproxen (**230**), 47% ee [136], from optically active carbamates **222** or **228**, respectively. The final step is a reductive removal of the carbamoyloxy group by Pd-catalyzed hydrogenolysis, which proceeds with inversion [138–140] but some erosion of the enantiomeric purity results.

The ee values have not been optimized. From today's knowledge improvements are possible by using a less polar solvent, not more than one equivalent of the complexing diamine, and keeping the deprotonation time as short as possible (~5 min). Polar solvents, such as THF, cause rapid racemization [133, 135, 141]. The same effect is observed when lithium is exchanged for potassium [133]. These results point to the importance of the incorporation of the lithium cation into a dense chelate complex which hampers the cation from changing the enantiotopic faces of the anion.

Deprotection of the tertiary benzyl *N,N*-diisopropylcarbamates is difficult. Acidic conditions lead to the destruction of the tertiary substrates. Occasionally, treatment with $LiAlH_4$ in refluxing THF [57, 136], or with a large excess of diisobutylaluminum hydride, liberates the carbinols, but ester and carbonyl groups are also reduced.

Therefore, we developed a carbamate group, which is cleavable under mild, alkaline conditions. The *N*-[2-(*tert*-butyldiphenylsilyloxy)-ethyl)]-*N*-isopropylcarbamoyl group (*Cbse*) [135, 137] in tertiary esters **232** can be removed by desilylation and subsequent neighboring group participation in the hydroxyalkylurethane **233** to give the alcohols **234**. Some examples of compounds synthesized by this method are collected [Eq. (63)] [137].

OTBDPS; *Cbse*; **231**; 1. *s*-BuLi 2. ElX; El OCbse Ar R; **232**; Bu_4NF; El O Ar R N OH; **233**; base; El OH Ar R; **234**; +

MeO_2C OH Ph CH_3 — **234a**, 89% 94% *ee*

HO CO_2Me Ph CH_3 — *ent*-**234a**, 76% 84% *ee*

HO Ph CH_3 — *ent*-**234b**, 81% 86% *ee*

(63)

HO Ph CH_3 — **234c**, 76% 92% *ee*

Me_2OC OH CH_3 — **234d**, 96% 76% *ee*

HO CO_2Me — **234e**, 87% 96% *ee*

OH OH — **234f**, 78% > 95% *ee*

From our results, obtained with chiral lithiobenzyl carbamates in the absence of any further chiral elements, such as a chiral complexing diamine, we suggested the following hypothesis for the stereochemical course of the reactions with electrophiles [142]: The preferred attack proceeds from the rear face with inversion of configuration. But, if a strong interaction in the transition state with the leaving group of the electrophile is possible, retention is favored. Examples for retention are found by the protonation reaction with Brønstedt acids, and, as well, in reactions with methyl esters and aliphatic aldehydes. The reaction with benzaldehyde [142] is an exception. We speculated that the origin is a π-π* interaction of the phenyl rings which is facilitated at the rear face.

However, such "generalizations" have a narrow basis and have to be considered with great care. These rules, deduced from open-chain benzylic carbamates must be modified, when cyclic carbamate esters are concerned. The lithiated indanyl carbamate **236** exhibits an enhanced tendency for front-side attack [Eq. (64)] [135, 143, 144].

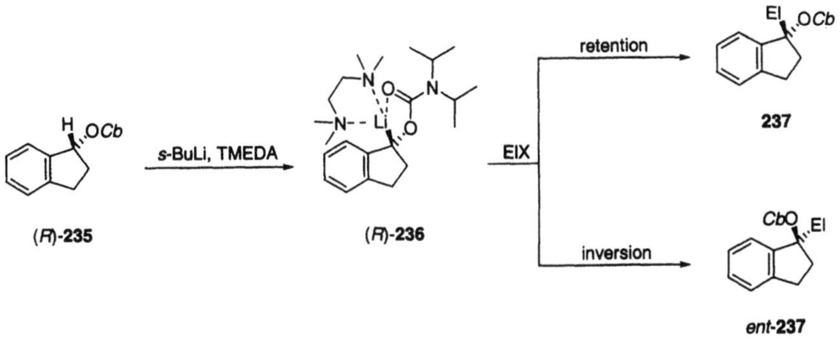

(64)

Table 10. Ratio of 237/*ent*-237 in dependence of the electrophile [Table to Eq. (64)]. [143]

Product	El	ElX	Yield [%]	237:*ent*237[a]	Course
235	H	HOMe	92	100:0	ret.
237a	MeOCO	$(MeO)_2CO$	85	99:1	ret.
237a	MeOCO	MeOCOCl	72	99:1	ret.
237a	MeOCO	i) CO_2; ii) CH_2N_2	82	99:1	ret.
237b[b]	MeSCS	i) CS_2; ii) MeI	80	99:1	inv.
237c	Me_3Si	Me_3SiCl	46	98:2	ret.
ent-237d	Me_3Sn	Me_3SnCl	74	25:75	inv.
ent-237e	Bu_3Sn	Bu_3SnCl	81	10:90	inv.

[a] Corrected to 100% ee in (*R*)-235.
[b] (*S*)-235 was the starting material.

The results are supported by Hammerschmidt et al. [144], who found partial retention (15:85) in the stannylation of the corresponding 2,4,6-tri-isopropylbenzoate **238a** when using a chiral trialkyltin bromide [Eq. (65)].

1. s-BuLi, TMEDA
2. (–)-MenSnMe$_2$Br
hexane, –78 °C

238a n = 1
b n = 2

239a
b

240a
b

(65)

Ar =

They also found, in the reaction of the lithiated indanyl carbamate (*S*)-**236** [Eq. (64) with tributyltin bromide and -triflate, products of opposite configuration [144]. This is in accordance with the assumption that the better complexing triflate group shifts the stereochemical course further towards retention. Why does the lithiated ring **236** behave so differently in comparison to the phenylethyl carbamate **216a** [Eq. (61)]? We calculated the ground state geometry of lithio carbamates [143]. The sum of the dihedral angles for the open-chain compound **216a** is 333.1°, whereas it was found to be 323.8° for the cyclic compound **236**. Ideal sp^2-hybridization counts to an angle sum of 360° whereas sp^3-hybridization requires a sum of 328°. Subsequently both components were forced in the calculations to planarity, requiring 10.0 kcal·mol^{-1} for **216a** and 14.1 kcal·mol^{-1} for **236**. It can be concluded that inversion for the cyclic lithium compound (which has to pass through a planar structure) is by around 4 kcal·mol^{-1} energetically more costly than for the open-chain compound. These calculations illustrate the importance of the geometrical situation as one of the major influences in the competition of inversion versus retention in the substitution step.

It should be added that the less strained 1-lithiotetralinyl carbamates show the "normal" sense of stereoselectivity [136].

An interesting approach to highly substituted α-lithiobenzyl carbamates consists in the carbolithiation of 1-arylalk-1-enyl carbamates (see also Normant, in this volume) [145, 146]. Lithiated benzyl *N,N*-diisopropylcarbamate was found to be configurationally stable on the microscopic scale in the Hoffmann test [147] but, so far, it has not been prepared in enantioenriched form free of a chiral ligand, see Sect. 3.2. Generally, the Hoffmann test can give valuable information on the configurational stability of carbanionic species – even if these are available only as racemates (see also Hodgson, Stent in this volume) – but one should take into account that, for synthesis on the macroscopic scale, enhanced stability may be necessary (see Beak et al, in this volume).

3.2
Chiral α-Oxybenzyllithium Compounds by Enantioselective Lithiation in the Presence of Chiral Ligands

When an achiral substrate **241** is deprotonated in the presence of a chiral ligand, such as (–)-sparteine, two diastereomeric ion pairs **242** and *epi*-**242** result; these differ in their energies and reactivity [Eq. (66)].

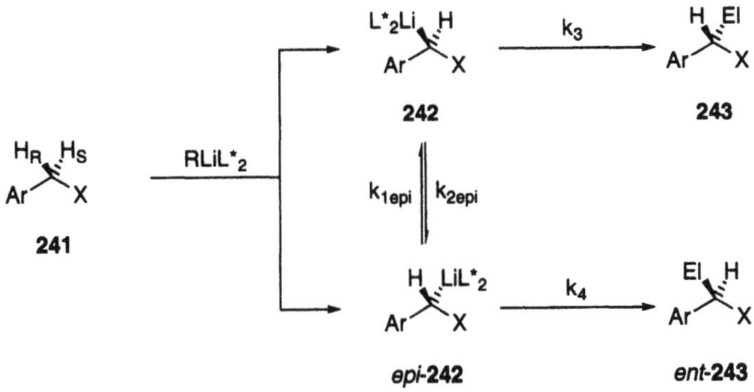

L*₂Li H / Ar X — **242** —k_3→ H El / Ar X — **243**

H_R H_S / Ar X — **241** —$RLiL^*_2$→ **242** ⇌ (k_{1epi}, k_{2epi}) *epi*-**242** (66)

H LiL*₂ / Ar X — *epi*-**242** —k_4→ El H / Ar X — *ent*-**243**

If these are configurationally stable, the situation is similar to that one encountered previously (Sect. 2); the enantiomeric ratio **243**/*ent*-**243** (inversion in the substitution step assumed) reflects the diastereomeric ratio **242**/*epi*-**242** after complete reaction. Configurational instability of the intermediates under the reaction conditions causes new problems, but also offers new possibilities [130] (s. a. Baek et al, in this volume). Two borderline cases and all situations between them are possible: If k_{1epi} and k_{2epi} are very small compared to the rates of substitution k_3 and k_4, after complete equilibration, the product ratio **243**/*ent*-**243** equals the ratio *epi*-**242**/**242** and reflects the difference of thermodynamic stabilities. On the other hand, if k_{1epi} and k_{2epi} are very large in comparison to the rates k_3 and k_4, the product ratio equals the quotient k_4/k_3. This is the typical situation of a dynamic kinetic resolution [148]. Since equilibration can be enforced by higher temperature and the substitution step carried out at low temperature interesting possibilities for the enhancement of selectivities arise. These have been explored predominantly by Beak for aryl- and amino-substituted benzyl anions (s. a. Baek et al, in this volume).

The first α-oxybenzyllithium compound, which offered these problems, was the lithiated benzyl carbamate **244** (Fig. 4), but it could not be tuned to useful selectivities [64, 149].

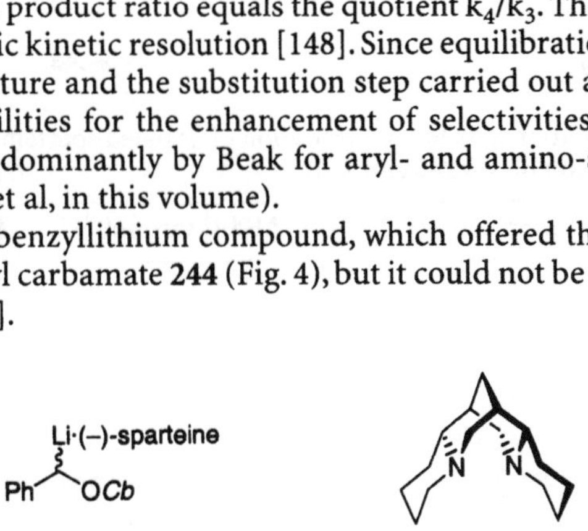

Li·(–)-sparteine / Ph OCb

244

144
(–)-α-isosparteine

Fig. 4.

The rate of epimerization of (–)-sparteine and (–)-α-isosparteine complexes **245/246**, derived from lithio-indenyl carbamates and, as well, their relative ratios in the equilibrium, were investigated NMR-spectroscopy in cooperation with Fraenkel [Eq. (67) and Table 11] [150]. The rates follow pseudo-first order. The activation energy for the epimerization process is increased by additional alkyl groups in the vicinity of the carbanionic center (**245a**, R=H: 13.5 kcal·mol^{-1}; **245b**, R=CH_3: >25 kcal·mol^{-1}). We were surprised to learn that the exchange of (–)-sparteine by the C_2-symmetrical diastereomer (–)-α-isosparteine (**144**) causes a significant decrease of the epimerization barriers (for **245b**: from >25 kcal·mol^{-1} to approx. 8 kcal·mol^{-1}). On the other hand, the equilibrium **245b**/*epi*-**245b** is shifted to the synthetically useful ratio 18:82 and this could be utilized in a silylation reaction [150].

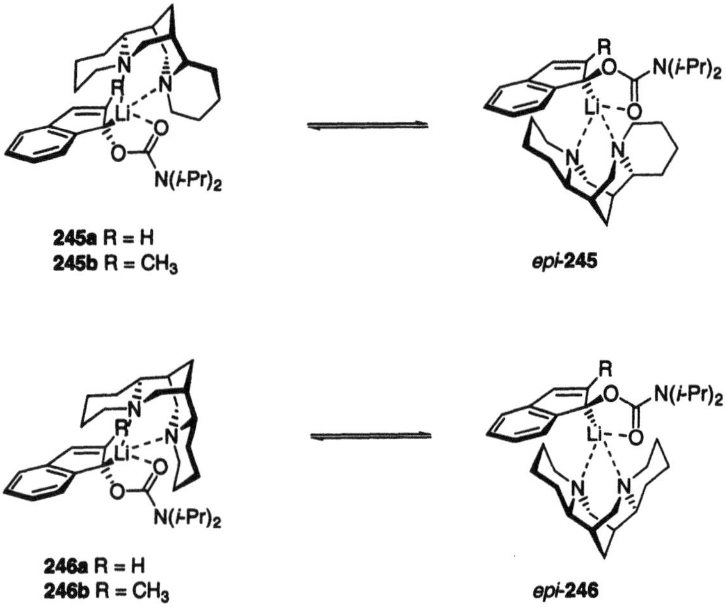

245a R = H
245b R = CH_3
epi-**245**

(67)

246a R = H
246b R = CH_3
epi-**246**

Table 11. Activation parameters for epimerization of lithiated sparteine- and isosparteine-complexed indenes (complexes **245** and **246**)

Compound	*d.r.*[a]	ΔH‡ [kcal·mol^{-1}]	ΔS‡ [cal·mol^{-1}·K^{-1}]
245a	55:45	13.5	-8
245b	60:40	>25	–[b]
246a	–[c]	<5	–[b]
246b	18:82	8.0–8.5	-25.1

[a] Isomer appearing at lower field at left.
[b] Line shape analysis not possible. [c] The diastereomeric ratio could no be determined due to fast epimerization.

Since (–)-sparteine is not C_2-symmetric, due to the chiral center at lithium, four diastereomers of each complex are possible [Eq. (68)]. The facile epimerization at the lithium atom could level differences arising from the configuration at the carbanionic center; semiempirical calculations point to this possibility [150, 151].

245a R = H
245b R = CH_3

(68)

A relatively high energetic difference between epimeric complexes **249** and *epi*-**249**, comprised of α-methoxybenzyllithium and (*S*,*S*)-Box-*i*-Pr (**248**), is the origin of the highly enantioselective carboxylation and hydroxyalkylation reported by Nakai et al. [Eq. (69)] [152, 153].

247 → *t*-BuLi, **248**, hexane, –78 °C → **249**; 1. CO_2 2. H_2O → **250**, > 95%, > 95% *ee*; RCHO → *anti*-**251** + *syn*-**251**, > 95% (R = Ph-C≡C), *anti* (> 98% *ee*):*syn* (84% *ee*) = 90:10

248

(69)

The deprotonation of **247** in hexane by *t*-BuLi/excess **248** at –78 °C for 1 h gave the best results: (*R*)-*O*-methylmandelic acid (**250**) with greater than 95% ee and the hydroxy ethers (several examples) with high *anti*-selectivity, and as well, very good ee. On the basis of control experiments, the authors assume a "dynamic thermodynamic resolution" as the origin. This mechanism implies an essentially complete conversion to one epimer **249** of unknown configuration at –78 °C and its stereospecific carbonyl addition.

4
Chiral 1-Oxy-2-Alkenyllithium Derivatives

4.1
Introduction

The first configurationally stable 1-oxy-2-alkenyllithium **253** was reported in 1986 by Hoppe and Krämer [Eq. (70)] [8]. It was generated by deprotonation of the enantioenriched allyl carbamate **252**, obtained from the corresponding alcohol via kinetic resolution through Sharpless epoxidation. More conveniently accessible are the 1-methyl derivatives **254** and analogues either from (*R*)- or (*S*)-lactaldehydes via Wittig olefination [154]. Kinetic resolution of *rac*-**254** during deprotonation is also possible [155–157].

H_3C R, CbO H → (*n*-BuLi, TMEDA, Et_2O, –78 °C) → H_3C R, O Li·TMEDA, $(i\text{-}Pr)_2N$ O → products (70)

68% *ee* → > 52% *ee*

252 R = $CH_2CH(CH_3)_2$ → **253**

254 R = CH_3 → **255**

Apart from the lithium carbanions, derived from secondary 2-alkenyl carbamates, no further types of configurationally stable α-oxyallyllithium derivatives have been reported in the last 15 years. One must conclude that the five-membered lithium chelate ring plays an important role for the stereochemical integrity. This structure has been nicely demonstrated by the X-ray structure analysis of a sparteine complex [158].

This section focuses on metallated, enantioenriched 2-alkenyl carbamates, their stereochemical features, and their use in homoaldol reactions. Much of the chemistry, often developed for racemates, has been summarized in previous reviews [6, 7, 64, 142, 159–165], and it can be easily applied to enantioenriched compounds.

The problems of configurational stability are similar to those encountered with the related benzyl compounds. The stereospecificity and the regioselectivity of the substitution step are even more complicated and it is due to the carbamoyloxy group that a useful regioselectivity is often observed. In Eq. (71), all stereochemical possibilities for a single enantiomer are summarized.

The lithium compound **256** can react from two conformations **256a** and **256b**. In an α-substitution, products **258** or *ent*-**258** are formed from both conformers, their ratio depends on the stereospecificity of the appropriate S_E reaction, which may occur with retention or inversion, depending on the electrophile. Electrophilic γ-substitution of **256a** and **256b** can lead, depending on the stereospecificity of the substitution step, to four different stereoisomers with two pairs of enantiomers (**257** and *ent*-**257**, **259** and *ent*-**259**); **257** and *ent*-**259** and as well, *ent*-**257** and **259** can be regarded as "pseudo-enantiomers". Since the latter pairs are diastereomers, in principle, these can be easily separated before deprotection.

Fortunately, reactions arising from conformer **256a** usually are favored by 1,3-allylic strain [166] and stereoelectronic effects. In particular, carbonyl addition reactions with covalently bound metal residue M proceed through a cyclic Zimmerman-Traxler transition state [167] in a formal *syn*-S_E'-reaction, leading to product **257** with high selectivity. The carbamoyloxy group is regarded as an important feature, holding the cation M in the α-position and avoiding the formation of regioisomers of type **258**.

(71)

4.2 Chiral, Non-Racemic 1-Oxy-2-Alkenyllithium Compounds by Sparteine-Mediated Deprotonation and Stereochemical Course of Substitution

The deprotonation of the secondary carbamate *rac*-**254** by *n*-butyllithium/(–)-sparteine (diethyl ether/hexane, –78 °C) is combined with an efficient kinetic resolution [Eq. (72)] [155, 168]. The lithium compound (*S*)-**256** is configurationally stable under these conditions and is formed with greater than 80% ee as estimated from trapping experiments. The less reactive enantiomer (*R*)-**254** is recovered with 41% yield and 80% ee.

(72)

Some trapping experiments were performed in order to elucidate their stereochemical course [Eq. (73)] [155, 157, 168]. The reaction with tributyltin chloride gave the γ-substitution products (*R*,*Z*)-**257a** (42%) and (*S*,*E*)-**259a** (6%), both with >80% ee, revealing that the stannylation proceeds as a clean *anti*-S_E' process from both conformations of the lithium compound (*S*)-**256** [168]. Carboxylation (followed by diazomethane esterification) yielded the α-adduct (*R*,*E*)-**258** in an invertive substitution [157]. Some of the γ-adduct, which has not been stereochemically assigned, was also found. The metal exchange reaction of these sparteine complexes with titanium tetraisopropoxide to form intermediate (*S*,*E*)-**260** proceeds α-regioselectively with inversion of the configuration; this was concluded from the result of carbonyl addition reactions (see below).

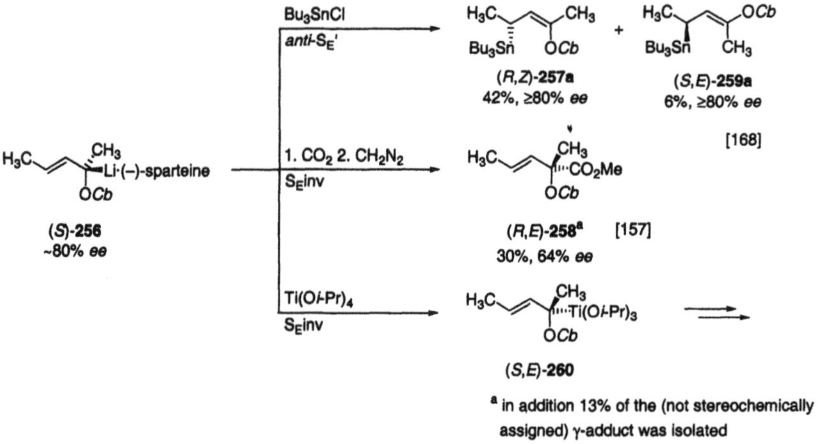

(73)

Additional experiments were performed with the corresponding lithium/TMEDA complexes (*S*)- and (*R*)-**255**. Most of the reactions take the same sense of stereospecificity, independent from the ligands at lithium. An exception is the triisopropoxytitanation, it proceeds with retention of configuration with the TMEDA complexes whereas the sparteine complexes react with inversion [169, 168]. Similar to results discussed in the benzyl section (Sect. 3), the interaction of the isopropoxy residue with the lithium cation may determine the reaction course. It seems that in the presence of the bulky (–)-sparteine as a ligand, such a suprafacial interaction does not contribute significantly. For the metal exchange with tris(diethylamino)titanium chloride inversion was observed, too [169].

As expected from the strong *anti*-S_E' preference in the stannylation, the stereoisomeric precursors (*R*,*E*)-**255** and (*S*,*Z*)-**255** lead to the identical stannane (*S*,*Z*)-**257b** as the major product [Eq. (74)] [168, 170].

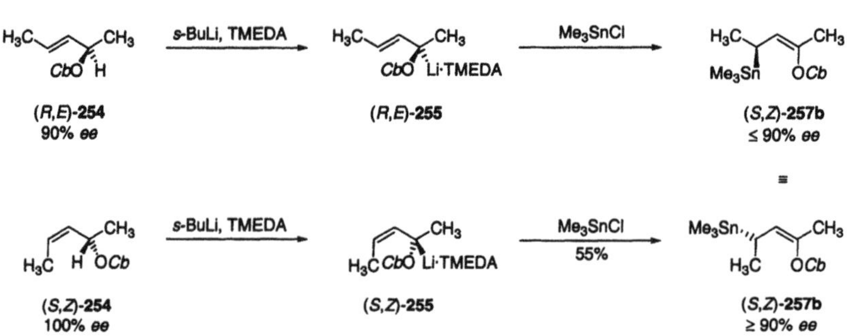

(74)

The stereochemical properties reported for the 3-penten-2-yl system could be confirmed for a number of further 1-methyl-2-alkenyl carbamates [171].

A very efficient approach to 1-aryl-substituted 1-lithio-2-alkenyl carbamates was very recently found by Seppi [172]. Two (from many more) examples are depicted in Eq. (75). On treating the (*Z*)-enol carbamate (*Z*)-**261** with *n*-butyllithium/(–)-sparteine, one of the enantiotopic γ-protons H_S or H_R is removed with complete selectivity, leading to a highly enriched, configurationally stable, five-membered chelate complex **262**. Trapping the reaction mixture with ketones or acid chlorides produces products **263** in essentially enantiomerically pure form, having the absolute configuration shown.

At the time of writing this manuscript, the configuration of intermediate **262** is still unknown; it cannot deduced on the basis of the stereochemical outcome of the homoaldol reactions, since it is not clear whether the often observed *syn*-S_E' course is retained with 1-aryl-substituted lithium compounds [this would imply (*R*)-configuration of **262**]. More evidence has accumulated for assuming the (*S*)-configuration. The great advantage of this approach when compared to the kinetic resolution strategy lies in the fact that the starting material is achiral and, hence, it can be completely converted into the chiral reagent **262**.

(*Z*)-**261** → [*n*-BuLi, (–)-sparteine, toluene, –78 °C] → **262** → [ElX] → **263** (75)

El = $(H_3C)_2$C-OH, 72%, > 95% *op*
El = $(H_3C)_3$CC=O, 83%, > 95% *ee*

The situation is more complicated when subjecting primary 2-alkenyl carbamates to a (–)-sparteine-mediated deprotonation. The butyllithium/(–)-sparteine base performs a quite efficient selection between the enantiotopic α-protons in carbamate **264** in favor of the (*S*)-configured complex (*S*)-**265**, which usually is preferred by the magnitude of 90:10 [Eq. (76)]. However, (*S*)- and (*R*)-**265** usually are configurationally unstable even at –78°C and do interconvert at a moderate rate. Thus, the kinetically achieved stereoselection is eroded by epimerization. In the very first investigated example, we were lucky that (*S*)-**265a** (R=CH_3) crystallizes from the solution under certain conditions, and according to a dynamic resolution [148], essentially all of the material is converted to (*S*)-**265a** [173]. Furthermore, trapping the solid by tetra(isopropoxy)titanium with complete inversion gave very efficiently intermediate (*R*)-**266**, which is stable in solution (see Sect. 4.3). Under optimized conditions, *d.r.* of 98:2 (for **265**) can be achieved [173]. In cooperation with Boche, we could obtain an X-ray crystal structure analysis of the γ-silyl derivative (*S*)-**265b** to secure structure and configuration [158].

It should be added here, that the 2,3-bond in (*E*)- and (*Z*)-lithioallyl carbamates is configurationally stable at the reaction temperature [174]. The γ-silyl derivatives are a notable exception. Even when performing the deprotonation with a (*Z*)-3-trimethylsilyl-2-propenyl carbamate, after few minutes at –78 °C, all of the material had been transformed to the lithiated (*E*)-isomer [175].

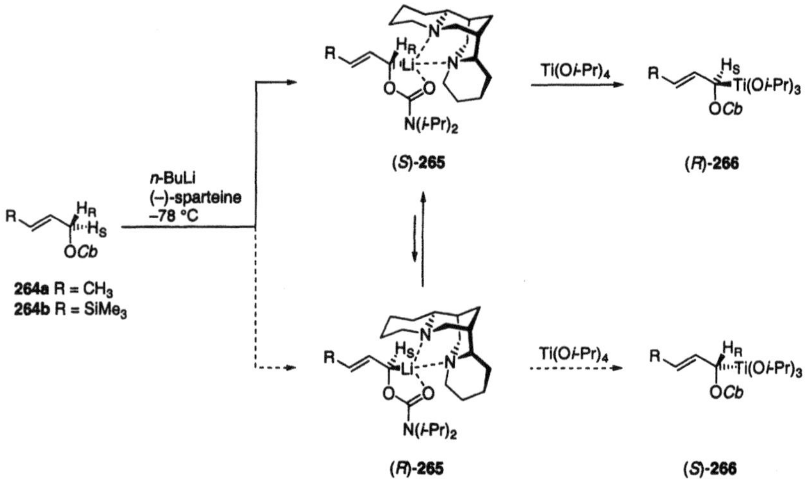

(76)

The cinnamyl carbamate **264c** was deprotonated under (–)-sparteine conditions and the reaction mixture was quenched after 30 min by several electrophiles [Eq. (77)] [93]. To our surprise, the silane **267a** (86% ee, *e.r.*=93:7) revealed its (*S*)-configuration by X-ray analysis. Assuming (*S*)-**265c** as the major intermediate, inversion to form silane (*R*)-**267a** was expected to occur.

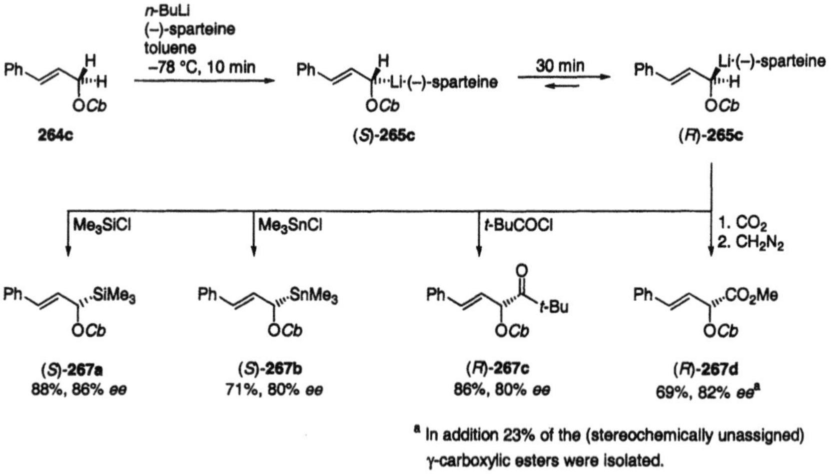

(77)

[a] In addition 23% of the (stereochemically unassigned) γ-carboxylic esters were isolated.

In a control experiment, **264c** was deprotonated under the same conditions but in the presence of chlorotrimethylsilane. Here, a nearly opposite ratio (*S*)-**267a**/(*R*)-**267a**, *e.r.*=21:79, 58% ee, was isolated [Eq. (78)]. It is obvious that, again, the *pro-S*-H in **264c** is removed preferentially to form intermediate (*S*)-**265c**, but it is rapidly epimerized to give (*R*)-**265c**, and the equilibrium lies close to 10:90. Consequently, silylation and stannylation by means of the chlorides proceeds with antarafacial S_E processes and the same is true for acylation by acid chlorides and carboxylation.

n-BuLi, (−)-sparteine, Me_3SiCl *in situ*, −78 °C, 29%: 264c → (S)-267a + (R)-267a; 21 : 79 (58% ee) (78)

Reaction with methyl iodide provided the (*S*)-configured major product (*S*)-**268** [Eq. (79)], which is the expected result of an *anti*-S_E' process from (*R*)-**265c** [93]. The ratio depends on the temperature and on the ratio **265**/methyl iodide. Without discussing the conclusions in detail, one must assume that a dynamic kinetic resolution does occur, the minor epimer being the more reactive one, thus, the degree of stereoselectivity is only low.

1. n-BuLi, (−)-sparteine 2. MeI, ~80%: 264c → (S)-268 + (R)-268; 58 to 78 : 42 to 22 (79)

In situ silylation of the lithium/(−)-sparteine complex (*S*)-**271** derived from the (*E,E*)-9-chloro-2,7-nonadienyl carbamate (*E,E*)-**269** gave the silanes (*S*)-**270** and (*R*)-**270** with an *e.r.* of 22:78 (56% ee); here also the *pro-S*-H is abstracted preferentially [Eq. (80)] [176, 177]. Even more of the initial chiral information is retrieved by an intramolecular allylic cycloalkylation of the intermediate (*S*)-**271** and by performing the deprotonation in toluene at −90°C [176, 177]. The *cis*-substituted 1,2-divinyl-cyclopentane **273** is formed with an *e.r.* of 90:10 (80% ee). Here, an *anti*-5_N reaction has to be assumed [177, 178].

(E,E)-269 → n-BuLi, (−)-sparteine, Me_3SiCl in situ, toluene, −78 °C, 70% → (R)-270, 56% ee

(E,E)-269 → n-BuLi, (−)-sparteine, toluene, −90 °C → (S)-271 ⇌ (R)-272

(S)-271 → [Li·(−)-sparteine, OCby, Cl]‡ → −LiCl → 273 (98% de) : ent-273 = 90 : 10 (90%) (80)

The isomer (*Z*,*Z*)-**269** furnished under identical conditions, surprisingly, the (1*S*,2*Z*,7*Z*)-cyclonona-2,7-dienyl carbamate **274** with 88% ee, which could be converted to the free alcohol **275** [Eq. (81)] [179]. The stereochemical outcome implies inversion of the configuration at the carbanionic center. The method could also be successfully applied to 5-oxy substituted dienes.

(*Z*,*Z*)-**269** → (*Z*,*Z*)-**271** → [TS]‡ → (–LiCl, 73%) **274** (88% *ee*) → (b, 95%) **275** (81)

a) 2.0 equiv *n*-BuLi/(–)-sparteine, toluene, –88 °C.
b) i) 2.0 equiv CH_3SO_3H, CH_3OH, 65 °C; ii) KOH/CH_3OH, 65 °C.

The remaining geometrically isomers (2*Z*,7*E*)- and (2*E*,7*Z*)-**269** [180] lead to planar-chiral (*E*,*Z*)-cyclononadienes (*M*,*R*)-**276** (80% ee) and (*M*,*R*)-**278**, respectively; some divinylcyclopentane **273** is formed from (*E*,*Z*)-**269** [Eq. (82)]. Both cyclononadienes come to equilibrium with the corresponding epimers (*P*,*R*)-**277** or (*P*,*R*)-**279** above 20°C by inversion of the chiral plane.

(*Z*,*E*)-**269** → (*Z*,*E*)-**271** → [TS]‡ → (–LiCl) (*M*,*R*)-**276** ⇌ (b) (*P*,*R*)-**277**, 82%, *d.r.* = 3:97, *e.r.* = 90:10

(*E*,*Z*)-**269** → (*E*,*Z*)-**271** → [TS]‡ → (–LiCl) (*M*,*R*)-**278** ⇌ (b) (*P*,*R*)-**279**, 57%, *d.r.* = 30:70, *e.r.* = 92:8; **273**, 25%, *d.r.* = 100:0, *e.r.* = 90.5:9.5 (82)

a) *n*-BuLi, (–)-sparteine, toluene, -88 °C.
b) Epimerization at 20-45 °C.

The S_N'-cyclization is also suitable for the construction of enantioenriched heterocycles as demonstrated by the synthesis of the divinylpyrrolidine **281** (90% ee) from the dialkenylamine **280** [Eq. (83)] [181]. The absolute configuration (3*R*,4*R*) (elucidated by anomalous X-ray diffraction) again results from an antarafacial process at the metal-bearing moiety during the formation of the five-membered ring by S_N'-substitution.

(83)

a) 2.2 Equiv *n*-BuLi/(–)-sparteine, toluene, –90 °C.

It was speculated that a higher substitution degree in the allylic system enhances the configurational stability of the lithium intermediate. In order to examine this assumption, an extensive study with (cycloalk-1-enyl)methyl *N*,*N*-diisopropylcarbamates **282** (R=H) in comparison with the appropriate 2′-methyl derivatives was undertaken. Out of the many examples, only a few can be discussed here [Eq. (84)] [182].

The half-life period of epimerization at –78 °C for the cyclopentenylmethyl derivative **283a** is close to 10 min, whereas the introduction of a 2′-methyl group (**283b**) enhances the configurational stability dramatically. Enhanced configurational stability is observed for the cyclohexenyl series **283c** and **283d.** No trace of epimerization was detected for **283d** during two hours. The 2′-methyl derivatives are, according to the efficiency of enantiotopos-differentiation and the high configurational stability, synthetically very useful chiral building blocks [182, 183] This is particularly true for their homoaldol reactions.

(84)

We realized another mode for the preparation of stereohomogeneous 1-lithio-2-alkenyl carbamates utilizing a chiral auxiliary and a rigid framework in

Table 12. [Table to Eq. (84)]

Product	Yield [%]	(*R*)-284:(*S*)-284	ee [%]	Time [min]
284a	54	67:33	34	10
284b	44	94.5:5.5	89	10
284c	78	77:23	54	30
284d	20	92.5:7.5	85	10
284d	88	92.5:7.5	85	120

the precursor **285**, which might offer another access to stereohomogeneous lithium carbanions [Eq. (85)] [184]. Deprotonation of **285** with *n*-butyllithium/TMEDA and quenching of the reaction mixture by different electrophiles led to the diastereomerically pure α-substitution products **287a–f.** In control experiments it was demonstrated that the (*R*)-lithium compound **286** is the kinetically and, as well, thermodynamically favored intermediate, which is substituted α-selectively under inversion. We assume that **286** adopts the configuration of the nine-membered ring chelate complex **286a**; it is, according to PM3 calculation, the most stable species in the system [184].

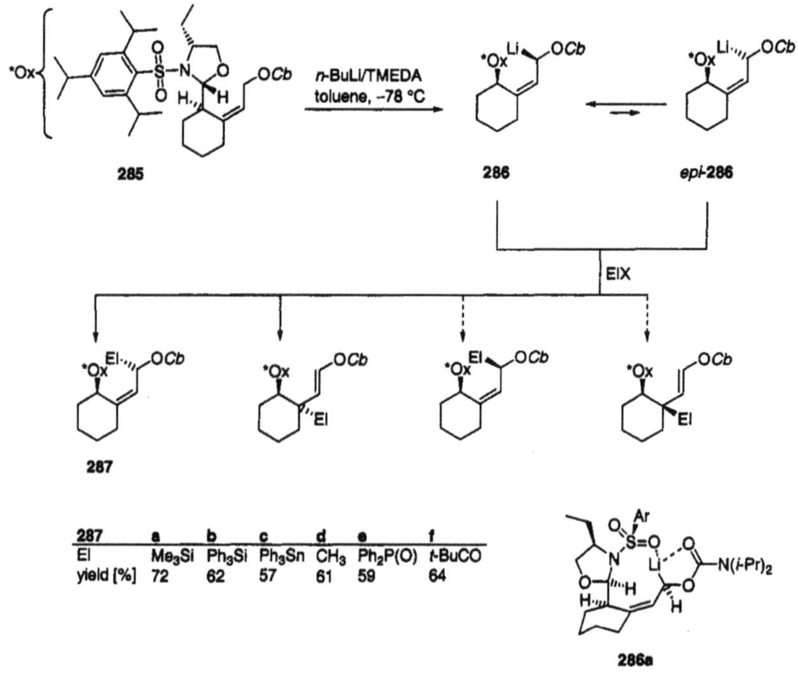
n-BuLi/TMEDA
toluene, –78 °C
285
286
epi-286
ElX
287
286a
287 a b c d e f
El Me3Si Ph3Si Ph3Sn CH3 Ph2P(O) t-BuCO
yield [%] 72 62 57 61 59 64

(85)

1-(*p*-Toluenesulfonyl)-2-alkenyl carbamates **288** have highly acidic α-protons [185]. The racemic carbanions **289** add to chiral aldehydes **290**, and diastereomerically pure alkenones **292** are finally isolated [Eq. (86)] [186–188]. The addition proceeds α-regioselectively to form the primary addition product **291.** A migration of the carbamoyloxy group takes place, and the lithium salt of the α-hydroxy sulfone looses lithium *p*-toluenesulfinate irreversibly, shifting the equilibrium completely towards the finals products. Clearly, only the Felkin-Ahn product **291** is involved in the product formation. The functionalized, stereochemically homogeneous enones **292** offer many possibilities for further useful transformations, as was demonstrated in the synthesis of the aminotriol **293** by means of standard transformations from **292b** [188].

(86)

292	R^1	R^2	R^3	X	Yield (%)
a	H	CH_3	CH_3	OBn	55
b	H	CH_3	$(CH_3)_2CHCH_2$	NBn_2	83
c	CH_3	CH_3	CH_3	OTBS	62

4.3
Addition to Aldehydes and Ketones; Enantioselective Homoaldol Reactions

4.3.1
Applying Racemic 1-Metallo-2-Alkenyl Carbamates

α-Metallated 2-alkenyl carbamates, in particular the titanium derivatives, add to aldehydes and form the corresponding (1*Z*)-3,4-*anti*-4-hydroxy-1-alkenyl carbamates with high simple diastereoselectivity [7, 64, 159, 189]. Since the reaction proceeds via a Zimmerman-Traxler transition state, an efficient reagent controlled chirality transfer is the result. Eq. (87) illustrates the reaction of the racemic metallo-carbamate **266a** with the enantioenriched aldehyde (*S*)-**295** [190, 191, 192]. The diastereomeric addition products **296** and **297** were formed in a 47:53 ratio; but when applying the racemic aldehyde *rac*-**295**, the ratio *rac*-**296** : *rac*-**297** shifted to 32:68. (*S*)-**266a** and (*S*)-**295**, respectively (*R*)-**266a** and (*R*)-**295** form the matched pair [72]. These combinations react preferentially when both compounds are present in racemic form. Since no interconversion between the enantiomers (*R*)- and (*S*)-**266a** is possible, these are configurationally stable under the reaction conditions on the time scale of the reactions. Thus, these undergo carbonyl addition independent from each other to complete conversion, although at different rates. If a rapid interconversion of the carbanionic reagent occurs, both ratios must be equal. The "Hoffmann test on configurational stability" [191, 193] is based on these facts, and the depicted example was an important precedent. *syn*- and *anti*-diol derivatives such as **296** and **297** have a different polarity on silica gel due to their different ability to undergo intramolecular hydrogen bridges; *syn*-**296** is the less polar one. This allows for an extremely facile separation of these diastereomers and a safe stereochemical assignment [194, 195]. As a consequence, this approach is attractive even if only racemic homoenolate reagents of type **266** are available.

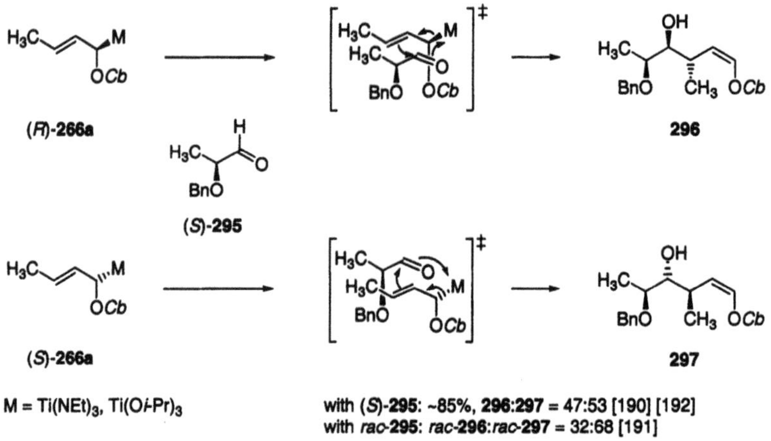

(*R*)-**266a** (*S*)-**295** **296**

(*S*)-**266a** **297** (87)

M = Ti(NEt)$_3$, Ti(O*i*-Pr)$_3$

with (*S*)-**295**: ~85%, **296**:**297** = 47:53 [190] [192]
with *rac*-**295**: *rac*-**296**:*rac*-**297** = 32:68 [191]

Fig. 5 collects examples for this strategy (only one diastereomer is shown).

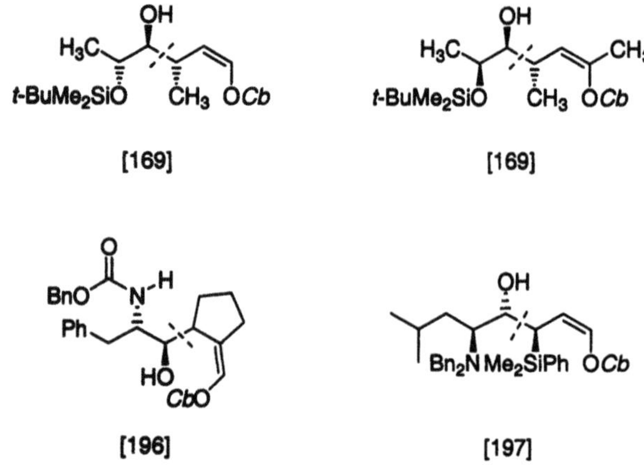

[169] [169]

[196] [197]

Fig. 5.

It has been demonstrated by Pancrazi, Ardisson et al., that an efficient kinetic resolution takes place when an excess of the racemic titanoalkenyl carbamate **266a** is allowed to react with the enantiopure ω-hydroxy-aldehyde **300** or alternatively the corresponding γ-lactol **299**, since the mismatched pair contributes to a lower extent to the product ratio [Eq. (88)] [198].

Under the best conditions, the ratio of the enantiomerically pure diastereomers 3,4-*anti*-4,5-*syn* (**301**) and 3,4-*anti*-4,5-*anti* (**302**) is close to 14:1. Surprisingly, approx. 9% of the *syn,syn*-diastereomer **303** were isolated. According to our own studies and those by Kocienski [199] 3,4-*syn*-diastereomers result when the starting (*E*)-crotyl carbamate is contaminated by the (*Z*)-isomer. The reasons which apply here are unknown. Extra base has to be used in order to neutralize the free hydroxy group. The pure *anti,anti*-addition product **305** was obtained with 85% yield from the reaction of the α-oxy-substituted titanate *rac*-**304** and lactol **299** [200]. Compound **305** is an intermediate in the asymmetric synthesis of tylosine.

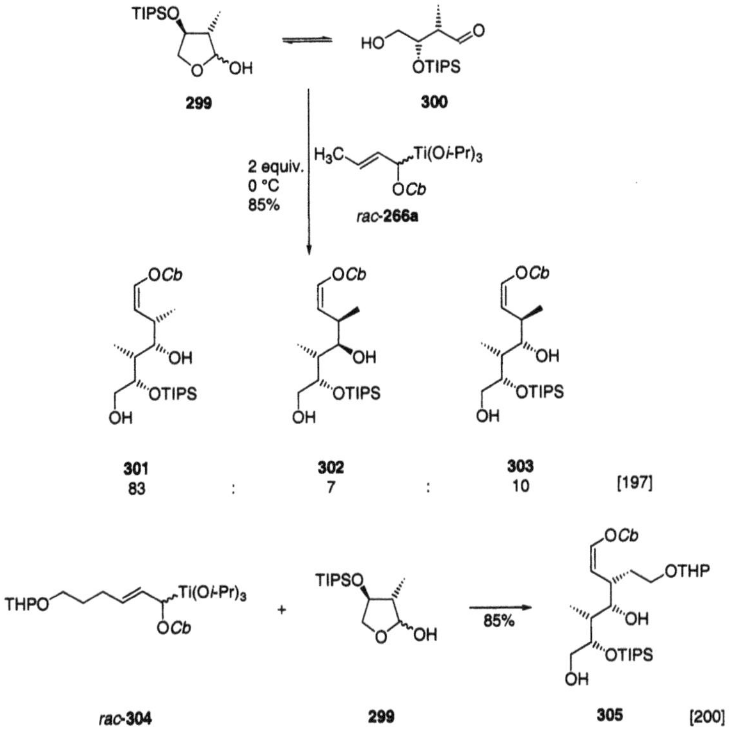

(88)

4.3.2
Applying Enantioenriched 1-Metallo-2-Alkenyl Carbamates

When employing enantioenriched 1-titano-2-alkenyl carbamates **266** in carbonyl addition, the selectivity depends on the enantiomeric purity which has been achieved in its preparation (see Sect. 4.2). The (*E*)-crotyl derivative (*R*)-**266a** has been employed several times [Eq. (89)] [201–203].

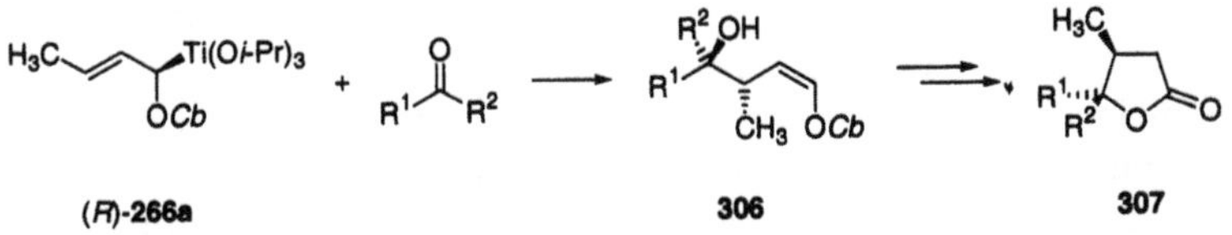

(89)

The optically active homoaldol products **306** are easily converted into γ-lactones **307** by four different pathways, which require an oxidation step (see Sect. 4.4). Applications in target synthesis include the natural products (+)-quer-

Table 13. Various optically active homoaldol products

306	R^1	R^2	Yield [%]	ee [%]	Yield 307 [%]	Ref.
a	$(CH_3)_2CH$	H	90	90[a]	89	[201]
b	CH_3	H	95	80[a]	–	[201]
c	$CH_3(CH_2)_3$	H	93	84[a]	90	[201]
d	CH_3CH_2	H	90	–[b]	–	[204, 205]
e	$(CH_3)_2C{=}CH{-}CH_2$	H	62	92	70	[202]
f	$H_2C{=}CCH_3$	H	78	86[a]	–	[203]
g	$H_2C{=}C$*i*-Pr	H	81	90	–	[203]
h	CH_3	CH_3	92	82	–	[201]

[a] Non-optimized preparation of **266a**.
[b] Racemic.

cus lactone (**307c**) [201] and (+)-eldanolide (**307e**) [202]. The reagent (*R*)-**266a** has also been applied in total syntheses of more complicated natural products: dihydroavermectin B_{1b} by Julia et al. [206], tylonolide [200] and *rac*-tylonolide [207] by Pancrazi, Ardisson et al., and jaspamide [208] and herboxidiene [209] by Kocienski et al. Some alkenyl carbamates **264** leading to configurationally unstable lithium intermediates could be subjected to asymmetric homoaldol reaction with less efficiency (Fig. 6); these reactions have not been optimized yet [158, 201].

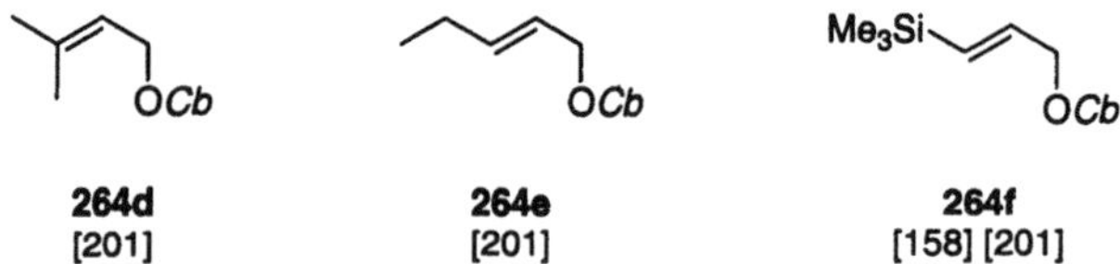

Fig. 6.

The titanium compounds, derived from the configurationally stable lithium/(–)-sparteine complexes **283b,d**, derived in turn from primary alkyl carbamates, undergo lithium-titanium exchange with chlorotriisopropoxytitanium to form the allyltitanates **308** [182]. These add to aldehydes providing the homoaldol products **309** with high stereoselectivity following the expected stereochemical course, as could be elucidated by several X-ray crystal structure analyses under anomalous dispersion. It is currently unknown why the yields are relatively low (21–35%), since we could not detect any side products besides traces of starting material **282**. The corresponding Li/TMEDA complexes, after titanation, deliver normal yields (71–79%). The homoaldol products are easily converted to enantioenriched bicyclic γ-lactones of type **310** [183].

Enantioenriched 3-(trialkylstannyl)alkenyl carbamates are accessible from the lithium compound (*S*)-**265a** in both enantiomeric forms [173]. These can be kept for several days in the refrigerator. Metal exchange with titanium tetrachloride in the presence of an aldehyde or ketone generates the highly reactive α-trichlorotitanium intermediate **312**, leading to homoaldol adducts **314** or *ent*-**314**, respectively. From the configuration of these adducts it is concluded that the

Li·(–)-sparteine OCb CH_3 — $ClTi(O\textit{i}\text{-}Pr)_3$ → $Ti(O\textit{i}\text{-}Pr)_3$ OCb CH_3 — RCHO →

283b, n = 0
283d, n = 1

308b, n = 0
308d, n = 1

(90)

309b, n = 0, R = 4-BrC_6H_4, 29%, 92% *ee*
309d, n = 1, R = 4-BrC_6H_4, 35%, 87% *ee*

309 → **310**

metal exchange proceeds as an *anti*-S_E' process, followed by the usual carbonyl addition via a Zimmerman-Traxler transition state, accompanied by a high degree of chirality transfer [Eq. (91)]. The stereochemical course is in agreement with results of Marshall [210, 211] and previous results, achieved with 1-substituted derivatives [168].

H_3C R_3Sn OCb — $TiCl_4$ (*anti*-S_E') → H_3C $TiCl_3$ OCb — R^LR^SCHO →

(*R*)-**311** (*R*)-**312**

(91)

313 → **314**

Table 14. Enantioselective Lewis acid catalyzed homoaldol reactions [Table to Eq. (91)]

Product (314)	Starting material (% ee)	R^L	R^S	Yield [%]	ee [%]	Ref.
a	(*R*)-**311a** (90)	$(CH_3)_2CH$	H	91	88	[173]
ent-**314a**	(*S*)-**311a** (82)	$(CH_3)_2CH$	H	82	82	[173]
a	(*R*)-**311b** (96)	$(CH_3)_2CH$	H	96	96	[173]
b	(*R*)-**311b** (80)	$(CH_3)_3C$	CH_3	80	74	[173]
c	(*R*)-**311b** (94)	$EtOCO(CH_2)_3$	CH_3	84	94	[173]
d	(*R*)-**311b** (94)	$EtOCO(CH_2)_3$	CH_3	91[a]	94	[173]
e	(*R*)-**311b**	C_2H_5	$PhCO(CH_2)_3$	67[b]	94	[212]

[a] Isolated as its corresponding lactone.
[b] 29% isolated as the corresponding dihydropyran from intramolecular condensation.

4.4 Synthetic Transformations and Applications of 4-Hydroxy-1-Alkenyl Carbamates

The vinyl carbamates of type **315**, although being enol esters, are very stable under acidic or basic conditions and can be handled, similar to usual alkenes, with-

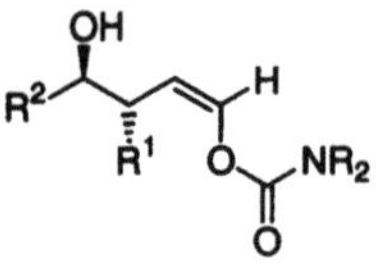

315

Fig. 7.

out problems [7]. Quite a number of synthetically useful transformations have been developed over the years. Five aspects can be addressed selectively by suitable reagents (Fig. 7)

- the enolic double bond,
- the C=O group of the carbamoyl moiety,
- the 4-hydroxy group,
- the highly acidic 1-H,
- the whole carbamoyl group by vinylic substitution.

Several of the reactions, mentioned in this chapter, proved to be successful for diastereomerically pure, but racemic substrates **315**. No reason is seen why racemization or epimerization should occur and therefore application to optically active substrates is possible without expecting difficulties. Of course, one must take into account possible sensitive functional groups present in the residues R^1 and R^2.

Atom C-1 in **315** is in the oxidation stage of an aldehyde, or a ketone in 1-alkylated products. The necessary solvolytic attack of the enolic double bond requires substoichiometric amounts of a catalyst such as mercuric acetate or palladium chloride and one equivalent of acid (e.g., methanesulfonic acid) for binding the liberated diisopropylamine. If the 4-hydroxy group is free, then when using methanol as solvent, anomeric lactol ethers **316** are formed with high yields [Eq. (92)].

Oxidation by the Grieco method [213] leads to γ-substituted γ-lactones [Eq. (92)] [174, 183, 196, 197, 214, 215].

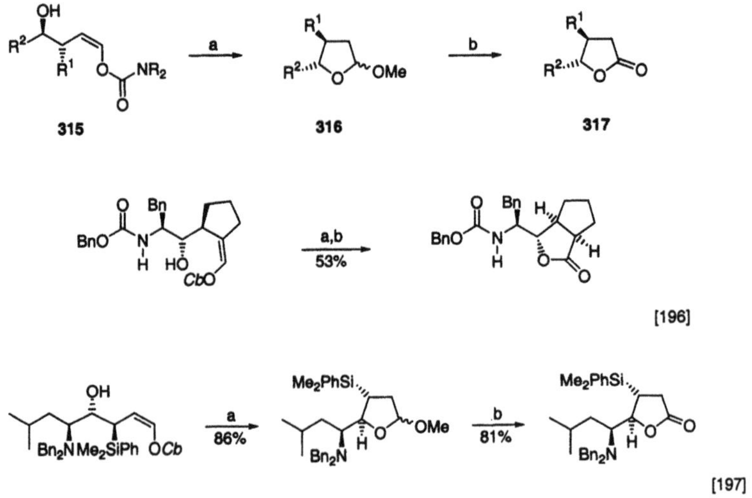

315 → (a) 316 → (b) 317

(a,b; 53%) [196]

(a, 86%; b, 81%) [197]

(92)

a) 0.1-0.25 equiv $Hg(OAc)_2$, 2 equiv $MeSO_3H$, MeOH.
b) m-CPBA, $BF_3 \cdot Et_2O$ or $MeSO_3H$, CH_2Cl_2.

Another approach for the deprotection of highly functionalized enol carbamates is based on hydroboration [Eq. (93)] [196, 197].

Cbo

a,b,c 48% [196]

b 75% d 63% [197]

(93)

a) 2-methoxypropene, cat. $POCl_3$, DMF.
b) i) $BH_3 \cdot Me_2S$ (2-4 equiv), 0–50°C; ii) 30% H_2O_2/NaOH.
c) PDC.
d) Jones reagent, acetone, r.t.

If Lewis acid-sensitive double bonds are located in the residues R^1 or R^2, an indirect oxidation of the vinylic hydrogen via lithiation/sulfenylation turned out to be a good solution [216, 217], and was applied during the final steps of (+)-eldanolide **321** synthesis [Eq. (94)] [216].

320 b 95% c 80% **321**

(94)

a **a** R = H; **b** R = $SiMe_3$ 92%

a) Me_3SiCl, Et_3N, CH_2Cl_2, 92%. b) i) *n*-BuLi/TMEDA, –78 °C; ii) MeSSMe. c) $MeSO_3H$, $MeOH/H_2O$, 40 °C. [216]

According to results published by Férézou et al. the *N,N*-diisopropylcarbamoyl group of homoaldol adducts can be directly attacked by "slim" nucleophiles such as lithium ethynylide or excess methyllithium [Eq. (95)] [218]. The TIPS ether **323** was treated with 3 equiv. of methyllithium to yield (via the *Z*-enolate **325**) the aldehyde **327**. With the free alcohol **322** and 4 equiv. of methyllithium the lactol **324** is obtained. Trapping of **325** by TBSCl gives rise to the synthetically valuable Z-silyl enol ether **326**.

322 R = H
323 R = Si(*i*-Pr)$_3$

a R = TIPS; **325**; + Me_2COLi + *i*-Pr_2NLi; b 42%; **327**

R = H a,b; c 84%

324 88%; **326**

(95)

a) 3 or 4 equiv. MeLi, THF, 0°C.
b) NH_4Cl/H_2O.
c) *t*-$BuMe_2SiCl$, TMEDA/THF.

[218]

Z-anti-4-Hydroxy-1-alkenyl carbamates **328**, when subjected to substrate-directed, vanadyl-catalyzed epoxidation [219, 220, 221], lead to diastereomerically pure epoxides of type **329** [Eq. (96)] [192, 203, 222]. These epoxides are highly reactive in the presence of Lewis or Brønsted acids to form β-hydroxylactol ethers **331**; in some cases the intermediate lactol carbamates **330** could be isolated [192]. However, most epoxides **329** survive purification by silica gel chromatography [192]. The asymmetric homoaldol reaction, coupled with directed epoxidation, and solvolysis rapidly leads to high stereochemical complexity. Some examples are collected in Eq. (96). The furanosides *ent*-**333** and *ent*-**332**, readily available from (*R*)-*O*-benzyllactaldehyde [192] have been employed in a short synthesis of the key intermediates of the Kinoshita rifamycin S synthesis [223, 224]. 1,5-Dienyl carbamates, obtained from 2-substituted enals, provide a facile access to branched carbohydrate analogues [203]. Epoxidation of the 1,2-double bond is followed by asymmetric attack at the 5,6-double bond; both oxirane rings in **334** can be opened separately under controlled conditions [203].

(96)

a) *t*-BuOOH, cat. $VO(acac)_2$ (1 mol%), CH_2Cl_2.
b) i) $MeSO_3H$ (1.0 equiv), MeOH, r.t.; ii) Et_3N.
c) i) HCl, H_2O/THF, 65 °C; ii) Et_3N.
d) i) *t*-BuOOH, cat. $VO(acac)_2$ (10 mol%), CH_2Cl_2; ii) *t*-BuOOH, $Ti(O\text{-}i\text{-}Pr)_4$, (+)-(*R*,*R*)-diisopropyl tartrate.
e) $ZnCl_2/Et_2O$, THF.

Strong nucleophiles applied under basic conditions attack epoxides **329** at C-2 with inversion of the configuration, leading to *trans,trans*-2,3,4-trisubstituted γ-lactols **335** which are easily oxidized to the corresponding γ-lactones **336** [Eq. (97)] [225].

The enol carbamates **337** can be regarded as weak enolate reagents. This type of reactivity is released by reactions with aldehydes, ketones, or the correspond-

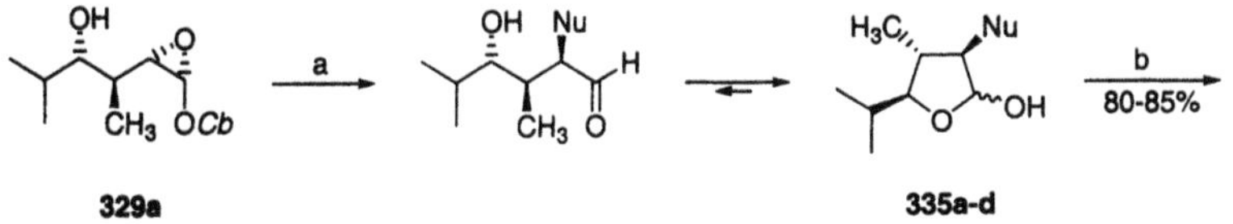

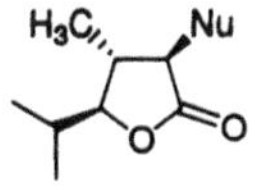

(97)

335	Nu	yield [%]
a	OAc	71
b	OPh	64
c	N_3	63
d	SPh	89

a) 2 equiv M^+Nu^-, DMF, 90 °C.
b) $CrO_3 \cdot 2$ pyridine. [225]

ing acetals under boron trifluoride catalysis [Eq. (98)] [226]. Stereochemically homogeneous, tri- or tetra-substituted 3-tetrahydrofuran-3-carbaldehydes or ketones **341** or **338** are isolated in high yields. Presumably, the (*E*)-oxonium ions **339** are formed via the attack at the free hydroxy group and undergo intramolecular Mukaiyama-type reactions to form the cations **340**, which are subsequently hydrolyzed to give the carbonyl compounds **341**. For cyclization, the intermediate **339** adopts the particular conformation which avoids 1,3-allylic strain [166]; this turns out to be a powerful tool for controlling the diastereoselectivity. Some examples are collected in Eq. (98). The 8-oxo derivatives **342** undergo an intramolecular version to form 8-oxabicyclo[3.2.1]octanes **343** [212].

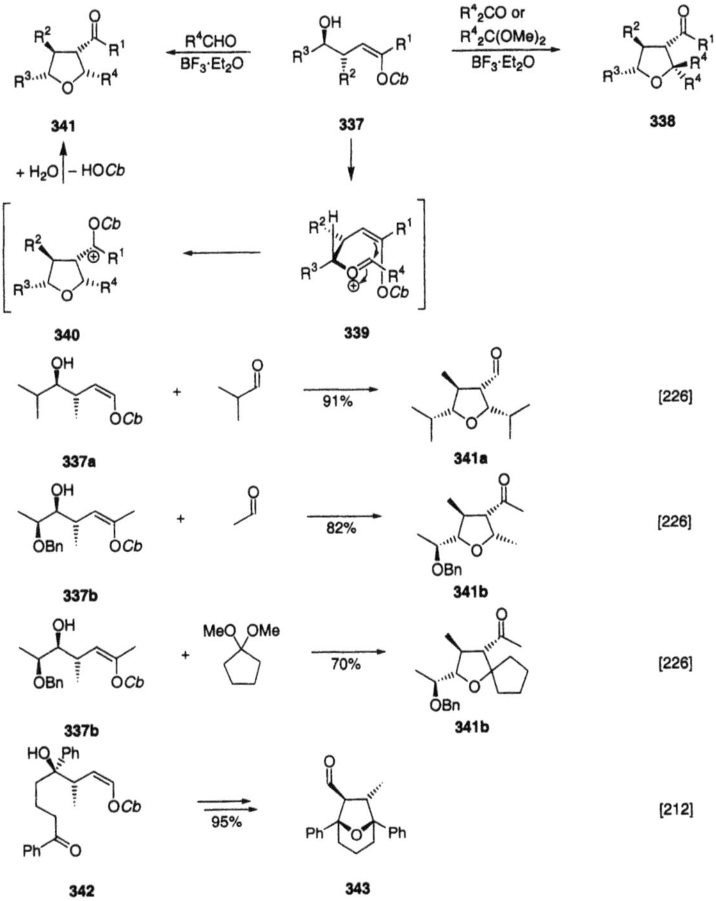

(98)

According to Eq. (99), stereochemically homogeneous 3-carbonyl-substituted tetrahydrofurans are constructed in a brick-box system by sequential homoaldol and aldol reaction. The metallated allyl carbamate serves as an equivalent for the chiral dianion **A**, which accepts two different aldehydes **B** and **C** in a highly controlled manner [226].

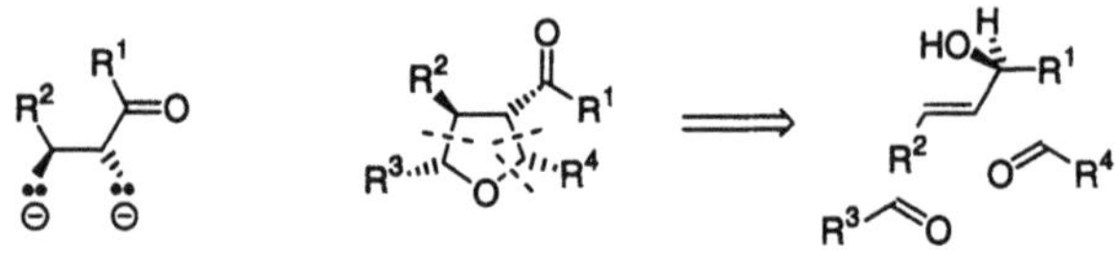

A 341 B C (99)

In comparison to other vinylic compounds [227], the vinyl proton in 1-alkenyl carbamates has a very high kinetic acidity [199, 228–230]. After protection of the 4-hydroxy group in the homoaldol products by silylation, deprotonation (*n*-BuLi, TMEDA, diethyl ether, or THF) is complete at −78 °C [Eq. (100)], and the resulting vinyllithium can be kept at this temperature without decomposition for several hours. Stannylation [199], silylation [229], methoxycarbonylation (with methyl chloroformate) [229], and alkylation (with propyl iodide/HMPA) [199] to form the carbamates **347** proceed without difficulties. Addition of aldehydes or ketones furnishes the alcoholates, such as **348**, which rearrange with migration of the carbamoyl group to give lithium enolates **350**; after aqueous workup, mixtures of alkenols **349** and ketones **351** are obtained [229]. Similar acyl migrations have been observed for simpler substrates by Snieckus and Sengupta [228].

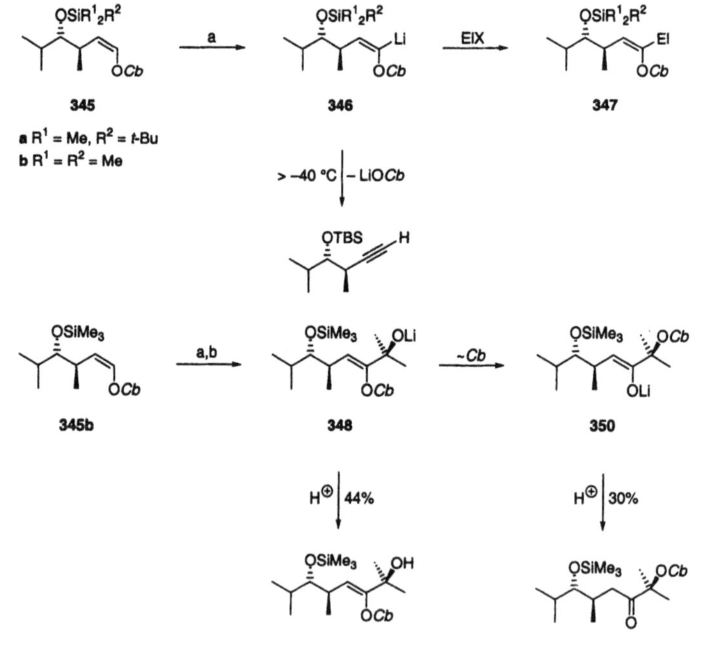

345 346 347

a R^1 = Me, R^2 = *t*-Bu
b R^1 = R^2 = Me

> −40 °C − LiOCb

345b 348 350 (100)

$H^{\oplus}$ 44% $H^{\oplus}$ 30%

349 351 [229]

a) BuLi/TMEDA, THF or Et_2O, −78 °C.
b) acetone.

Table 15. [Table to Eq. (100)]

347	R^1	R^2	El	Yield [%]	Ref.
a	Me	*t*-Bu	Me_3Sn	95	[199]
b	Me	*t*-Bu	$CH_2CH_2CH_3$	56	[199]
c	Me	Me	Me_3Si	93	[229]
d	Me	Me	CO_2Me	69	[229]

When warming these vinyllithiums of type **346** to temperatures above –40°C, a Fritsch-Buttenberg-Wiechell rearrangement, which includes the α-elimination of lithium carbamate and the migration (with retention of configuration) of the alkyl group attached at C-2 to C-1, provides 1-alkynes [199].

A number of further useful reactions, involving organometallic reagents, have been developed – mainly by Kocienski et al. – for converting 4-hydroxy-1-alkenyl carbamates into various alkenes [Eq. (101)] [199, 231–234]. For each reaction type, only one example is depicted.

OH [199] ← a, 70% — OR, OCb **345a** — b → [OTBS, R, Cu(CN), OCb]$^{2\ominus}$ 2 Li$^{\oplus}$ — 63-90% → OTBS, R [119]

d, 66% → OMe [234]; c, 50-80% → OTBS, Ar, SPh [232]

(101)

a) BuMgBr, (dppe)$NiCl_2$.
b) i) *n*-BuLi/TMEDA,; ii) Me_3SnCl; iii) RCu(CN)Li, –78 °C → 0 °C.
c) i) *t*-BuLi, > –40 °C; ii) $PhSSO_2Ph$, –10 °C → r.t.; iii) Bu_3SnH, Pd(0); iv) *n*-BuLi, THF, –78 °C; v) $ZnBr_2$/THF, → r.t.; vi) ArI, Pd(0).
d) i) 0.2 equiv EtMgBr, 0.1 equiv $NiCl_2(PPh_3)_2$; ii) 3.0 equiv $CH_3(CH_2)_3C{\equiv}C\text{-}MgBr$.

These reactions involve metallate rearrangements [231], migratory insertion and transition metal-catalyzed vinylic substitution reactions. They also perform well in applications in natural product synthesis [204, 205, 209, 233].

Many useful synthetic possibilities arise from application of ring-closing olefin metathesis (RCM) to unsaturated homoaldol products and their derivatives by means of the Grubbs catalyst **354** [235–237]. Eq. (102) collects some examples [217].

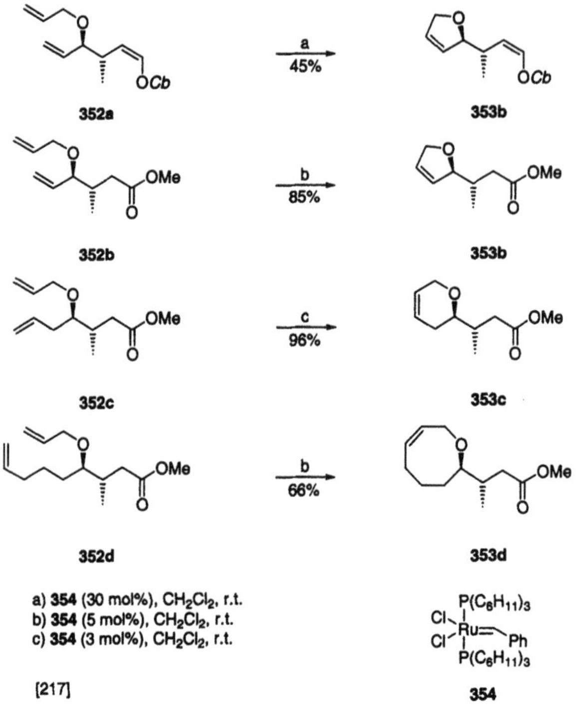

a) **354** (30 mol%), CH_2Cl_2, r.t.
b) **354** (5 mol%), CH_2Cl_2, r.t.
c) **354** (3 mol%), CH_2Cl_2, r.t.

$Cl_2Ru(=CHPh)(P(C_6H_{11})_3)_2$ **354**

[217]

(102)

The *N*,*N*-diisopropylcarbamoyloxy group, as seen in the allylic ether **352a**, interferes with the catalyst; high amounts of **354** are required. The metathesis reaction proceeds more smoothly when the vinyl carbamate is converted into an ester (see subsequent examples).

5 Metallated 2-Alkynyl Carbamates

2-Alkynyl carbamates **355** are acidic enough to be rapidly deprotonated by *n*-butyllithium/TMEDA in diethyl ether [Eq. (103)] [238, 239]. The resulting lithium compounds *rac*-**356** possess a propargylic structure; this was demonstrated in NMR investigations by Reich [240]. After lithium-titanium exchange, the titanium intermediates *rac*-**357** undergo carbonyl addition with high simple diastereoselectivity to form the *syn*-4-hydroxy-1,2-alkadienyl carbamates *rac*-**358**.

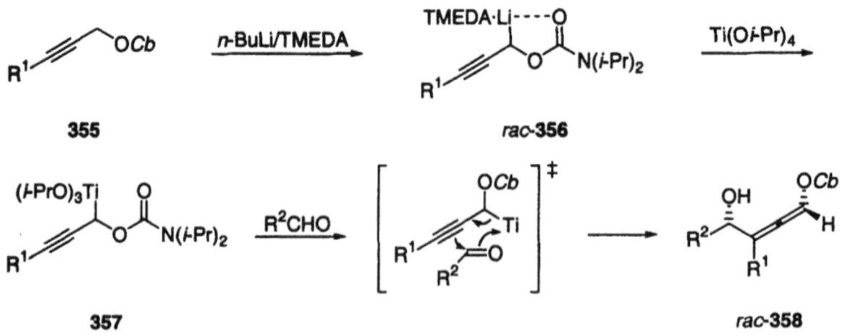

(103)

Some interesting elimination and rearrangement reactions for the allenic alcohols *rac*-**358** have been developed [241–243].

A hint to asymmetric deprotonation and a certain configurational stability of the resulting lithiated primary 2-alkynyl alkenyl ether, can be taken from a report by Marshall et al. [244]. In an extensive study, enantioenriched secondary alkynyl carbamates (*R*)-**359** were deprotonated (*n*-BuLi/TMEDA, hexane, 20 min at −78 °C) and added to achiral aldehydes to form mixtures of the adducts *syn*- and *anti*-**363** [Eq. (104)] [9]. It turned out that the ion pairs, derived from the 3-pentyn-2-yl derivative (*R*)-**359c**, underwent complete racemization. The *tert*-butyl substituted alkynyl carbamate (*R*)-**359a** retained its chiral information completely, resulting in a complete chirality transfer (*c.t.*=100%). It was shown that, in both diastereomers, the allene moiety has the identical sense of axial chirality. With the cyclohexyl derivative (*R*)-**359b**, some erosion of chiral information was observed. The low simple diastereoselectivity of the aldehyde addition could be improved by metal exchange by addition of titanium tetraisopropoxide [9]. With our present-day knowledge it is assumed that the lithium titanium exchange was incomplete under the applied reaction conditions and chlorotriisopropoxytitanium is expected to be a better-suited reagent [245].

(104)

a) *n*-BuLi/TMEDA, Et_2O, 20 min, −78 °C.
b) i) + R^2CHO; ii) aqueous workup.

Table 16. [Table to Eq. (104)]

363	R^1	R^2	*c.t.* [%]	Ref.
a	*t*-Bu	CH_3	100	[9]
b	*c*-C_6H_{11}	*t*-Bu	73	[9]
c	CH_3	$CH(CH_3)_2$	0	[9]

Hydrolysis of (*R*)-**360a** furnished the allene **361** (of unknown configuration) with complete chirality transfer [9].

The deprotonation of primary 2-alkynyl carbamates **364** by the sparteine method leads to complexes **365** which are not configurationally stable in solution (diethyl ether, toluene, pentane) even at –78 °C [245]. For intermediates (*S*)-**365a** and (*S*)-**365b** conditions could be found for the kinetic resolution by preferential crystallization of one of the diastereomers. (*S*)-**365a,b** are transmetallated by rapid addition of chlorotitanium tri(isopropoxide) from the solid state with complete inversion of the configuration; subsequent addition of an aldehyde leads to the highly enantioenriched, diastereomerically pure allenyl carbinol **367** with greater than 93% ee [245]. Protonation of the titanium intermediates (*S*)-**366a,b** furnishes the enantioenriched allenyl carbamates (*M*)-**368a,b** (80%, 84% ee, and 86%, 88% ee, respectively) [245].

(105)

a) *s*-BuLi/(–)-sparteine, pentane, –78 °C. b) Crystallization of (*S*)-**365** or **b**.
c) $ClTi(O\text{-}i\text{-}Pr)_3$, –78 °C. d) HOAc. e) i) R^2CHO, –78 °C; ii) HOAc.

Allene **368a** is deprotonated by *n*-butyllithium/TMEDA adjacent to the carbamoyl group, leading to the configurationally stable ion pair **369a** with axial chirality (see Clayden in this volume)] [Eq. (106)] [246]. By addition to α,β-enones to yield the alkoxides **370a**, an interesting series of subsequent rearrangements is triggered [246].

OCb
H
H
s-BuLi/TMEDA
toluene, –78 °C
OCb
Li·TMEDA
H

368a

369a

(106)

CbO
OLi
R^1
R^2-CH=CH-CO-R^1
H
R^2
[246]

370a

Transient enantioenriched 1-lithio-1-alkoxyalkynides have been generated via lithiodestannylation by Nakai for enantioselective Wittig rearrangements [247] (s. a. Hodgson et al, in this volume).

References

1. For the sake of simplicity, the 1-oxy-substituted lithium compounds such as **2** are drawn in a simplified form although it is known that the C-Li-bond has only a small covalent portion (if at all) and a high tendency for the formation of dimers or tetramers exists. Furthermore, the lithium cation is coordinatively saturated by Lewis-basic ligands; see: Lambert C, Schleyer PvR (1993) Carbanionen – polare Organometall-Verbindungen. In: Hanack M (ed) Houben-Weyl, Carbanionen, vol. E 19 d, 4th edn. Thieme-Verlag, Stuttgart, p 1
2. Still WC, Sreekumar C (1980) J Am Chem Soc 102:1201
3. Jephcote VJ, Pratt AJ, Thomas EJ (1989) J Chem Soc Perkin Trans 1 1529
4. The term "chiral carbanion" or "chiral lithium reagent" is used in this article only for those reagents that bear the negative charge directly at a stereogenic carbon atom (or moiety). Oxy-substituted enolates and related species are not covered since the sp^2-hybridized enolate moiety is not chiral
5. Evans DA, Andrews GC, Buckwalter B (1974) J Am Chem Soc 96:5560
6. Ahlbrecht H, Beyer U (1999) Synthesis 365
7. Hoppe D (1984) Angew Chem 96:930; Angew Chem Int Ed Engl 23:932
8. Hoppe D, Krämer T (1986) Angew Chem 98:171; Angew Chem Int Ed Engl 25:160
9. Dreller S, Dyrbusch M, Hoppe D (1991) Synlett 397
10. Chong JM, Mar EK (1991) Tetrahedron Lett 32:5683
11. Chan PCM, Chong JM (1988) J Org Chem 53:5584
12. Tomooka K, Igarashi T, Nakai T (1994) Tetrahedron Lett 35:1913
13. Matteson DS, Tripathy PB, Sarkar A, Sadhu KM (1989) J Am Chem Soc 111:4399
14. Marshall JA, Gung WY (1989) Tetrahedron 45:1043
15. Marshall JA (1992) Chemtracts, Org Chem 5:75
16. Sawyer JS, Kucerovy A, Macdonald TL, McGarvey GJ (1988) J Am Chem Soc 110:842
17. McGarvey GJ, Kimura M (1982) J Org Chem 47:5420
18. Stork G, Manabe K, Liu L (1998) J Am Chem Soc 120:1337
19. Yamada JI, Abe H, Yamamoto Y (1990) J Am Chem Soc 112:6118
20. Yamamoto Y (1991) Chemtracts, Org Chem 4:255
21. Chong JM, Mar EK (1989) Tetrahedron 45:7709
22. Smyj RP, Chong JM (2001) Org Lett 3:2903
23. Chan PCM, Chong JM (1990) Tetrahedron Lett 31:1985

24. Chong JM, Mar EK (1990) Tetrahedron Lett 31:1981
25. Linderman RJ, Griedel BD (1990) J Org Chem 55:5428
26. Linderman RJ, Griedel BD (1991) J Org Chem 56:5491
27. Hutchinson DK, Fuchs PL (1987) J Am Chem Soc 109:4930
28. Christopher JA, Kocienski PJ, Procter MJ (1998) Synlett 425
29. Christopher JA, Kocienski PJ, Kuhl A, Bell R (2000) Synlett 463
30. Screttas CG, Micha-Screttas M (1978) J Org Chem 43:1064
31. Cohen T, Daniewski WM, Weisenfeld RB (1978) Tetrahedron Lett 4665
32. Cohen T, Bhupathy M (1989) Acc Chem Res 22:152
33. Cohen T, Lin MT (1984) J Am Chem Soc 106:1130
34. Rychnovsky SD, Mickus DE (1989) Tetrahedron Lett 30:3011
35. Lancelin JM, Morin-Allory L, Sinaÿ P (1984) J Chem Soc Chem Commun 355
36. Fernandez-Mayoralas A, Marra A, Trumtel M, Veyrières A, Sinaÿ P (1989) Tetrahedron Lett 30:2537
37. Pedretti V, Veyrières A, Sinaÿ P (1990) Tetrahedron 46:77
38. Beau JM, Sinaÿ P (1985) Tetrahedron Lett 26:6185
39. Beau JM, Sinaÿ P (1985) Tetrahedron Lett 26:6189
40. Beau JM, Sinaÿ P (1985) Tetrahedron Lett 26:6193
41. Frey O, Hoffmann M, Kessler H (1995) Angew Chem 107:2194; Angew Chem Int Ed 34:2026
42. Hoffmann M, Kessler H (1995) Tetrahedron Lett 35:6067
43. Rychnovsky SD, Mickus DE (1989) Tetrahedron Lett 30:3011
44. Rychnovsky SD, Buckmelter AJ, Dahanukar VH, Skalitzky DJ (1999) J Org Chem 64:6849
45. Rychnovsky SD, Zeller S, Skalitzky DJ, Griesgraber G (1990) J Org Chem 55:5550
46. Rychnovsky SD (1995) Chem Rev 95:2021
47. Rychnovsky SD, Griesgraber G, Schlegel R (1994) J Am Chem Soc 116:2623
48. Rychnovsky SD, Griegraber G, Kim J (1994) J Am Chem Soc 116:2621
49. Rychnovsky SD, Hoye RC (1994) J Am Chem Soc 116:1753
50. Richardson TI, Rychnovsky SD (1996) J Org Chem 61:4219
51. Rychnovsky SD, Khire UR, Yang G (1997) J Am Chem Soc 119:2058
52. Richardson TL, Rychnovsky SD (1997) J Am Chem Soc 119:12360
53. Ley SV, Jones P (1998) Chemtracts, Org Chem 11:1005
54. Schneider C (1998) Angew Chem 110:1445; Angew Chem Int Ed Engl 37:1375
55. Yamauchi Y, Katagiri T, Uneyama K (2002) Org Lett 4:173
56. Hoppe D, Paetow M, Hintze F (1993) Angew Chem 105:430; Angew Chem Int Ed Engl 32:394
57. Paetow M, Kotthaus M, Grehl M, Fröhlich R, Hoppe D (1994) Synlett 1034
58. Schlosser M (2002) Organoalkali chemistry. In: Schlosser M (ed) Organometallics in synthesis, 2nd edn. Wiley, Chichester, p 1
59. Schlosser M (1992) Superbases as powerful tools in organic synthesis. In: Scheffold R (ed) Modern synthetic methods, vol. 6, Springer, Berlin, p 227
60. Mordini A (1992) Superbases and their use in organic synthesis. In: Snieckus V (ed) Advances in carbanion chemistry, vol. 1, Jai Press Inc., Greenwich, Connecticut, p 1
61. Lochmann L (2000) Eur J Inorg Chem 1115
62. Hoppe D, Hintze F, Tebben P (1990) Angew Chem 102:1457; Angew Chem Int Ed Engl 29:1422
63. Hoppe D, Hintze F, Tebben P, Paetow M, Ahrens H, Schwerdtfeger J, Sommerfeld P, Haller J, Guarnieri W, Kolczewski S, Hense T, Hoppe I (1994) Pure Appl. Chem 66:1479
64. Hoppe D, Hense T (1997) Angew Chem 109:2376; Angew Chem Int Ed Engl 36:2282
65. Haller J, Hense T, Hoppe D (1993) Synlett 726
66. Haller J, Hense T, Hoppe D (1996) Liebigs Ann Chem 489
67. Schwerdtfeger J, Hoppe D (1992) Angew Chem 104:1547; Angew Chem Int Ed Engl 31:1505
68. Schwerdtfeger J, Kolczewski S, Weber B, Fröhlich R, Hoppe D (1999) Synthesis 1573
69. O'Donnell MJ, Delgado F, Hostettler C, Schwesinger R (1998) Tetrahedron Lett 39:8775, and previous work cited therein

70. Boie C, Hoppe D (1997) Synthesis 176
71. Weber B, Kolczewski S, Fröhlich R, Hoppe D (1999) Synthesis 1593
72. Masamune S, Choy W, Petersen JS, Sita LR (1985) Angew Chem 97:1; Angew Chem Int Ed Engl 24:1
73. Weber B, Schwerdtfeger J, Fröhlich R, Göhrt A, Hoppe D (1999) Synthesis 1915
74. Woltering MJ, Fröhlich R, Wibbeling B, Hoppe D (1998) Synlett 797
75. Guarnieri W, Grehl M, Hoppe D (1994) Angew Chem 106:1815; Angew Chem Int Ed Engl 33:1734
76. Guarnieri W, Sendzik M, Fröhlich R, Hoppe D (1998) Synthesis 1274
77. Sendzik M, Guarnieri W, Hoppe D (1998) Synthesis 1287
78. Ahrens H, Paetow M, Hoppe D (1992) Tetrahedron Lett 33:5327
79. Helmke H, Hoppe D (1995) Synlett 978
80. Helmke H (1995) Dissertation, University of Münster
81. Hoppe D (1995) (-)-Sparteine. In: Paquette LA (ed), Encyclopedia of reagents for organic synthesis, vol 7, 1st edn. Wiley, Chichester, p 4662
82. Aggarwal VK (1994) Angew Chem 106:185; Angew Chem Int Ed Engl 33:175
83. Sigma No S2126, Aldrich No 41,531-6
84. Ebner T, Eichelbaum M, Fischer P, Meese CO (1989) Arch Pharm 322:399
85. Würthwein E-U, Behrens K, Hoppe D (1999) Chem Eur J 5:3459
86. Hintze F, Hoppe D (1992) Synthesis 1216
87. Hoppe I, Marsch M, Harms K, Boche G, Hoppe D (1995) Angew Chem 107:2328; Angew Chem Int Ed Eng 34:2158
88. Hintze F (1993) Dissertation, University of Kiel
89. Rein KS, Chen Z-H, Perumal PT, Echegoyen L, Gawley RE (1991) Tetrahedron Lett 32:1941 and references therein
90. Papillon JPN, Taylor RJK (2002) Org Lett 4:119
91. Tomooka K, Shimizu H, Nakai T (2001) J Organomet Chem 624:364
92. Behrens K, Fröhlich R, Meyer O, Hoppe D (1998) Eur J Org Chem 2397
93. Hoppe I, unpublished results
94. Boie C (1996) Dissertation, University of Münster
95. Sommerfeld P, Hoppe D (1992) Synlett 764
96. Paetow M, Ahrens H, Hoppe D (1992) Tetrahedron Lett 33:5323
97. Ahrens H (1994) Dissertation, University of Münster
98. Christoph G, Hoppe D (2002) Org Lett 4:2189
99. Hoppe D, Ahrens H, Guarnieri W, Helmke H, Kolczewski S (1996) Pure Appl Chem 68:613
100. For a more recent example, combined with chiral amplification, see Sato I, Omiya D, Saito T, Soai K (2000) J Am Chem Soc 122:11739
101. Würthwein E-U, unpublished results
102. For calculations on the (-)-sparteine-mediated *N*-Boc-pyrrolidines, see Wiberg KB, Bailey WF (2000) Angew Chem 112:2211; Angew Chem Int Ed 39:2127
103. Paetow M (1993) Dissertation, University of Kiel
104. van Bebber J (1997) Dissertation, University of Münster
105. Tomooka K, Shimizu H, Inoue T, Shibata H, Nakai T (1999) Chem Lett 759
106. Boche G, Bosold F, Lohrenz JCW, Opel A, Zulauf P (1993) Chem Ber 126:1873
107. For a similar reaction of -(dibenzylamino)alkyl carbamates, see ref. [71]
108. Kotthaus M (1997) Dissertation, University of Münster
109. Marek I (1999) J Chem Soc Perkin Trans 1 535
110. Woltering MJ, Fröhlich R, Hoppe D (1997) Angew Chem 109:1804; Angew Chem Int Ed Engl 36:1764
111. Tomooka K, Komine N, Sasaki T, Shimizu H, Nakai T (1998) Tetrahedron Lett 39:9715
112. Hoppe D, Woltering MJ, Oestreich M, Fröhlich R (1999) Helv Chim Acta 82:1860
113. Oestreich M, Fröhlich R, Hoppe D (1998) Tetrahedron Lett 39:1745
114. Oestreich M, Fröhlich R, Hoppe D (1999) J Org Chem 64:8616
115. Oestreich M, Hoppe D (1999) Tetrahedron Lett 40:1881, 3283
116. Gralla G, Wibbeling B, Hoppe D (2002) Org Lett 4:2193

117. Christoph G (2002) Dissertation, University of Münster
118. Seebach D, Prelog V (1982) Angew Chem 94:696; Angew Chem Int Ed Engl 21:654
119. Hense T, Hoppe D (1997) Synthesis 1394
120. van Bebber J, Ahrens H, Fröhlich R, Hoppe D (1999) Chem Eur J 5:1905
121. Laqua H, Fröhlich R, Wibbeling B, Hoppe D (2001) J Organomet Chem 624:96
122. For another example of different reactivity of diastereomers in the cyclocarbolithiation see ref. [111]
123. Hodgson DM, Lee GP (1996) J Chem Soc Chem Commun 1015
124. Hodgson DM, Robinson LA (1999) J Chem Soc Chem Commun 309
125. Hodgson DM, Cameron ID (2001) Org Lett 3:441
126. Hodgson DM, Gras E (2002) Angew Chem 114:2482; Angew Chem Int 41:2376
127. Streitwieser Jr. A, Juaristi E, Nebenzahl LL (1980) Equilibrium carbon acidities in solution. In: Buncel E, Durst T (eds) Comprehensive carbanion chemistry, vol. 5A, Elsevier, Amsterdam, p 323
128. Ruhland T, Hoffmann RW, Schade S, Boche G (1995) Chem Ber 128:551
129. Beak P, Basu A, Gallagher DJ, Park YS, Thayumanavan S (1996) Acc Chem Res 29:552
130. Basu A, Thayumanavan S (2002) Angew Chem 114:740; Angew Chem Int 41:717
131. Hoppe D, Carstens A, Krämer T (1990) Angew Chem 102:1455; Angew Chem Int Ed Engl 29:1424
132. Hammerschmidt F, Hanninger A (1995) Chem Ber 128:1069
133. Carstens A, Hoppe D (1994) Tetrahedron 50:6097
134. Some experiments were carried out with (*S*)-**214a**, the results have been transferred for the sake of simplicity to the enantiomer (*R*)-**214b**
135. Derwing C (1995) Dissertation, University of Münster
136. Carstens A (1993) Dissertation, University of Kiel
137. Derwing C, Hoppe D (1996) Synthesis 149
138. Brewster JH, Braden WE Jr (1964) Chem Ind 1759
139. Mitsui S, Imaizumi S, Senda Y, Konno K (1964) Chem Ind 233
140. Mitsui S, Senda Y, Konno K (1963) Chem Ind 1354
141. Zhang P, Gawley RE (1993) J Org Chem 58:3223
142. Hoppe D (1997) Generation and reactions of chiral, mesomerically stabilized lithium carbanions. In: Nair V, Kumar S (eds) New horizons in organic synthesis. New Age International Publishers, New Delhi, p 130
143. Derwing C, Frank H, Hoppe D (1999) Eur J Org Chem 3519
144. Hammerschmidt F, Hanninger A, Simov BP, Völlenkle H, Werner A (1999) Eur J Org Chem 3511
145. Peters JG, Seppi M, Fröhlich R, Wibbeling B, Hoppe D (2002) Synthesis 381
146. Superchi S, Sotomayor N, Miao G, Joseph B, Campbell MG, Snieckus V (1996) Tetrahedron Lett 37:6061
147. Hoffmann RW, Rühl T, Harbach J (1992) Liebigs Ann Chem 725
148. Caddick S, Jenkins K (1996) Chem Soc Rev:447
149. Retzow S (1993) Dissertation, University of Kiel
150. Heinl T, Retzow S, Hoppe D, Fraenkel G, Chow A (1999) Chem Eur J 5:3464
151. For calculations on a higher level concerning the problem, see: Wiberg KB, Bailey WF (2000) J Mol Struct (special issue) 556:239
152. Komine N, Wang L-F, Tomooka K, Nakai T (1999) Tetrahedron Lett 40:6809
153. Tomooka K, Wang L-F, Komine N, Nakai T (1999) Tetrahedron Lett 40:6813
154. Schwark J-R, Hoppe D (1990) Synthesis 291
155. Zschage O, Schwark J-R, Hoppe D (1990) Angew Chem 102:336; Angew Chem Int Ed Engl 29:296
156. Kunz H, Waldmann H (1990) Chemtracts Org Chem 3:421
157. Zschage O, Hoppe D (1992) Tetrahedron 48:8389
158. Marsch M, Harms K, Zschage O, Hoppe D, Boche G (1991) Angew Chem 103:338; Angew Chem Int Ed Engl 30:321
159. Hoppe D, Krämer T, Schwark JR, Zschage O (1990) Pure Appl Chem 62:1999

160. Hoppe D, Zschage O (1990) Chiral metallated carbamates: Tools for new strategies in asymmetric synthesis. In: Dötz KH, Hoffmann RW (eds) Organic synthesis via organometallics, 1st edn. Vieweg, Braunschweig, p 267
161. Hoppe D (1986) Synthesis of enantiomerically pure unnatural compounds via nonbiomimetic homoaldol reactions. In: Schneider MP (ed) Enzymes as catalysts in organic synthesis. D. Reidel, Dordrecht, p 177
162. Hoppe D (1995) Formation of C-C bonds by addition of allyl-type organometallic compounds to carbonyl compounds. General aspects. In: Helmchen G, Hoffmann RW, Mulzer J, Schaumann E (eds) Houben-Weyl, Stereoselective synthesis, vol. E 21b, 4th edn. Thieme-Verlag, Stuttgart, p 1357
163. Hoppe D (1995) Formation of C-C bonds by addition of allyl-type organometallic compounds to carbonyl compounds. Allyl alkali metal reagents (M=Li, Na, K). In: Helmchen G, Hoffmann RW, Mulzer J, Schaumann E (eds) Houben-Weyl, Stereoselective synthesis, vol. E 21b, 4th edn. Thieme-Verlag, Stuttgart, p 1379
164. Hoppe D (1995) Formation of C-C bonds by addition of allyl-type organometallic compounds to carbonyl compounds. Allyltitanium and allylzirconium reagents. In: Helmchen G, Hoffmann RW, Mulzer J, Schaumann E (eds) Houben-Weyl, Stereoselective synthesis, vol. E 21b, 4th edn. Thieme-Verlag, Stuttgart, p 1551
165. Hoppe D (1995) (*E*)-1-(*N*,*N*-Diisopropylcarbamoyloxy)crotyllithium. In: Paquette LA (ed) Encyclopedia of reagents for organic synthesis, vol. 3, 1st edn. Wiley, Chichester, p 1927
166. Hoffmann RW (1989) Chem Rev 89:1841
167. Zimmermann HE, Traxler MD (1957) J Am Chem Soc 79:1920
168. Zschage O, Schwark J-R, Krämer T, Hoppe D (1992) Tetrahedron 48:8377
169. Krämer T, Hoppe D (1987) Tetrahedron Lett 28:5149
170. Unfortunately, the assignments in the preliminary communication (Krämer T, Schwark J-R, Hoppe D (1989) Tetrahedron Lett 30:7037) are incorrect
171. Schwark JR (1991) Dissertation, University of Kiel
172. Seppi M (2001) Dissertation, University of Münster
173. Paulsen H, Graeve C, Hoppe D (1996) Synthesis 141
174. Hoppe D, Hanko R, Brönneke A, Lichtenberg F, van Hülsen E (1985) Chem Ber 118:2822
175. van Hülsen E. Hoppe D (1985) Tetrahedron Lett 26:411
176. Deiters A, Hoppe D (1999) Angew Chem 111:529; Angew Chem Int Ed 38:546
177. Deiters A, Hoppe D (2001) J Org Chem 66:2842
178. In the preliminary communication [176] the intermediacy of (*R*)-**272** and retention at C-1 were proposed. However, in the subsequent experiments, all evidence points to (*S*)-**271** as the decisive intermediate
179. Deiters A, Fröhlich R, Hoppe D (2000) Angew Chem 112:2189; Angew Chem Int Ed 39:2105
180. Deiters A, Mück-Lichtenfeld C, Fröhlich R, Hoppe D (2002) Chem Eur J 8:1833; (2000) Org Lett 2:2415
181. Deiters A, Wibbeling B, Hoppe D (2001) Adv Synth Catal 343:181
182. Özlügedik M, Kristensen J, Wibbeling B, Fröhlich R, Hoppe D (2002) Eur J Org Chem:414
183. Özlügedik M, Hoppe D (2002) Synthesis, in preparation
184. Heimbach D, Fröhlich R, Wibbeling B, Hoppe D (2000) Synlett 950
185. Reggelin M, Tebben P, Hoppe D (1989) Tetrahedron Lett 30:2915
186. Tebben P, Reggelin M, Hoppe D (1989) Tetrahedron Lett 30:2919
187. We had to correct the originally assumed opposite relative configurations
188. Hoppe D, Tebben P, Reggelin M, Bolte M (1997) Synthesis 183
189. Hanko R, Hoppe D (1982) Angew Chem 94:378; Angew Chem Int Ed Engl 21:372
190. Hoppe D, Tarara G, Wilckens M, Jones PG, Schmidt D, Stezowski JJ (1987) Angew Chem 99:1079; Angew Chem Int Ed Engl 26:1034
191. Hoffmann RW, Lanz J, Metternich R, Tarara G, Hoppe D (1987) Angew Chem 99:1196; Angew Chem Int Ed Engl 26:1145
192. Hoppe D, Tarara G, Wilckens M (1989) Synthesis 83

193. Hirsch R, Hoffmann RW (1992) Chem Ber 125:975
194. Hoffmann RW, Weidmann U (1985) Chem Ber 118:3980
195. Landmann B, Hoffmann RW (1987) Chem Ber 120:331
196. Hanko R, Rabe K, Dally R, Hoppe D (1991) Angew Chem 103:1725; Angew Chem Int Ed Engl 30:1690
197. Rehders F, Hoppe D (1992) Synthesis 859
198. Berque I, Le Ménez P, Razon P, Pancrazi A, Ardisson J, Neuman A, Prangé T, Brion J-D (1998) Synlett 1132
199. Kocienski P, Dixon NJ (1989) Synlett 52
200. Berque I, Le Ménez P, Razon P, Pancrazi A, Ardisson J, Brion J-D (1998) Synlett 1135
201. Zschage O, Hoppe D (1992) Tetrahedron 48:5657
202. Paulsen H, Hoppe D (1992) Tetrahedron 48:5667
203. Peschke B, Lüßmann J, Dyrbusch M, Hoppe D (1992) Chem Ber 125:1421
204. Le Ménez P, Fargeas V, Poisson J, Ardisson J, Lallemand J-Y, Pancrazi A (1994) Tetrahedron Lett 35:7767
205. Le Menez P, Firmo N, Fargeas V, Ardisson J, Pancrazi A (1994) Synlett 995
206. Férézou JP, Julia M, Khourzom R, Pancrazi A, Robert P (1991) Synlett 611
207. Le Ménez P, Fargeas V, Berque I, Poisson J, Ardisson J, Lallemand J-Y, Pancrazi A (1995) J Org Chem 60:3592
208. Ashworth P. Broadbelt B. Jankowski P. Pimm A. Kocienski P (1995) Synthesis 199
209. Smith ND, Kocienski PJ, Street SDA (1996) Synthesis 652
210. Marshall JA (1996) Chem Rev 96:31
211. Marshall JA (1998) Chemtracts, Org Chem 11:697
212. Paulsen H, Graeve C, Fröhlich R, Hoppe D (1996) Synthesis 145
213. Grieco PA, Oguri T, Yokoyama Y (1978) Tetrahedron Lett 419
214. Hoppe D, Brönneke A (1983) Tetrahedron Lett 24:1687
215. Zschage O, Hoppe D (1992) Tetrahedron 48:5657
216. Paulsen H, Hoppe D (1992) Tetrahedron 48:5667
217. Prasad KRK, Hoppe D (2000) Synlett 1067
218. Madec D, Henryon V, Férézou J-P (1999) Tetrahedron Lett 40:8103
219. Sharpless KB, Verhoeven TR (1979) Aldrichimica Acta 12:63
220. Rossiter BE, Verhoeven TR, Sharpless KB (1979) Tetrahedron Lett 4733
221. Mihelich ED (1979) Tetrahedron Lett 4729
222. Hoppe D, Lüßmann J, Jones PG, Schmidt D, Sheldrick GM (1986) Tetrahedron Lett 27:3591
223. Tarara G, Hoppe D (1989) Synthesis 89
224. Nakata M, Toshima K, Kai T, Kinoshita M (1985) Bull Chem Soc Jpn 58:3457 and previous publications
225. Lüßmann J, Hoppe D, Jones PG, Fittschen C, Sheldrick GM (1986) Tetrahedron Lett 27:3595
226. Hoppe D, Krämer T, Freire Erdbrügger C, Egert E (1989) Tetrahedron Lett 30:1233
227. Braun M (1998) Angew Chem 110:445; Angew Chem Int Ed 37:430
228. Sengupta S, Snieckus V (1990) J Org Chem 55:5680
229. Peschke B (1991) Dissertation, University of Kiel
230. Howarth JA, Owton WM, Percy JM (1994) Synlett 503
231. Kocienski P, Barber C (1990) Pure Appl Chem 62:1933
232. Pimm A, Kocienski P, Street SDA (1992) Synlett 886
233. Hareau-Vittini G, Kocienski PJ (1995) Synlett 893
234. Madec D, Pujol S, Henryon V, Férézou JP (1995) Synlett 435
235. Sanford MS, Love JA, Grubbs RH (2001) J Am Chem Soc 123:6543
236. Grubbs RH, Miller SJ, Fu GC (1995) Acc Chem Res 28:446
237. Schuster M, Blechert S (1997) Angew Chem 109:2124; Angew Chem Int Ed Engl 36:2036
238. Hoppe D, Riemenschneider C (1983) Angew Chem 95:64; Angew Chem Int Ed Engl 22:54
239. Hoppe D, Gonschorrek C, Schmidt D, Egert E (1987) Tetrahedron 43:2457
240. Reich HJ, Holladay JE (1995) J Am Chem Soc 117:8470

241. Hoppe D, Gonschorrek C (1987) Tetrahedron Lett 28:785
242. Egert E, Beck H, Schmidt D, Gonschorrek C, Hoppe D (1987) Tetrahedron Lett 28:789
243. Hoppe D, Gonschorrek C, Egert E, Schmidt D (1985) Angew Chem 97:706; Angew Chem Int Ed Engl 24:700
244. Marshall JA, Lebreton J (1987) Tetrahedron Lett 28:3323
245. Schultz-Fademrecht C, Wibbeling B, Fröhlich R, Hoppe D (2001) Org Lett 3:1221
246. Schultz-Fademrecht C, Tius MA, Grimme S, Wibbeling B, Hoppe D (2002) Angew Chem 115:1610; Angew Chem Int Ed 41:1532
247. Tomooka K, Komine N, Nakai T (1997) Synlett 1045

Topics Organomet Chem (2003) 5: 139–176
DOI 10.1007/b10335

Enantioselective Synthesis by Lithiation Adjacent to Nitrogen and Electrophile Incorporation

Peter Beak, Timothy A. Johnson, Dwight D. Kim, Sung H. Lim

Department of Chemistry, University of Illinois at Urbana-Champaign, Urbana, IL 61801, USA. *E-mail: beak@scs.uiuc.edu*

The formation of asymmetric carbon-carbon and carbon-heteroatom bonds by lithiation-substitution at a carbon adjacent to nitrogen can be accomplished by deprotonations or destannylations, followed by reaction with electrophiles. Applications of these sequences for amine elaboration are summarized for reactions controlled by chiral ligands and chiral auxiliaries. Notable features include syntheses of alkaloids, the ability to make both enantiomers with a single chiral ligand and diastereoselective and enantioselective conjugate additions with benzylic and allylic lithiation intermediates. The sequences are classified, where information is available, in terms of stereocontrol in the lithiation or substitution step.

Keywords. Amine elaboration, Asymmetric deprotonation, Dynamic thermodynamic resolution, Conjugate addition

1
Introduction

Asymmetric carbon-carbon and carbon-heteroatom bond formation at a carbon adjacent to nitrogen can be accomplished by the lithiation-substitution sequence shown for the conversion of **1** to **2** to **3** [1, 2]. The combination of a directing group Z on nitrogen and an enantioenriched ligand which directs a deprotonation of **1** (Y=H) to provide an enantioenriched carbanionic intermediate **2** has been used in a number of asymmetric syntheses. The fact that the ligand can induce enantioselectivity in the reaction of **2** is an emerging methodology. The tin-lithium exchange approach, whether from a racemic or enantioenriched **1** ($Y=SnR_3$), has been known for some time (Scheme 1).

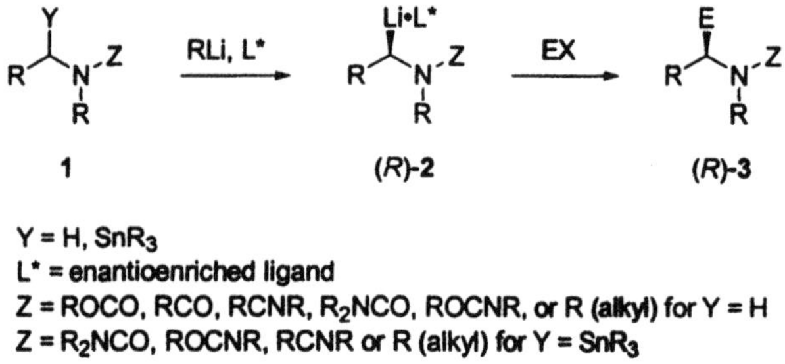

Scheme 1.

We will present principles and recent developments with illustrative applications which guide this chemistry. Our focus will be on representative results in which a chiral auxiliary or a chiral ligand controls the formation of a diastereomerically enriched organolithium intermediate, which provides an enantioenriched product. A few cases in which a preexisting stereogenic center controls the stereoselectivity will be noted. Classic asymmetric syntheses via carbanions which have an anion stabilizing group, typically a carbonyl group, on the carbon bearing the nitrogen and those in which chemical or enzymatic resolutions follow formation of a racemic product obtained by lithiation-substitution will not be covered.

1.1
Reaction Pathways

The general reaction pathways for conversion of an amine derivative **1** to an enantioenriched product **3** in a lithiation-substitution sequence are shown in Scheme 2. In the deprotonation pathway, **1** (Y=H) is achiral. In the first step, formation of a complex between the organolithium base, chiral ligand and the substrate is supported by structural and kinetic studies, investigations of isotope effects and calculations. The formation of a complex is not obligatory for directed metallations, but is definitively established in a closely related directed metallation and has been used to rationalize these cases [3–6]. Tin-lithium exchange is considered to proceed via an ate complex and can give a racemic or enantioenriched intermediate. For the tin-lithium exchange protocol, **1** ($Y=SnR_3$) can be racemic or enantioenriched. The enantiodetermining step in a sequence can be either the formation of the carbanionic intermediate or its subsequent reaction with an electrophile. In the case of stereocontrol by an asymmetric deprotonation or reaction of an enantioenriched tin precursor, the carbanionic intermediate **2** must be formed highly stereoselectively, maintain its configuration, and react highly stereoselectively with the electrophile to afford a highly enantioenriched product. If enantioselectivity is established to be an asymmetric substitution, there are two possibilities. The epimeric organolithium diastereoisomers, (*R*)-**2** and (*S*)-**2**, can be in rapid equilibrium with respect to the rate at which they react with an electrophile in a dynamic kinetic resolution. Alternatively, the rate of interconversion of the epimeric intermediates can be slow with respect to the rate of reaction with the electrophile in which case the reaction pathway is a dynamic thermodynamic resolution [7].

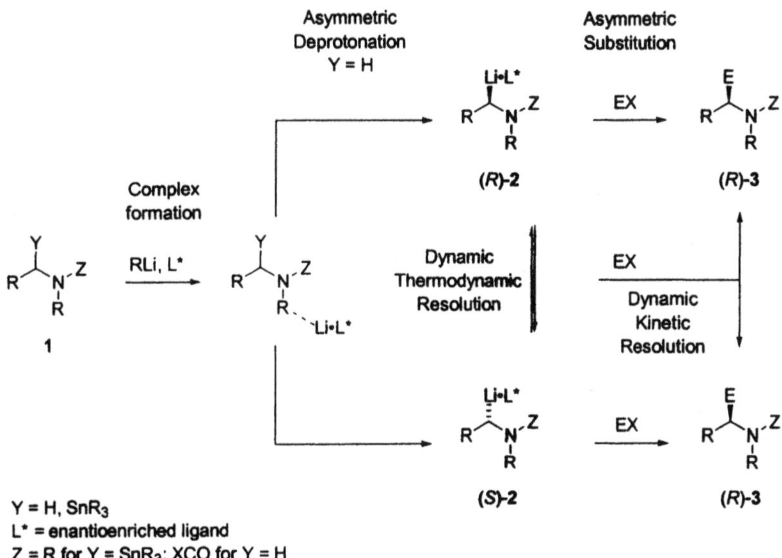

Scheme 2.

The lithiation-substitution of *N*-Boc-*N*-(*p*-methoxyphenyl)-benzylamine (**4**) in the presence of (−)-sparteine (**5**) is illustrative of an asymmetric deprotonation (Scheme 3). Lithiation of **4** with *n*-BuLi in the presence of (−)-sparteine followed by electrophilic incorporation with methyl triflate provides (*S*)-**6** with an enantiomeric ratio of 97:3. Generation of the epimeric organolithium intermediates by tin-lithium exchange from (*S*)-**7** in the presence of (−)-sparteine provides the epimeric product (*R*)-**6** with a 95:5 er. Additional proof that the enantioselectivity is not established in the methylation is the demonstration that generation of the racemic organolithium species in the presence of (−)-sparteine by tin-lithium exchange of the racemic tin precursor followed by methyl triflate provides a racemic product [8].

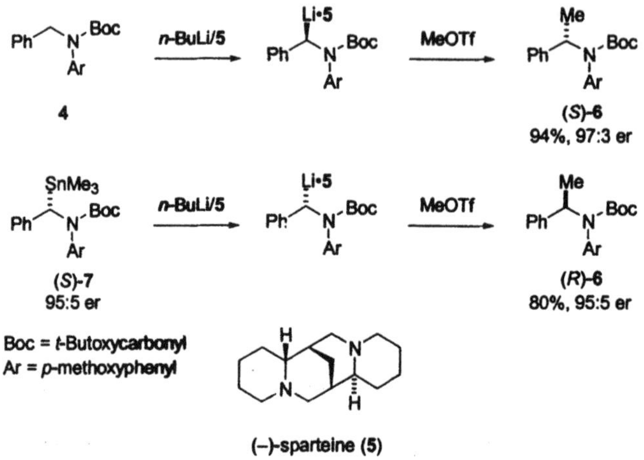

Scheme 3.

Enantioenrichment can be established as an asymmetric substitution in the reaction of the carbanionic intermediate **2** with the electrophile. This reaction pathway can be established by generation of a racemic organolithium intermediate prior to addition of the chiral ligand and the electrophile and observation of an enantioenriched product. In these asymmetric substitutions the influence of the chiral ligand is demonstrated through thermodynamic and/or kinetic effects. The lithiation-substitution of *N*-pivalolyl-*o*-ethylaniline (**8**) is illustrative of the former. Generation of **9** with *sec*-BuLi followed by addition of (–)-sparteine prior to chlorotrimethylsilane (TMSCl) provides (*R*)-**11** with enantioenrichments which are dependent on reaction conditions. As shown in Scheme 4, if the initial reaction is kept at –78 °C, a modest yield and very little enantioenrichment is observed. However, if the reaction mixture is warmed to –25 °C after the addition of (–)-sparteine and cooled back to –78 °C before addition of the electrophile, the yield improves and the enantioenrichment is enhanced to 90:10. It is notable that when the initial temperature is kept at –78 °C and 0.1 equivalent of the electrophile TMSCl is used, the enantiomeric ratio is 91:9. Finally, if a cycling sequence of a warm/cool process is combined with nine sequential additions of 0.1 equivalents of the electrophile, the yield is enhanced and the enantiomeric ratio is improved to 99:1 [9–11]. These results are clearly inconsistent with a dynamic kinetic resolution.

A reaction profile which provides an understanding of these observations is a dynamic thermodynamic resolution as described in Fig. 1 [7, 9–11]. The initial reaction of **9** with (–)-sparteine carried out at –78 °C produces effectively a one-to-one mixture of epimers at the carbanionic carbon as non-equilibrating diastereomeric complexes. The reaction with one equivalent of the electrophile then captures all of the carbanionic species to give an essentially racemic product. When the reaction is warmed to –25 °C prior to the addition of the electrophile, equilibration occurs and the diastereomeric complexes achieve their thermodynamic equilibrium. Rapid cooling to –78 °C maintains this ratio and subsequent

Temp. 1 (°C)	Temp. 2 (°C)	eq TMSCl	Yield (%)	er
−78	−78	1	52	56:44
−78	−25, −78	1	72	90:10
−78	−78	0.1		91:9
−78	−25, −78	0.1[a]	70[a]	99:1

[a]Cycle repeated nine times

Piv = $(CH_3)_3C(O)$

Scheme 4.

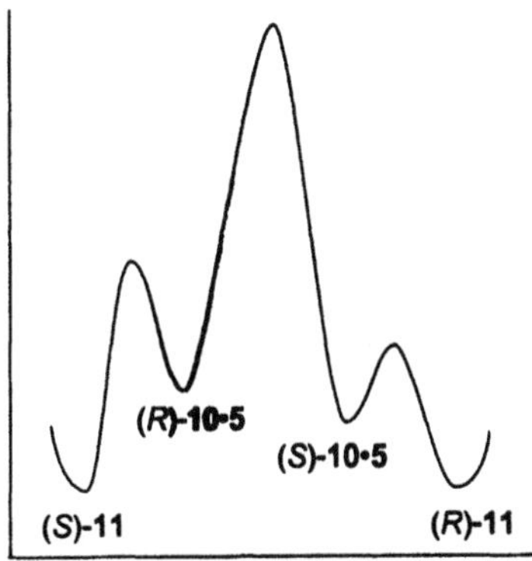

Fig. 1. Energy diagram for a dynamic thermodynamic resolution

reaction of 1 equivalent of the electrophile then reflects the population of the equilibrated species with a significant improvement in the enantiomeric ratio. The fact that 0.1 equivalent of the electrophile also produces an improved enantiomeric ratio reflects the fact that the activation energies for the reaction of each of the diastereomeric complexes with TMSCl are different. In this case the more stable diastereomer is the more reactive.

Understanding the reaction profile opens the way for the experiment in which sequential additions 0.1 equivalent of the electrophile are followed by thermal re-equilibrations of the organolithium intermediate. The results show the increased population of the more stable diastereoisomer ((*S*)-**10**•**5**) and the significant improvement of the enantiomeric ratio. With understanding of the reaction profile, an essentially racemic reaction can be converted to a reaction which gives a highly enantioenriched product. Dynamic thermodynamic resolution is not limited to this case, nor for that matter to this type of reaction, but is much more widely applicable. It clearly offers other strategies for improvement of enantiomeric ratios and for preparations of each enantiomer in a reaction sequence [7,9–11].

Dynamic kinetic resolution is a well-recognized pathway for enantioenrichment of epimeric species. In a dynamic kinetic resolution the equilibration of the diastereomeric species is fast with respect to the rate of reaction of the electrophile and the enantiomeric ratio is determined only by the difference in activation energies for reaction of each diastereomer with an electrophile [12].

The direct deprotonative approaches with an enantioenriched ligand from a prochiral reactant are synthetically most convenient; however, the tin-lithium exchange is able to provide more diverse intermediates. Both approaches have been used to achieve asymmetric bond formations at carbons adjacent to nitrogen.

2 Alkyl Systems

2.1 Chiral Auxiliary Mediated Reactions

When there is a heteroatom as a second substituent on the anionic carbon bearing nitrogen, asymmetric reactions provide an enantioenriched formyl anion synthetic equivalent. Gawley has used the chiral auxiliary approach in a deprotonation-substitution sequence to make highly diastereoenriched **13** and **14**, and converted the separated diastereoisomers to the enantiomeric diols **18** and **19** (Scheme 5). A number of cases are reported. Structural, kinetic and computational studies were carried out and found to be consistent with reaction via a complex species [13].

In an extensive investigation, Seebach has developed a deprotonative chiral auxiliary approach with an oxazolidinone to provide a reagent for enantioselective formylation of aldehydes and ketones [14–16]. Lithiation-substitution of **20** gives a diasteromeric mixture of **21**, as representative examples, with the major diastereoisomer formed in drs greater than 70:30, and up to 95:5 in most cases. The separated diastereoisomers were converted to highly enantioenriched products via the hemiaminal and hydrolysis, as shown in the representative examples in Scheme 6. Additions to imine derivatives were also found to be possible in this approach [14–16].

Syntheses of enantioenriched acyclic alkyl α-lithioamine derivatives have been accomplished by tin-lithium exchanges, usually from highly enantioenriched precursors. The chiral centers have been synthesized by three different approaches and the enantioenriched α-lithioamines were found to have good configurational stability, although this is very much a function of the solvent, temperature and ligand [17,18].

In early work, Pearson reported an extensive study of imidazolidin-2-ones **22** and oxazolidin-2-ones **24** as chiral auxiliaries to provide diastereoisomeric pre-

Scheme 5.

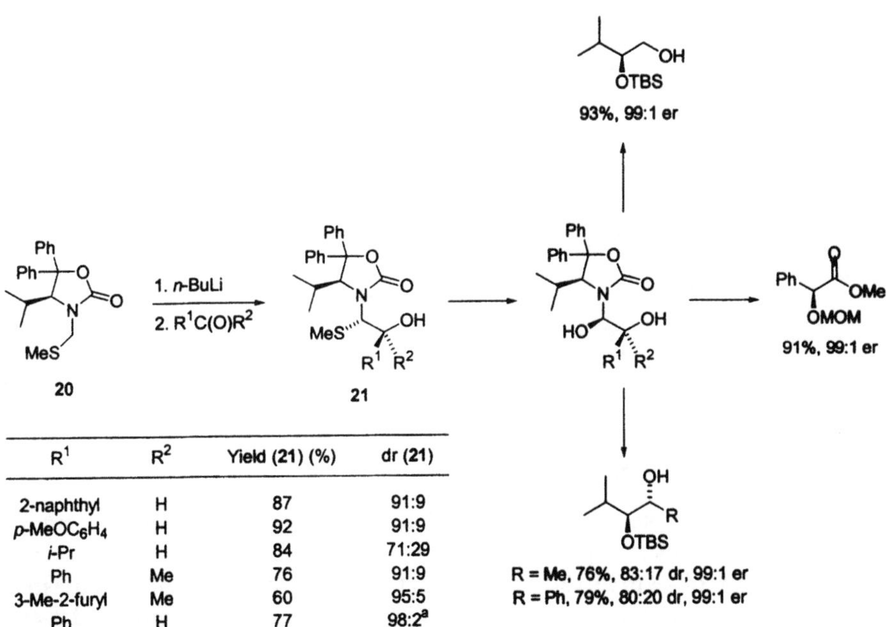

R^1	R^2	Yield (21) (%)	dr (21)
2-naphthyl	H	87	91:9
p-$MeOC_6H_4$	H	92	91:9
i-Pr	H	84	71:29
Ph	Me	76	91:9
3-Me-2-furyl	Me	60	95:5
Ph	H	77	98:2[a]

[a]Protected as MOM ether

Scheme 6.

cursors for tin-lithium exchange (Scheme 7) [19, 20]. The tin compounds were prepared by displacement reactions and the separated diastereoisomers were subjected to tin-lithium exchange and electrophilic substitution. Good yields were obtained as shown for representative cases. An important finding was that epimeric tin derivatives can lead to the same diastereoisomeric product ratios, due to equilibration of the lithiated intermediates.

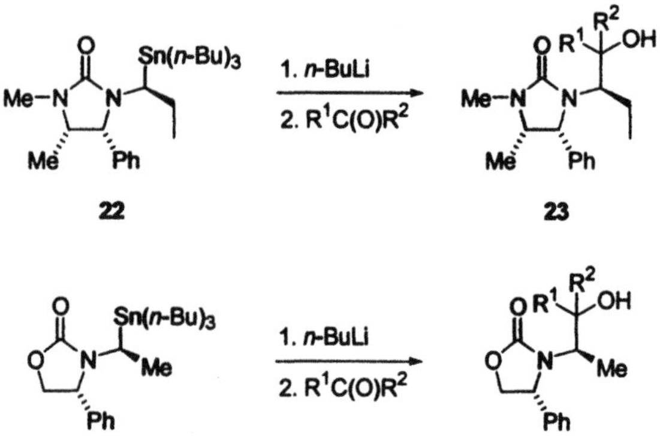

1. n-BuLi
2. $R^1C(O)R^2$

22 → 23

1. n-BuLi
2. $R^1C(O)R^2$

24 → 25

Substrate	R^1	R^2	dr (23)
22	$(CH_2)_5$	-	-
22	OEt	Cl	92:8
22	Ph	H	87:13
24	n-Pr	H	83:17 (25)

Scheme 7.

A recent development of this approach by Jeajean et al. is a synthesis of highly enantioenriched amino acids from the oxazolidinone **26**, as illustrated in Eq. (1) for the preparation of **27** with drs of 96:4 [21].

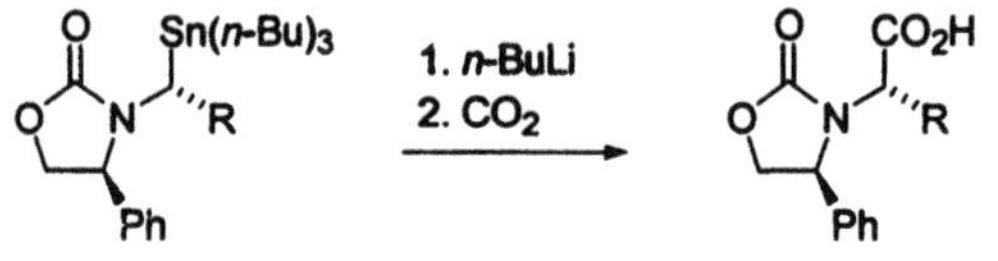

1. n-BuLi
2. CO_2

26 → 27 (1)

R = CH_2CH_2SMe, CH_2CH_2SBn, Me

Nakai and coworkers have reported that a mixture of diastereomeric tin precursors **28** can be used to provide highly enantioenriched products, as shown for the conversion of **28** to **29** (Scheme 8). Furthermore, reduction of **29** is diastereoselective to afford (following chiral auxiliary removal) enantioenriched β-amino alcohols, **30** [22]. In another report, Nakai has described the conversion of the diastereomeric organolithium intermediates from tin-lithium exchange of **28** to copper species which can be used for 1,4 additions to α,β-unsaturated aldehydes

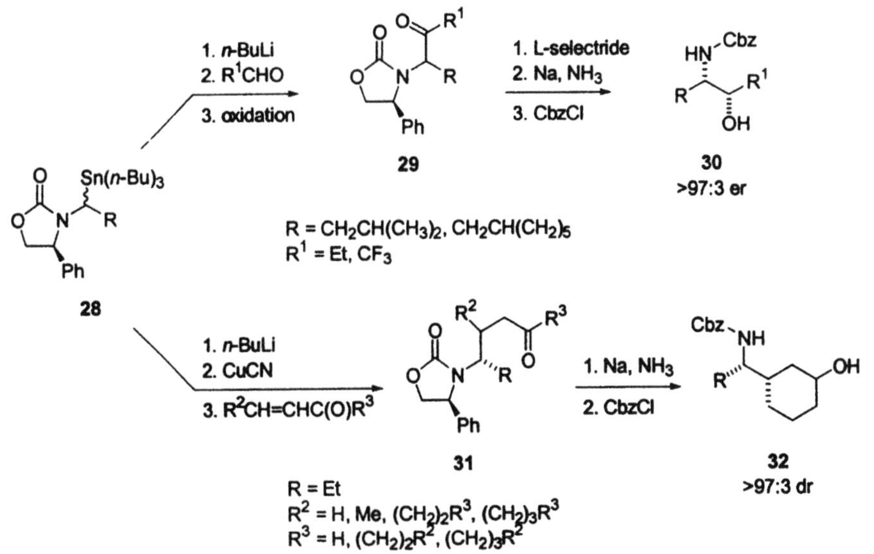

1. n-BuLi
2. R^1CHO
3. oxidation

1. L-selectride
2. Na, NH_3
3. CbzCl

28 29 30 >97:3 er

R = $CH_2CH(CH_3)_2$, $CH_2CH(CH_2)_5$
R^1 = Et, CF_3

1. n-BuLi
2. CuCN
3. $R^2CH{=}CHC(O)R^3$

1. Na, NH_3
2. CbzCl

31 32 >97:3 dr

R = Et
R^2 = H, Me, $(CH_2)_2R^3$, $(CH_2)_3R^3$
R^3 = H, $(CH_2)_2R^2$, $(CH_2)_3R^2$

Scheme 8.

and ketones to provide **31** [23]. The conversion to the amino alcohol **32** from the cyclohexanone adduct illustrates the methodology for the preparation of **32**.

A chiral auxiliary approach to control a carbolithiation from a tin precursor has been used by Coldham and coworkers [24–26]. The lithiation-cyclization of **33** produced **34** with a dr of 79:21 in the presence of (–)-sparteine, and a ratio of 74:26 in the absence of the ligand, Eq. (2).

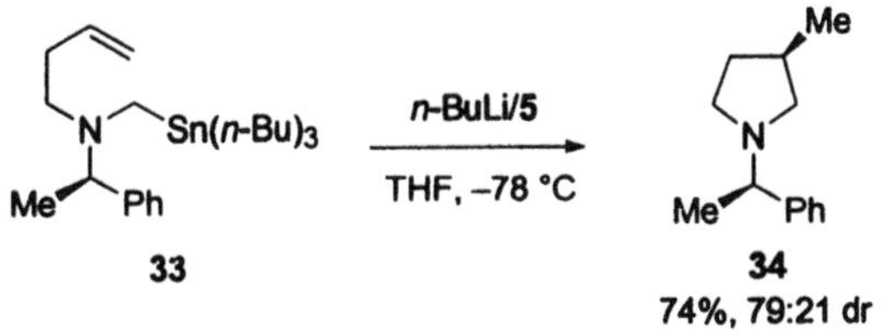

n-BuLi/5, THF, –78 °C

33 → 34 74%, 79:21 dr (2)

Gawley has used the chiral auxiliary approach to prepare highly diastereoenriched 2-tri n-butyltin-substituted piperidines from **35** [27]. Replacement of the chiral auxiliary on nitrogen by a methyl group and tin-lithium exchange provides **36** which on alkylation and acylation gives **37** with high enantioenrichments by retentive and invertive pathways (Scheme 9). This work along with the studies of corresponding pyrrolidine derivatives (vide infra), showed the *N*-alkyl organolithium species to be configurationally more stable than the *N*-Boc analogues [17].

Chong showed that acylstannanes can be asymmetrically reduced and subjected to subsequent displacement by carbamates to afford **38** with high stereointegrity [28, 29].

Tin-lithium exchange of **38** followed by carboxylation provided the expected carboxylic acids **39** in good yield with ers up to 94:6 depending on the reaction

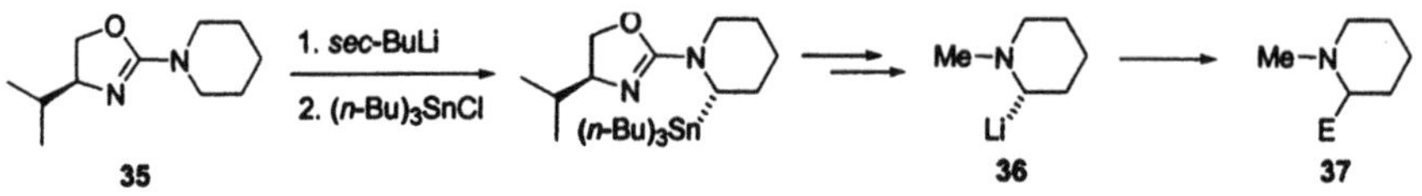

E	Yield (%)	er	Pathway
$(CH_2)_3Ph$	75	99:1	inversion
$(CH_2)_{11}OH$	78	95:5	inversion
CO_2Me	76	99:1	retention
$COH(CH_2)_5$	79	99:1	retention

Scheme 9.

conditions. The same sequence carried out with **40**, followed by reduction afforded **41** in good yield with an er of 97:3 (Scheme 10). However, if the *n*-pentyl group is replaced by an *i*-Pr group, the configurational stability of the intermediates is reduced.

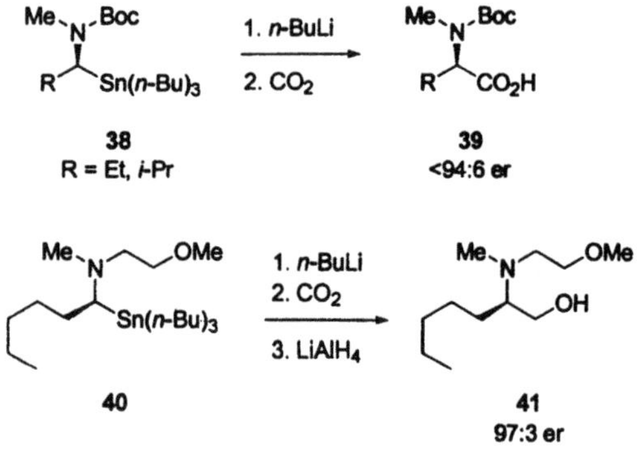

Scheme 10.

2.2 Chiral Ligand Mediated Reactions

The prototypical case for asymmetric deprotonation of a cyclic *N*-Boc amine followed by substitution, is the reaction of *N*-Boc-pyrrolidine (**42**) with *sec*-BuLi/(–)-sparteine which has been reported from our laboratory. The organolithium intermediate (*S*)-**43** reacts with electrophiles to provide (*R*) or (*S*)-**44** with significant enantioenrichment (Scheme 11). Enantioselectivity is established in the deprotonation step [30, 31], the structure of the organolithium base has been determined and the kinetics are consistent with formation of a complex in the deprotonation [32, 33]. Theoretical calculations are consistent with the selective removal of the pro-*S* hydrogen in an *N*-Boc-pyrrolidine alkyllithium complex in which the major factor in preference for removal of the pro-*S* hydrogen is relief of steric interactions on proceeding from the complex to the transition state for deprotonation [3–6]. Efforts to design chiral ligands which would be more effective than

(–)-sparteine in this deprotonation have revealed that the basic bispidine core structure is a favored structural motif [34, 35]. Although other ligands appear to be more effective at binding to the organolithium and have a higher asymmetry, we speculate that the advantage of (–)-sparteine lies in its intermediate binding capacity. The ligand needs to be firmly bound to be present in the critical transition state. However, the binding must not be so strong as to prevent reaction with a substrate relative to the ligand free background reaction. Further studies established that there was an optimal bond distance and bond angle relationship between the activating carbamate group and the proton to be removed [36].

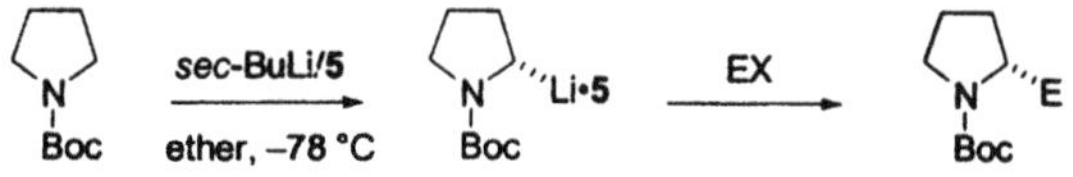

42 → (*sec*-BuLi/5, ether, –78 °C) → (*S*)-43 → (EX) → (*R*) or (*S*)-44

E	Yield (%)	er
Me	88	97:3
$SiMe_3$	71	97:3
$Sn(n\text{-}Bu)_3$	83	98:2
CO_2H	55	94:6

Scheme 11.

An interesting feature of the C_2 character of *N*-Boc-pyrrolidine is that sequential substitution by the same electrophile leads to significant improvement in enantioselectivity as illustrated for the conversion of **42** to (*S,S*)-**45**, Eq. (3).

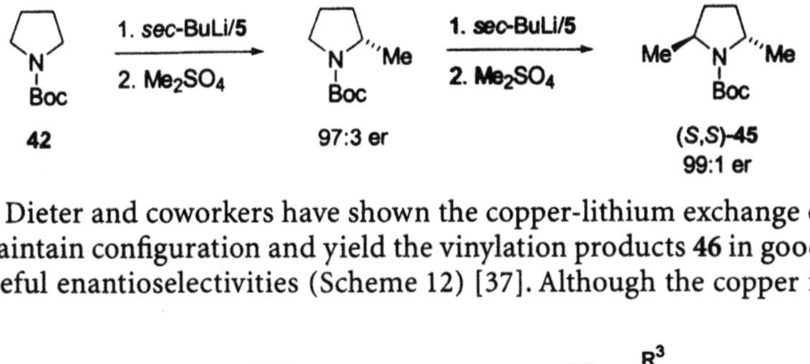

42 → (1. *sec*-BuLi/5; 2. Me_2SO_4) → 97:3 er → (1. *sec*-BuLi/5; 2. Me_2SO_4) → (*S,S*)-45, 99:1 er (3)

Dieter and coworkers have shown the copper-lithium exchange of (*S*)-**43** can maintain configuration and yield the vinylation products **46** in good yields with useful enantioselectivities (Scheme 12) [37]. Although the copper intermediate

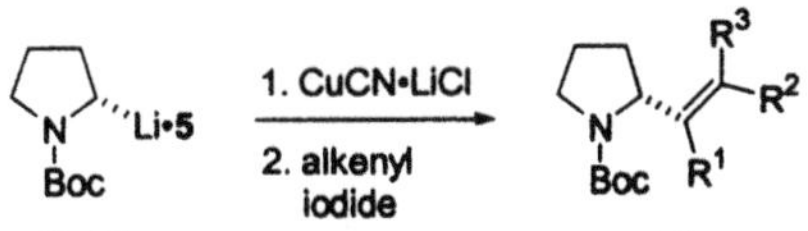

(*S*)-43 → (1. CuCN•LiCl; 2. alkenyl iodide) → 46

R^1	R^2	R^3	Yield (%)	er
H	H	Ph	95	93:7
H	CO_2Et	H	89	95:5

Scheme 12.

adds to vinyl halides, racemization is observed when reaction conditions, such as temperature and solvent, are varied.

Extensions to related systems have been reported. Treatment of **47** with *sec*-BuLi and (–)-sparteine in cumene, followed by reactions with electrophiles gives products **48** with variable yields but often good enantioselectivities (Scheme 13). The regioselectivity of the lithiation depends on solvent with substitution being observed at the 7-position, in some cases [38].

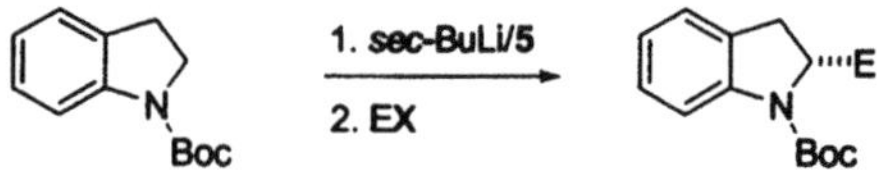

1. sec-BuLi/5
2. EX

47 → **48**

E	Yield (%)	er
Me	52	98:2
$CH_2CH=CH_2$	15	99:1
$SiMe_3$	57	97:3
$SnMe_3$	70	99:1
CH(OH)Ph	74 (82:18 dr)	91:9-98:2

Scheme 13.

Coldham has reported an application of this approach to provide enantioenriched derivatives of 1,2-diamines, as demonstrated by the conversion of **49** to **50** (Scheme 14) [39]. The substituted imidazoline intermediate was obtained with an er of 92:8 on reaction with benzophenone while reaction with allyl bromide gave racemic product. An interesting aspect of Coldham's study is the fact that the yield in the initial lithiation appears to be related to the rotational conformer of the Boc group which provides a complex which orients the base to remove the pro-*S* proton selectively.

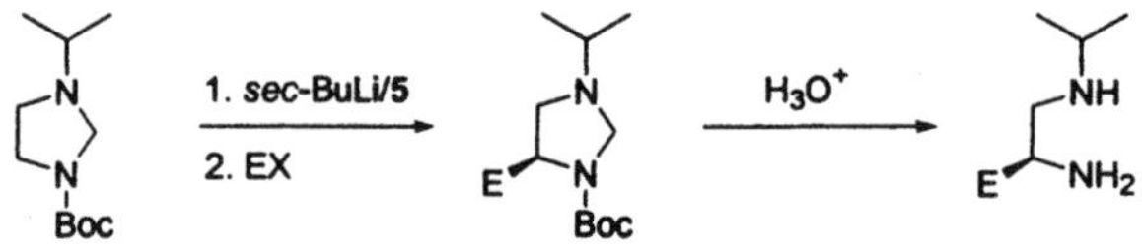

1. sec-BuLi/5
2. EX
H_3O^+

49 → **50**

E	Yield (%)	er
Me	44	92:8
$CH_2CH=CH_2$	40	50:50
$SiMe_3$	40	93:7
$SnMe_3$	40	94:6
$COH(Ph)_2$	50	92:8

Scheme 14.

An asymmetric deprotonation of **51** has been reported by Kise and Yoshida to give a lithiated intermediate which reacts with benzaldehyde to give **52** with a

95:5 er, and **53** with a 94:6 er. If exchange is carried out with $MgBr_2$, the diastereoselectivity improves from 46:54 to 90:10, albeit with a slight erosion in enantioselectivity, Eq. (4) [40].

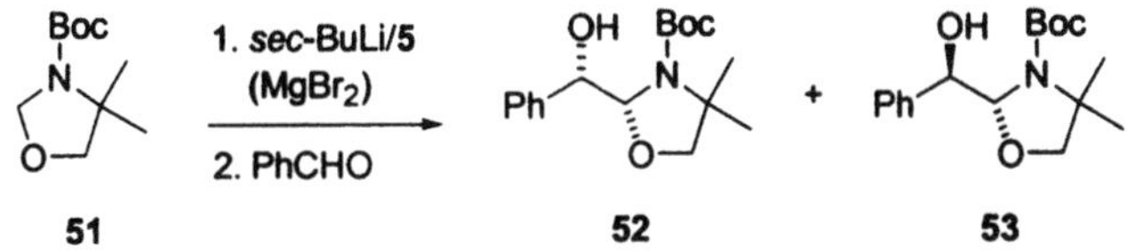

1. *sec*-BuLi/5 ($MgBr_2$)
2. PhCHO

51 → **52** + **53** (4)

Efforts to extend the direct asymmetric lithiation and substitution from *N*-Boc-pyrrolidine to *N*-Boc-piperidine have not been successful. Although a low yield of an enantioenriched product can be obtained from *N*-Boc-piperidine, competing reactions intervene, a result which is consistent with calculations [41].

Meyers has used the tri *n*-butylstannyl-*N*-Boc-pyrrolidine **54**, obtained by asymmetric deprotonation and stannylation of *N*-Boc-pyrrolidine (vide supra), to generate **55**, for investigation of its configuration stability. Gawley used the approach to **57** for a similar study, as well as for the preparation of a number of enantioenriched derivatives **58** (Scheme 15) [42–45]. The general conclusion is that α-lithio-*N*-alkyl derivatives show greater configurational stability than do α-lithio-*N*-formamidine and α-lithio-*N*-Boc derivatives. The configurational stabilities, however, are very much a function of ligand, solvent and temperature [17, 18].

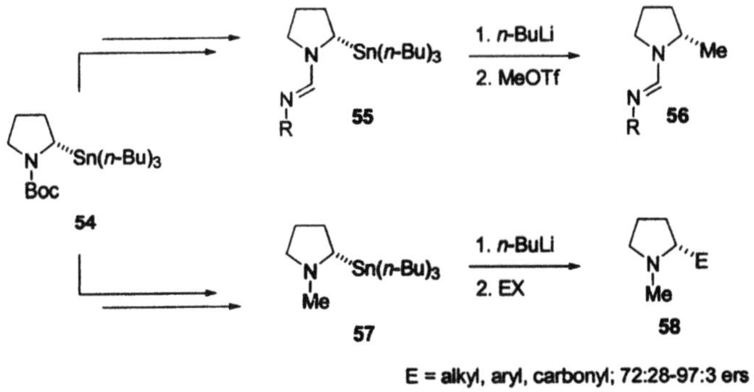

E = alkyl, aryl, carbonyl; 72:28-97:3 ers

Scheme 15.

Two interesting invertive [2,3] sigmatropic rearrangements of α-lithioamines from tin derivatives of pyrrolidine, which have been reported by Gawley, are shown for the conversions of **59** to **60** and **61** to **62** (Scheme 16) [46]. An isotopic substitution showed that the rearrangement of **59** has a [1, 2] sigmatropic component.

Intramolecular carbolithiations from tin derivatives to afford 4-, 5-, or 6-membered rings have been reported by Coldham and coworkers [24–26]. The reaction is shown for the conversion of **63** to **64** and **65** (Scheme 17). The cyclization is more selective for four- and five-membered ring formation than for the

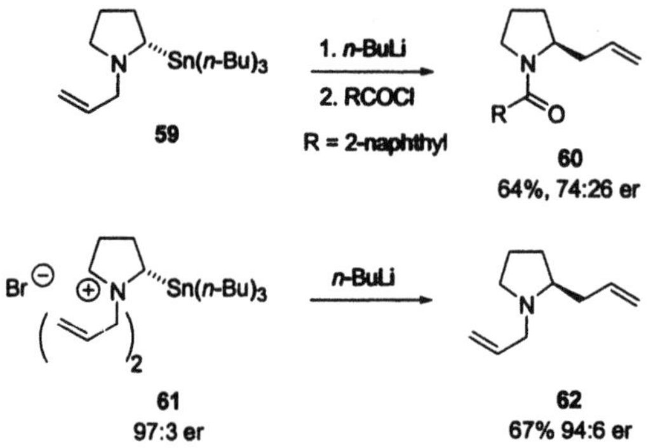

Scheme 16.

six-membered ring formation in the unsubstituted cases. In the later case, substitution of a thiophenyl group at the double bond increases the rate of cyclization relative to the rate of epimerization. This careful study measured the rate of epimerization of the intermediate organolithium.

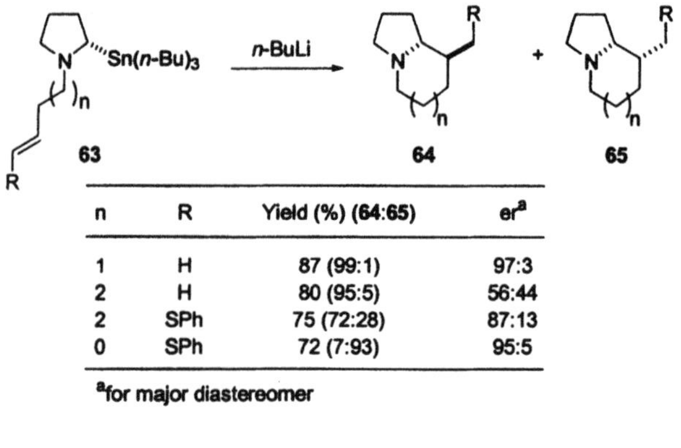

n	R	Yield (%) (64:65)	er[a]
1	H	87 (99:1)	97:3
2	H	80 (95:5)	56:44
2	SPh	75 (72:28)	87:13
0	SPh	72 (7:93)	95:5

[a]for major diastereomer

Scheme 17.

2.3 Asymmetric Center Mediated Reactions

Although it is outside the scope of this review, it is important to note that diastereoselective lithiation-substitutions of enantioenriched substrates which have a permanent chiral center can be very useful. A representative example which indicates the flexibility of the approach is the conversion of (*S*)-**66** to the *cis*- and *trans*-2,6-disubstituted piperidine natural products, **67** and **68** (Scheme 18) [47, 48]. The key to the synthesis of both diastereomers is initial formation of *trans*-products which, in case of formylation, can be equilibrated to a more stable *cis*-isomer. The ready availability of either (*S*) or (*R*)-**66** makes this procedure of general value for stereocontrolled synthesis of highly enantioenriched *cis*- and *trans*-disubstituted piperidines. It should be noted there is a large number of diastereoselective reactions involving α-lithioamine synthetic equivalents that would be of significant synthetic value if the reactions were carried out with enantioenriched starting materials.

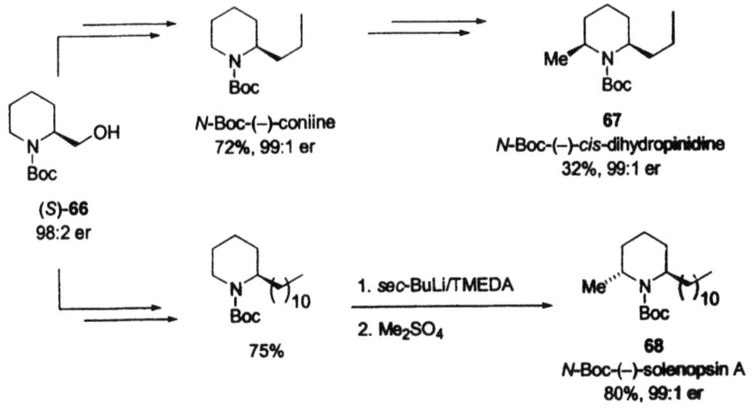

(*S*)-**66** 98:2 er

N-Boc-(–)-coniine 72%, 99:1 er

67 *N*-Boc-(–)-*cis*-dihydropinidine 32%, 99:1 er

75%

1. *sec*-BuLi/TMEDA 2. Me_2SO_4

68 *N*-Boc-(–)-solenopsin A 80%, 99:1 er

Scheme 18.

3 Benzylic Systems

3.1 Chiral Auxiliary Mediated Reactions

In 1983 Meyers and coworkers reported their seminal work on the stereoselective carbon-carbon bond forming reactions derived from dipole stabilized carbanions [49]. In the preliminary communication, enantioselective alkylations were performed using tetrahydroisoquinoline formamidines derived from (1*S*, 2*S*)-(+)-2-amino-1-phenyl 1,3-propanediol [50]. The lithiation and substitution of the isoquinoline formamidine **69** proceeded with various electrophiles to provide (*S*)-α-substituted isoquinolines in moderate yields with drs greater than 96:4. The hydrolytic removal of the formamidine auxiliary led to the highly enantioenriched isoquinolines **70** (Scheme 19). Using appropriate starting materials, several naturally occurring isoquinoline alkaloids, including reticuline [51], isopavine [52], aporphines [53], dextromethorphan [54], homolaudansine [55], and laudansosine [55], have been prepared with high enantiomeric ratios by this approach.

The use of oxazolines as chiral auxiliaries in the asymmetric syntheses of isoquinoline alkaloids was investigated by Gawley and coworkers [56, 57]. Lithiation and substitution of **71** resulted in isoquinolines **72**, which were obtained in high yields with ers of 90:10 to 97:3 (Scheme 20). The products **72** obtained by this route were enantiomeric with **70**. The rationale is that the isopropyl group in the two different auxiliaries have the opposite orientation leading to different

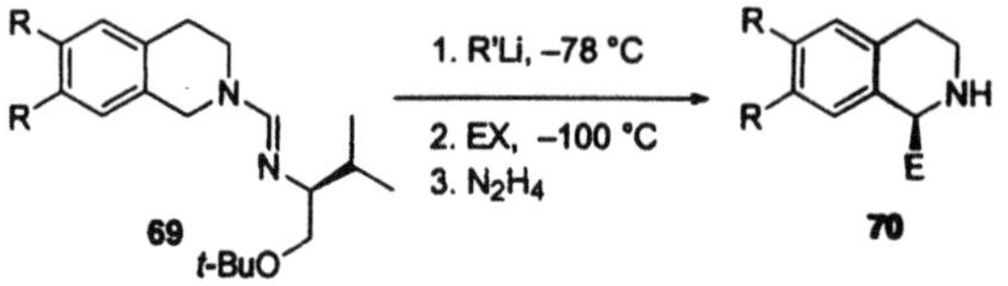

1. R'Li, –78 °C
2. EX, –100 °C
3. N_2H_4

69 → 70

R	E	Yield (%)	er
H	Me	50	99:1
H	$CH_2CH{=}CH_2$	63	98:2
OMe	Me	61	97:3

Scheme 19.

enantiomers. The methodology has been utilized in the asymmetric syntheses of several alkaloids, including *R*-laudanosine [58], 9-*R*-*O*-methylflavinatine [58], (–)-egenine [59, 60], (–)-corytensine [60], and (–)-bicuculline [60].

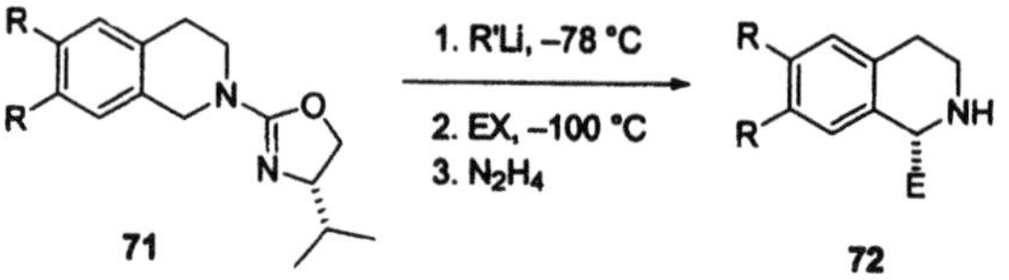

1. R'Li, –78 °C
2. EX, –100 °C
3. N_2H_4

71 → 72

R	E	Yield (%)	er
H	Me	84	95:5
H	Bn	88	91:9
OMe	Me	93	90:10
OMe	CH_2Ar	84	97:3

Ar = 3,4-$(CH_3O)_2C_6H_3$-

Scheme 20.

Meyers and coworkers have shown that the simple modification of replacing the *t*-butoxy group with the smaller methoxy group allowed deprotonation of the remaining benzylic proton in 1-substituted tetrahydroisoquinolines [61]. Subsequent alkylation of the resulting carbanion afforded α,α-disubstituted isoquinolines **74** in high yields of 85–90% with good diastereoselectivities of up to 93:7 (Scheme 21). The stereochemical course of the reactions has been summarized [62–65].

Seebach and coworkers demonstrated that tetrahydroisoquinoline pivalamides **75** containing a stereogenic center at the 3-position can be used in diastereoselective alkylations [66]. After dilithiation of the acid with 2 equivalents of *t*-BuLi, the resulting organolithium was quenched with alkyl halides, such as MeI and benzyl bromide, to afford products **76** in ca. 80% yield, Eq. (5). Oxidative decarboxylation followed by amine deprotection provided **77** with ers greater than 97:3.

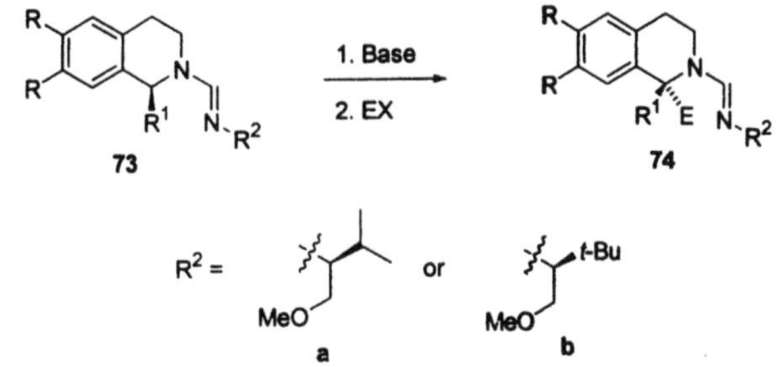

R	Isoquinoline	R^1	E	Base	Yield (%)	dr
OMe	73a	Bn	Me	*n*-BuLi	90	88:12
OMe	73b	Me	Bn	*s*-BuLi	88	93:7
H	73b	Et	Bn	*n*-BuLi	87	91:9
H	73b	Bn	Et	*n*-BuLi	85	90:10

Scheme 21.

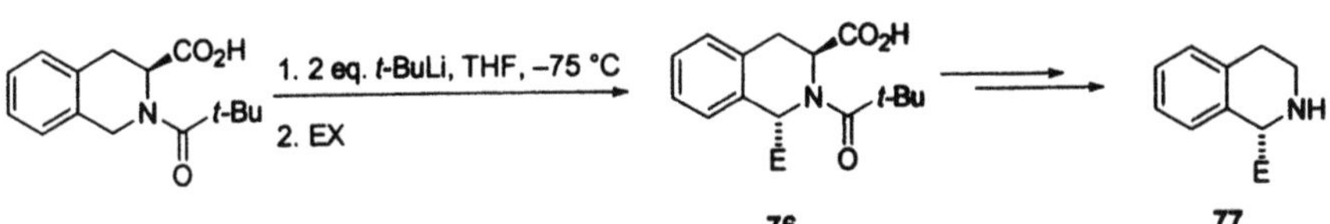

1. 2 eq. *t*-BuLi, THF, –75 °C 2. EX (5)

75 → 76 → 77 >97:3 er

Quirion has reported the first application of chiral carboxylic acids as activating agents and chiral inducers for α-lithiation of tetrahydroisoquinolines [67]. The diastereoselective lithiation and alkylation of acid-derived amides **78** provided **79** in 41–57% yields with moderate drs of up to 92:8 (Scheme 22). The chiral auxiliary can be removed under refluxing basic conditions (KOH/MeOH) in 60–70% yields.

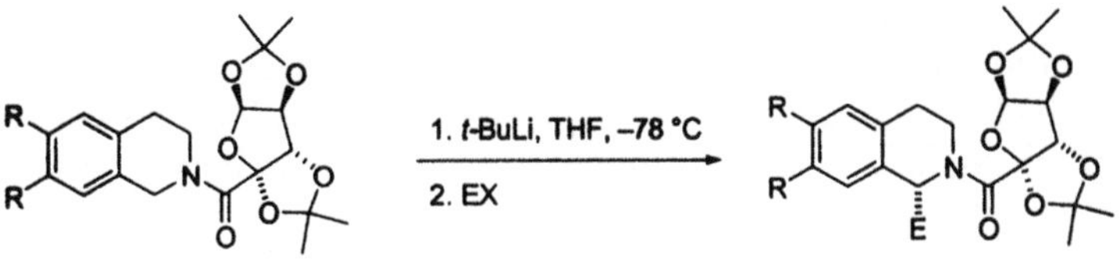

1. *t*-BuLi, THF, –78 °C 2. EX

78 → 79

R	E	Yield (%)	dr
H	Me	53	91:9
H	Bn	50	92:8
H	Bn	41	83:17
OMe	Me	55	89:11
OMe	Bn	57	86:14

Scheme 22.

The formamidine methodology was further extended by Meyers to asymmetric lithiation-substitution of tetrahydrobenzazepine formamidines **80** to provide the corresponding α-alkylated benzazepines **82** in good yields with enantioselectivities of up to 95:5 er (Scheme 23) [68]. Furthermore, a series of enantioenriched 1,1-dialkyltetrahydrobenzazepines **83** were prepared via a second lithiation, in good yields and with enantioselectivities of up to 99:1 er [69]. The diastereocontrol in the alkylation of the 1-substituted 2-benzazepines was significantly influenced by the nature of the E^1 group at C-1, although introduction of methyl iodide in the second step generally proceeded with high stereoselectivity.

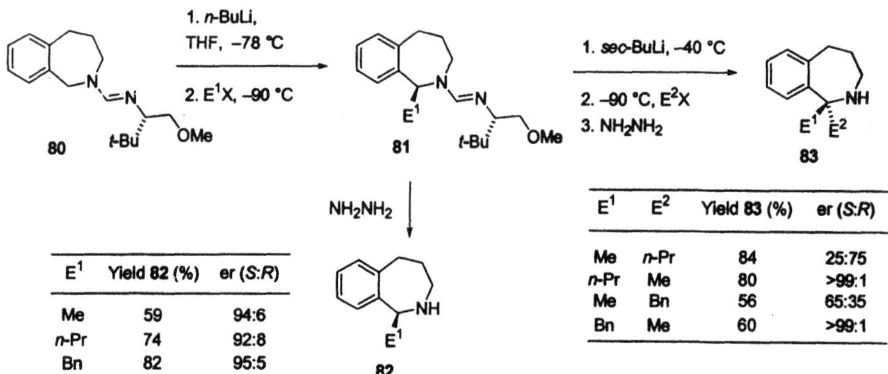

E^1	Yield **82** (%)	er (*S*:*R*)
Me	59	94:6
n-Pr	74	92:8
Bn	82	95:5

E^1	E^2	Yield **83** (%)	er (*S*:*R*)
Me	*n*-Pr	84	25:75
n-Pr	Me	80	>99:1
Me	Bn	56	65:35
Bn	Me	60	>99:1

Scheme 23.

An extension of the formamidine mediated lithiation-substitution methodology to the β-carboline system was reported by Meyers and coworkers [70]. The indole nitrogen was protected using chloromethyl methyl ether, prior to lithiation of the β-carboline formamidine **84** with *n*-BuLi and alkylation with selected electrophiles. Removal of the chiral auxiliary and the protecting group provided **85** in moderate yields with high enantioselectivity, Eq. (6). Subsequent synthetic transformations led to tetracyclic or pentacyclic ring systems and several indole alkaloids, including deplancheine [70], (–)-yohimbone [71], the corynantheine family of alkaloids [72], and yohimban derivatives have been prepared in high enantiomeric purity [73].

1. KH, $ClCH_2OMe$
2. *n*-BuLi, EX
3. NH_2NH_2, H^+

84 → **85** (6)

E = Me, 46%, 99:1 er

The recent use of *N*-indole-substituted gulonic amides by Quirion and coworkers demonstrates diastereoselective lithiation and alkylation of **86** to afford a

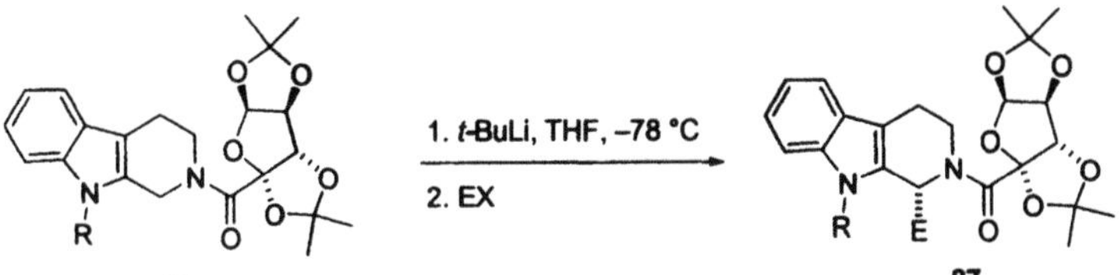

R	E	Yield (%)	dr
Me	Me	51	91:9
Me	$CH_2CH{=}CH_2$	41	63:37
CH_2OEt	Me	40	68:32
CH_2OEt	$CH_2CH{=}CH_2$	40	56:44

Scheme 24.

variety of indole alkaloids **87** (Scheme 24) [67]. Quirion established that the diastereoselectivity is determined during the alkylation step.

Recently, Clayden et al. reported a dearomatizing cyclization reaction of lithiated 1-naphthamides with a phenylglycinol-derived chiral auxiliary (Scheme 25) [74]. Amides **88** were dilithiated with *t*-BuLi which readily cyclized to afford **89** in moderate yields with diastereoselectivities up to 10:1.

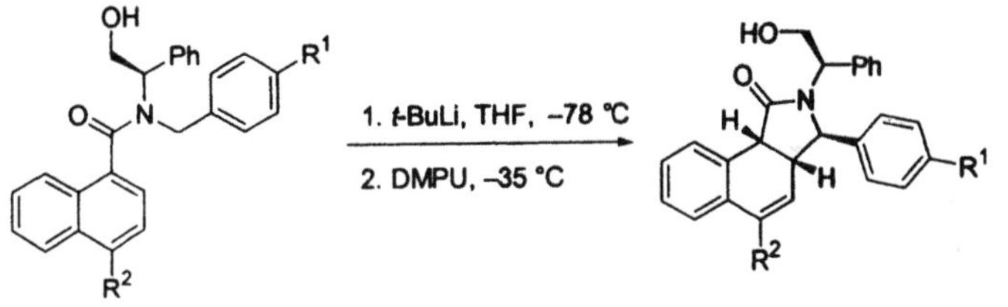

Scheme 25.

Gawley and coworkers developed a methodology for the asymmetric syntheses of primary amines via acyclic dipole-stabilized organolithium intermediates [75]. In the case of the aminooxazoline **90**, 4:1 diastereoselectivity was observed, but the selectivity was improved significantly with oxazolidinone **92**, as shown in Scheme 26. The chiral oxazolidinone was subsequently removed to provide the highly enantioenriched primary amines **94**.

Beak et al. reported that chiral homoenolate equivalents can be formed by dilithiation of an amide and highly diastereoselective reaction with electrophiles to provide the benzylically substituted products [76]. Treatment of (*S*)-*N*-(1'-phenylethyl)-3-phenylpropionamide **95** with 2.2 equivalents of *sec*-BuLi and TMEDA followed by addition of an electrophile affords the alkylated products **96** in 46–55% yield with 90:10–94:6 drs (Scheme 27).

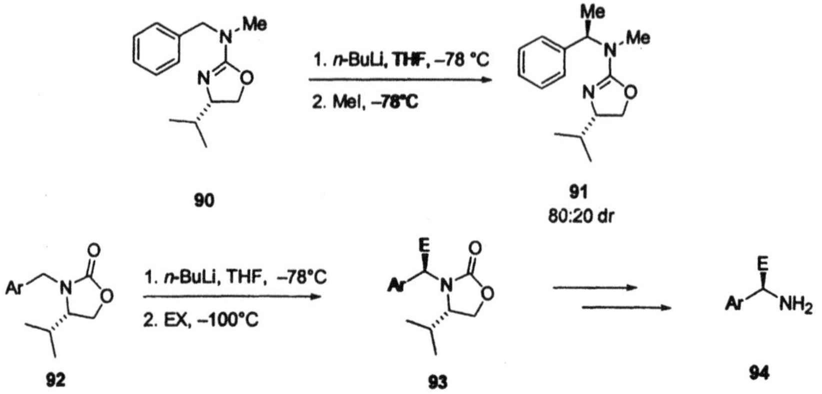

1. n-BuLi, THF, –78 °C
2. MeI, –78 °C

90 → 91 (80:20 dr)

1. n-BuLi, THF, –78 °C
2. EX, –100 °C

92 → 93 → 94

Ar	E	Yield 93 (%)	dr
Ph	Me	75	>99:1
Ph	Et	74	>99:1
α-naphthyl	Me	92	>99:1

Scheme 26.

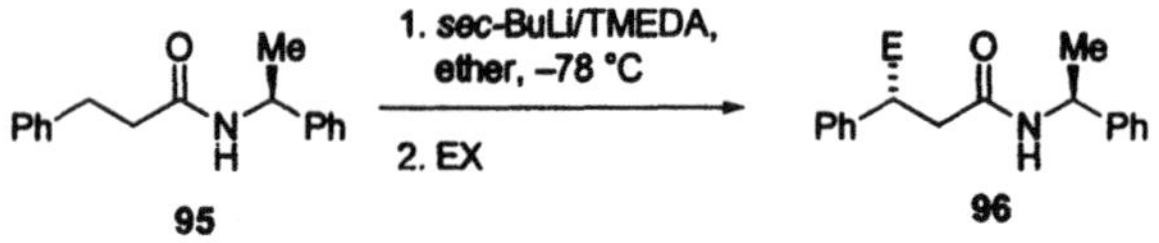

1. sec-BuLi/TMEDA, ether, –78 °C
2. EX

95 → 96

E	Yield (%)	dr
Me	50	91:9
Bn	55	90:10
$SiMe_3$	46	94:6

Scheme 27.

3.2 Chiral Ligand Mediated Reactions

Use of a chiral proton source, a chiral base or base/chiral ligand complex circumvents the problem of incorporation and removal of a chiral auxiliary. Simpkins and coworkers opened the possibility of enantioselective protonation as a method for the asymmetric syntheses of 1-substituted tetrahydroisoquinolines [77]. Using the chiral amine **98** as a proton source, deracemization of **97** proceeded in up to 93:7 er, alleviating the requirement for a chiral auxiliary (Scheme 28).

In 1993 Beak and Du reported the direct benzylic lithiation-substitution of *N*-methyl-3-phenylpropionamide **100** (Scheme 29). In the presence of (–)-sparteine (**5**), alkylation of the lithiated intermediate with a number of electrophiles afforded the *N*-methyl-3-phenyl-3-substituted-propionamides **102** in

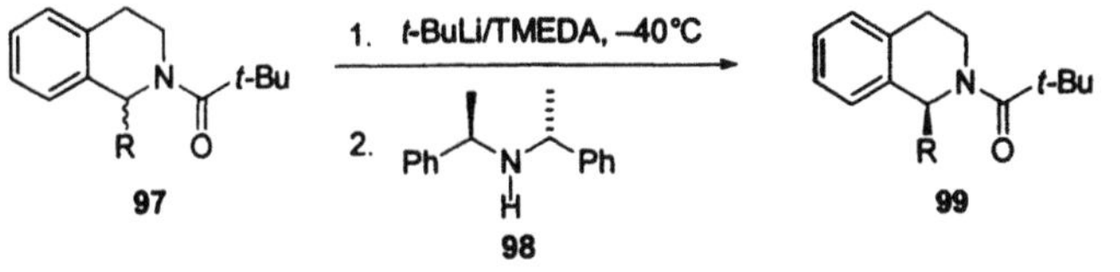

1. t-BuLi/TMEDA, –40°C
2. 98

97 → 99

R	Yield (%)	er
Me	95	93:7
$CH_2CH{=}CH_2$	99	77:23
Bn	83	73:27

Scheme 28.

yields ranging from 63–86%, with high ers ranging from 82:18 to 97:3 [78]. The asymmetric lithiation-substitution of the *N*-methylamide **100** by *sec*-BuLi/**5** has been shown to proceed through a pathway of asymmetric substitution. The enantioselectivity of the reaction at –78 °C arises primarily from a dynamic thermodynamic resolution, although a contribution from a dynamic kinetic resolution process is possible. The synthetic utility of the β-lithiation of β-aryl secondary amides was demonstrated by syntheses of enantioenriched β-aryl-substituted amides, acids and lactones [79].

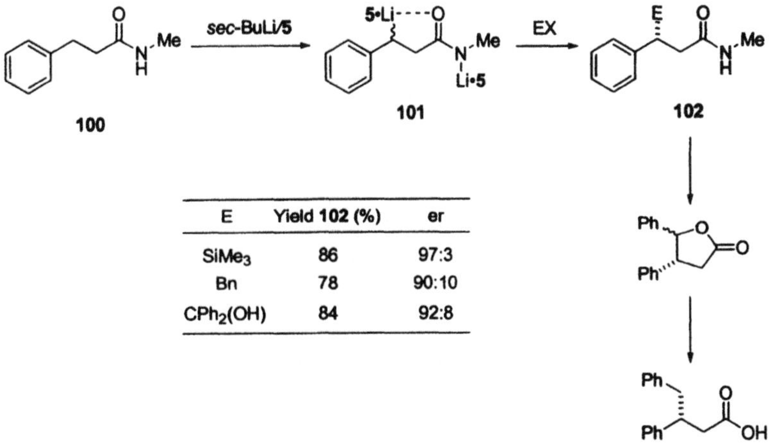

100 → (sec-BuLi/5) → 101 → (EX) → 102

E	Yield **102** (%)	er
$SiMe_3$	86	97:3
Bn	78	90:10
$CPh_2(OH)$	84	92:8

Scheme 29.

Beak and Basu have studied the lateral benzylic lithiation of *N*-pivaloyl-*o*-ethylaniline **103** to give the (–)-sparteine/organolithium complex **104** [9, 10]. Subsequent electrophilic substitution afforded the products **105** in good yields with enantiomeric ratios ranging from 88:12 to 95:5 (Scheme 30). Higher enantioselectivities were obtained by allowing the diastereomeric complexes to equilibrate at –25 °C prior to cooling to –78 °C. Extensive work has been conducted on determining the reaction pathway of this reaction sequence (vide supra).

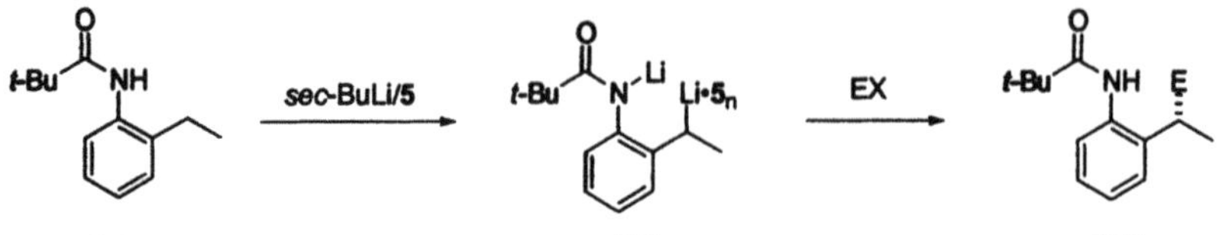

E	Yield 105 (%)	er
$SiMe_3$	72	95:5
$COH(CH_2)_5$	80	88:12
$CH_2CH=CH_2$	67	91:9

Scheme 30.

Schlosser [80] and Voyer [81] reported that *N*-Boc-*N*-methylbenzylamine **106** can be deprotonated with *sec*-BuLi in the presence of (–)-sparteine. The resulting organolithium can be trapped with electrophiles to provide α-substituted benzylamines **107** with high enantioselectivities (Scheme 31). Schlosser and coworkers showed that the reaction pathway of *N*-methyl-*N*-Boc benzylamines **107** is an asymmetric deprotonation followed by racemization and asymmetric substitution, and provided rationalization of solvent effects in terms of ion paired species [80]. The enantioselectivity of this reaction sequence is highly dependent on the solvent, electrophile, and reaction time.

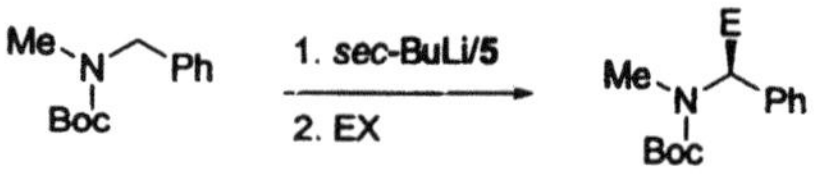

E	solvent	er (*R*:*S*)
CO_2H	Hexane	90:10
CO_2H	THF	8:92
Me	Hexane	13:87
Me	THF	90:10

Scheme 31.

In 1996 Beak and coworkers reported highly enantioselective syntheses of both mono- and disubstituted *N*-Boc-*N*-(*p*-methoxyphenyl)-benzylamine via (–)-sparteine mediated lithiation-substitution sequences [82]. Using alkyl triflates, carbonyl compounds and imines as electrophiles, the lithiated intermediate **108** afforded mono-substituted products in good yields. The subsequent oxidative deprotection of the *p*-methoxyphenyl group with ceric ammonium nitrate (CAN) provided *N*-Boc-benzylamines **109** and **110** with consistently high ers of 97:3 (Scheme 32). The mechanism of this highly enantioselective reaction has been established to be an asymmetric deprotonation through tin-lithium exchange and NMR investigations. The configuration of the stable benzyllithium

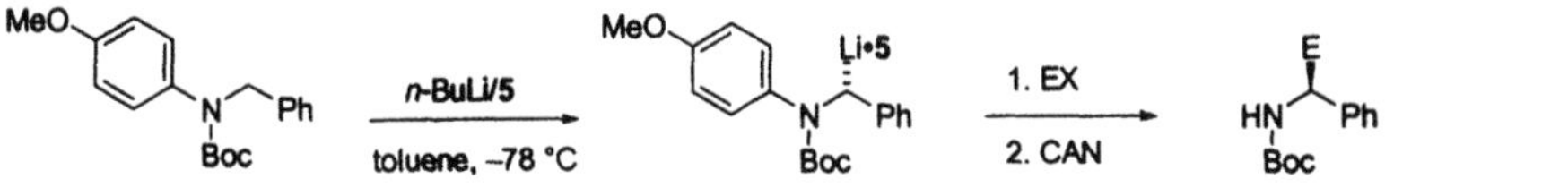

4 108 109, E = Me, 81%, 97:3 er; 110, E = Bn, 73%, 98:2 er

Scheme 32.

(R)-**108** has been assigned by derivatizing with trimethyltin chloride and then by X-ray analysis of this derivative.

Further lithiation of **111** followed by reaction with allyl triflate as the electrophile provided highly enantioenriched benzylamines **112** with quaternary centers [82]. The enantiomer of **112** can be prepared from **113** by a similar reaction sequence (Scheme 33).

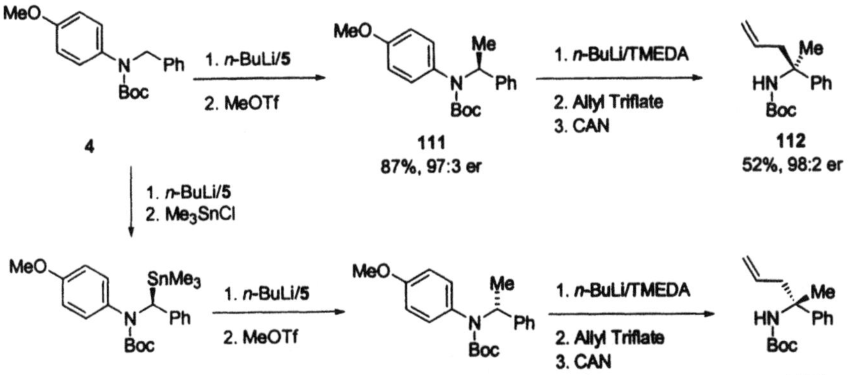

4; 111, 87%, 97:3 er; 112, 52%, 98:2 er; 113, 95:5 er; *ent*-111, 81%, 95:5 er; *ent*-112, 43%, 99:1 er

Scheme 33.

The synthetic utility was demonstrated by access to either enantiomer of highly enantioenriched natural and unnatural α-, β-, γ-arylamino acids and esters from **4** by convenient lithiation substitution sequences (Scheme 34) [83].

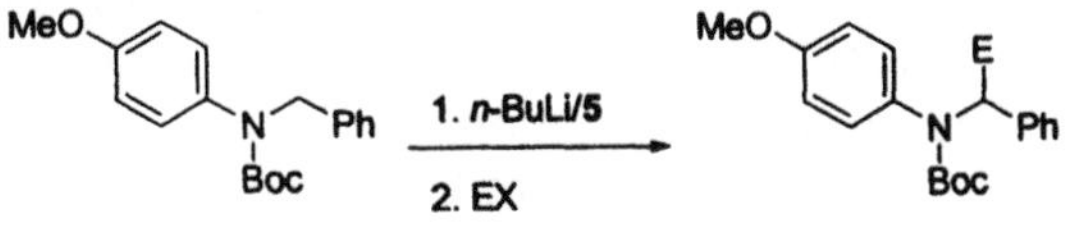

4 114

E	Yield (%)	er
CO_2H	95	96:4 (*R*)
CO_2Me	83	93:7 (*S*)
CH_2CO_2H	76	93:7 (*S*)
$CH_2CH_2CO_2H$	55	97:3 (*S*)

Scheme 34.

Electrophilic substitution of **108** with CO_2 afforded the phenylglycine derivative in 95% yield with an er of 96:4. Interestingly, the opposite enantiomer of the phenylglycine derivative was accessed using methyl chloroformate as the electrophile, which provided the corresponding product with an er of 93:7.

Additional synthetic applications of this methodology are illustrated by enantioselective syntheses of 4-phenyl-β-lactams [84]. Highly enantioenriched β-phenylamino acid derivative **114** can be transformed into β-lactam **115** in good yields with 94:6 er, Eq. (7).

1. $SOCl_2$, MeOH
2. *t*-BuMgCl

114 → **115** 86%, 94:6 er (7)

Conjugate additions of the configurationally stable organolithium species **108** to Michael acceptors provide highly diastereo- and enantioenriched 1,4-addition products **116** in good yields (Scheme 35) [85, 86]. When cyclic enones such as 2-cyclohexen-1-one were used as Michael acceptors in conjugate additions of the benzylic organolithum species, TMSCl was required to improve reaction yields. Reaction of 2-cyclohexen-1-one gave conjugate addition product **116** in 82% yield with an er of 96:4. When more activated olefins were used as electrophiles, generally TMSCl was not necessary, and the products were obtained in good yields.

The synthetic utility of this methodology is further demonstrated by hydrolysis and cyclization of Michael adduct **117** to the stereochemically pure butyrolactam **118** in 79% yield (Scheme 36). The absolute configuration of **118** was assigned by X-ray crystallography of the alkylated derivative **119**.

Beak and coworkers also presented a convenient methodology for enantioselective syntheses of (*S*)-2-aryl-Boc-pyrrolidines **121** by intramolecular lithiation-substitution of arylmethyl-3-chloropropyl-Boc-amines [87]. Treatment of **120** with *sec*-BuLi/**5** at –78°C led to the corresponding cyclized products **121** in moderate to good yields with high enantioselectivities of 96:4–98:2 (Scheme 37). The mechanism of the stereoinduction was established to be an enantioselective deprotonation by the *sec*-BuLi/**5** complex to yield an enantioenriched lithiated species which undergoes rapid cyclization.

In 1998 Simpkins et al. reported that the metallation and alkylation of *N*-methylisoindoline-borane complex **122** with *n*-BuLi/**5** provided products **123** in up to 94:6 er (Scheme 38) [88,89]. The mechanism of the reaction was established to involve an asymmetric deprotonation to give an organolithium which is configurationally stable at nitrogen but stereochemically labile with respect to the C-Li bond.

Very recently, Clayden has used the chiral base (*N*)-lithio-(*R*,*R*)-bis-(1-phenylethyl)-amine for an enantioselective deprotonation cyclization of a benzyl benzamide similar to the conversion in Scheme 25 [90].

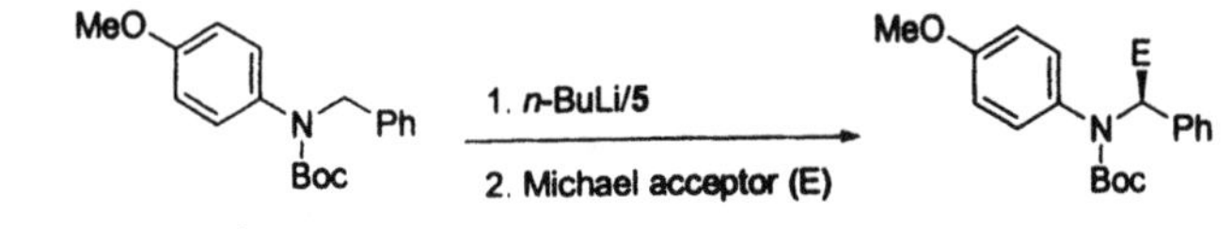

1. n-BuLi/5
2. Michael acceptor (E)

4 → **116**

Michael acceptor (E)	Product		Yield (%)	dr	er[a]
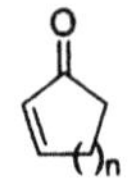 cyclic enone (b)	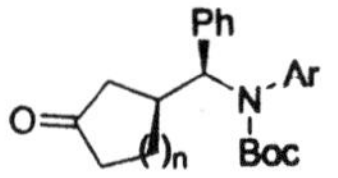	n = 1	86	>99:1	96:4
		n = 2	82	>99:1	96:4
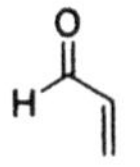 acrolein (b)	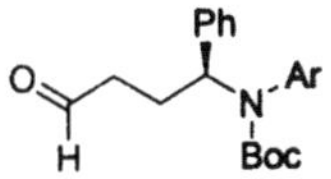		72	-	97:3
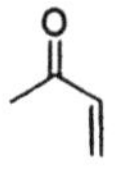 methyl vinyl ketone (b)	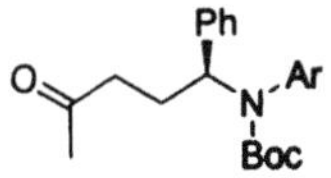		63	-	97:3
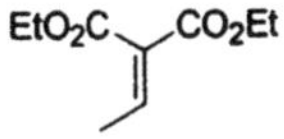 EtO_2C, CO_2Et	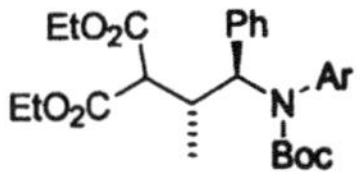		92	80:20	97:3
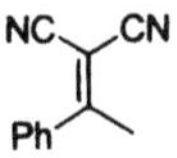 NC, CN, Ph	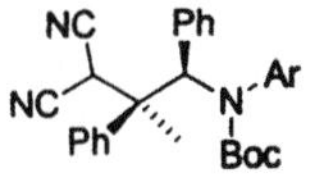		91	92:8	95:5
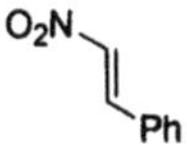 O_2N, Ph	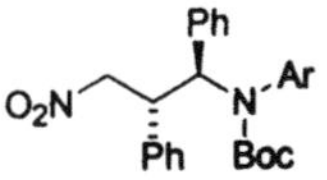		80	90:10	97:3
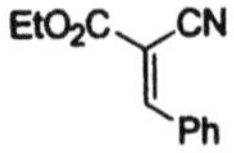 EtO_2C, CN, Ph	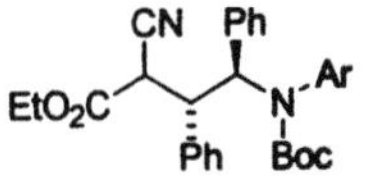		85	75:25	>99:1[c]

[a]major diastereomer
[b]with TMSCl present
[c]following hydrolysis and decarboxylation

Scheme 35.

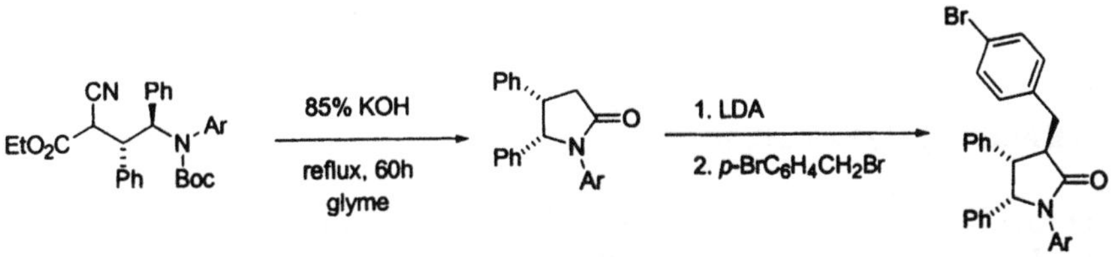

117 → 85% KOH, reflux, 60h, glyme → **118** 79%, >99:1 dr, >99:1 er → 1. LDA 2. p-$BrC_6H_4CH_2Br$ → **119** 71%, 99:1 dr

Scheme 36.

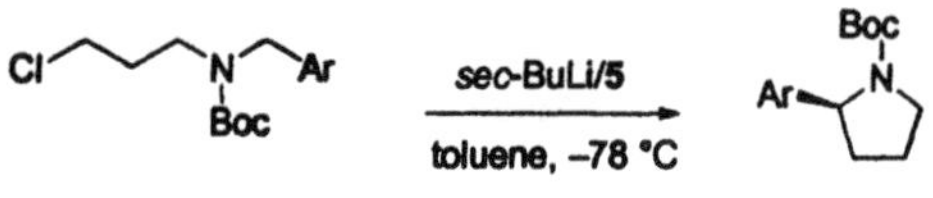

120 → (S)-121 (sec-BuLi/5, toluene, –78 °C)

Ar	Yield (%)	er
Ph	72	98:2
1-naphthyl	68	96:4
3-thienyl	52	96:4
3-furyl	21	98:2

Scheme 37.

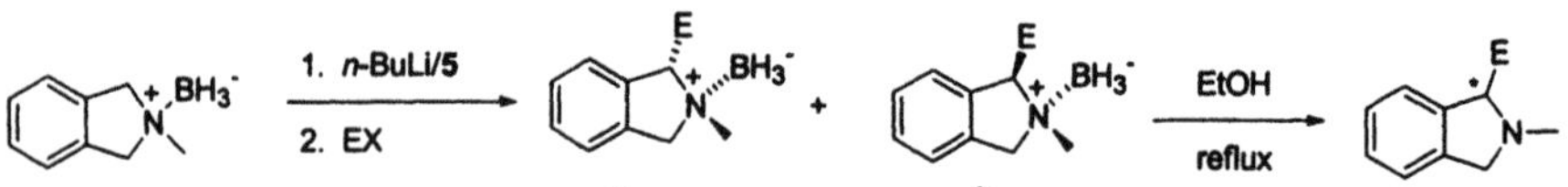

122 → 123 (a + b) (1. n-BuLi/5; 2. EX) → 124 (EtOH, reflux)

E	Yield 123 (%)	dr (a:b)	er[a]
$SnBu_3$	95	1:20	94:6
$CH_2CH{=}CH_2$	65	1:3	93:7
Me	70	1:2	91:9
Et	71	10:1	87:13
C_5H_{11}	65	20:1	78:22

[a]for major diastereomer

Scheme 38.

4 Allylic Systems

A valuable feature of reactions of allylic amino organolithiums relative to their alkyl analogues is that the delocalized anion can lead to substitution at either the α- or γ-position to nitrogen. Consequently, these compounds can serve as both chiral α- and γ-lithioamine synthetic equivalents, as well as chiral β-homoenolate synthetic equivalents [91], in substitution reactions.

4.1 Chiral Auxiliary Mediated Reactions

The first lithiation and substitution of an allylic amine to provide enantioenriched products was reported in 1980 by Ahlbrect and coworkers [92, 93]. Deprotonation of cinnamylamine **125** and substitution with electrophiles provided γ-substituted enamines **126**, which were hydrolyzed to the corresponding aldehydes. The aldehydes were isolated in good yield with high enantiomeric ratios when methyl *t*-butyl ether (MTBE) was used as the solvent (Scheme 39).

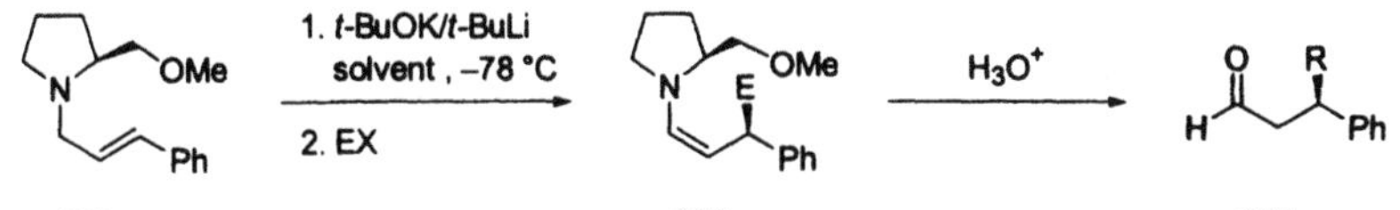

E	Solvent	Yield 127 (%)	er 127
Me	THF	-	69:31
Me	pet. ether	27	82:18
Me	MTBE	80	90:10
$CH_2CH=CH_2$	MTBE	63	90:10
$SiMe_3$	MTBE	68	94:6

Scheme 39.

The authors extended the scope of this methodology by utilizing 1-substituted cinnamylamines **128** [94]. Hydrolysis of the resulting enamines **129** provided direct access to β-substituted ketones **130** (Scheme 40). The absolute configurations of the products could be controlled by modification of solvent and temperature with the *R* products obtained in MTBE at 0°C, and *S* configured products obtained in THF/HMPA at -78°C.

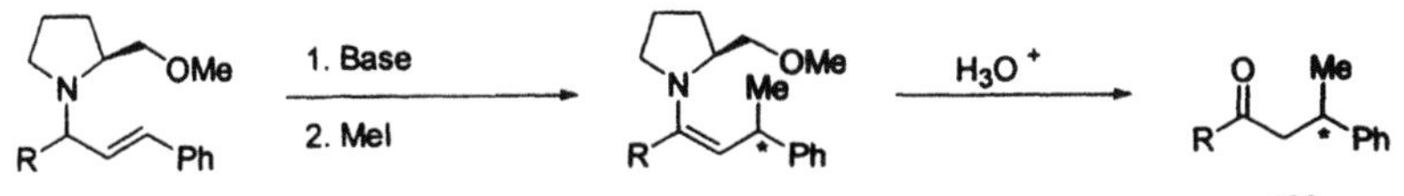

R	Solvent	Base	Temp. (° C)	Yield 130 (%)	er 130	Config.
Me	MTBE	*t*-BuLi/*t*-BuOK	0	78	77:23	*R*
Me	THF/HMPA	*t*-BuLi/*t*-BuOK	–78	78	65:35	*S*
Ph	MTBE	*n*-BuLi	0	78	88:12	*R*
Ph	THF/HMPA	*n*-BuLi	–78	89	80:20	*S*

Scheme 40.

In addition to investigating substitutions of allylic organolithiums with alkyl halides as electrophiles, aldehyde electrophiles provided access to homoaldol products resulting from hydrolysis of the enamine (Scheme 41) [95]. While *syn*/*anti* selectivity was generally low, each diastereomer was highly enantioenriched. The stereochemical path of aldehyde substitution was assigned to occur with retention of configuration.

In addition to the auxiliary mediated additions of allylamino organolithiums to aldehydes described above, the chiral *N*-allylurea **132** has also been shown to add to aldehydes and ketones to provide highly diastereoenriched homoaldol products (Scheme 42) [96]. The reaction involves deprotonation with *n*-BuLi, followed by transmetallation with bis(diethylamino)titanium chloride and sub-

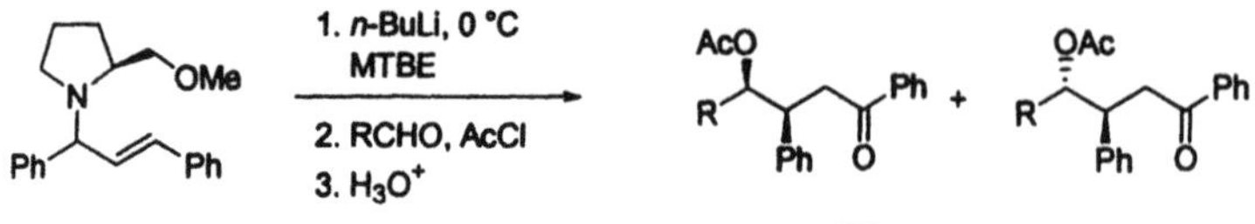

1. *n*-BuLi, 0 °C MTBE
2. RCHO, AcCl
3. H_3O^+

128 → *syn*-131 + *anti*-131

R	Yield (%)	*syn*:*anti*	er (*syn*-131)	er (*anti*-131)
Ph	80	59:41	>95:5	>95:5
t-Bu	73	84:16	>95:5	>95:5
Cy	72	63:37	>95:5	>95:5
Et	66	56:44	>95:5	>95:5

Scheme 41.

stitution with the carbonyl compounds. The high diastereoselectivity was rationalized by a chair-like transition state and the diastereoenriched products were easily converted to enantioenriched substituted lactones by methanolysis and oxidation.

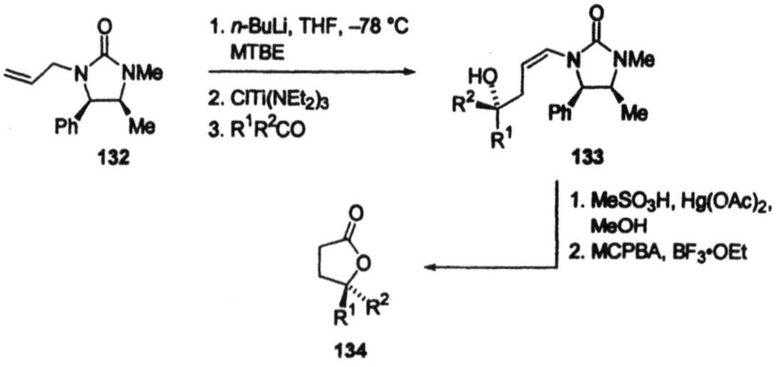

1. *n*-BuLi, THF, –78 °C MTBE
2. $ClTi(NEt_2)_3$
3. R^1R^2CO

132 → 133

1. $MeSO_3H$, $Hg(OAc)_2$, MeOH
2. MCPBA, $BF_3{\cdot}OEt$

→ 134

R^1	R^2	Yield 133 (%)	dr	Yield 134 (%)
C_8H_{17}	H	96	94:6	99
Et	H	94	96:4	79
i-Pr	H	95	96:4	-
i-Pr	Me	93	98:2	97

Scheme 42.

Chiral cyclic allylic amines have also been utilized in lithiation-substitution reactions to access substitution α to nitrogen. Meyers and coworkers have demonstrated that lithiation of chiral formamidine-protected pyrroline **135** and tetrahydropyridine **136** followed by substitution with electrophiles provides both the α and γ-substituted heterocycles (Scheme 43) [97].

Removal of the formamidine auxiliary provided the α-substituted heterocycle **139**, while the γ-substituted product is destroyed. Subsequent hydrogenation

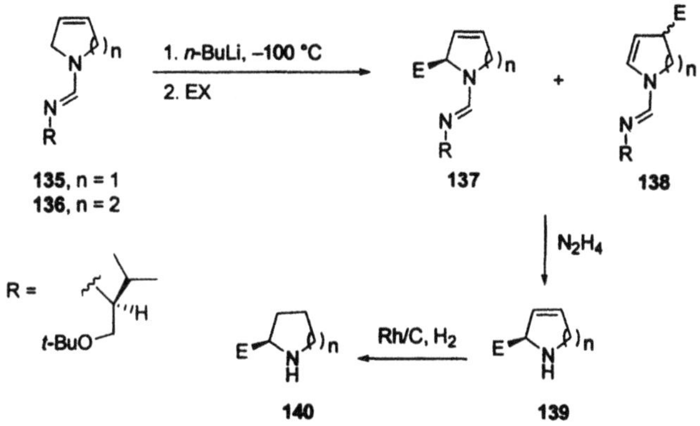

Substrate	E	ratio 137/138	Yield 139 (%)	Yield 140 (%)	er
135	$(CH_2)_6CH_3$	92:8	64	78	98:2
135	$(CH_2)_3Ph$	92:8	76	81	97:3
136	Bn	66:33	48	70	98:2
136	$(CH_2)_6CH_3$	70:30	71	77	96:4
136	$(CH_2)_3Ph$	66:34	69	77	95:5

Scheme 43.

provided access to 2-substituted pyrrolidines and piperidines. The authors have utilized this approach in the synthesis of the antibiotic (+)-anisomycin [98], the benzomorphan (+)-metacozine [97], as well as other morphinans [54]. While the ratios of **137/138** were not as high for the tetrahydropyridine as the pyrroline, the removal of the minor γ-substituted product during the auxiliary cleavage makes this a synthetically useful approach.

Rein and coworkers have utilized a chiral oxazoline auxiliary in a similar lithiation-substitution of the tetrahydropyridine system [99]. The authors, however, report poor α/γ ratios on this substrate, with the γ-substituted product being the major product.

4.2 Chiral Ligand Mediated Reactions

A chiral ligand mediated approach to lithiation-substitutions of allylic amines has also been well developed. Weisenburger and Beak demonstrated that lithiation of doubly protected allylic amines **141** in the presence of the chiral ligand (–)-sparteine (**5**), and substitution with a variety of electrophiles provided highly enantioenriched enecarbamate products **142** (Scheme 44) [100]. The authors demonstrated that the intermediate organolithium could be viewed as either an aldehyde β-homoenolate or γ-lithioamine synthetic equivalent by hydrolysis or reduction and deprotection of the enecarbamates, respectively.

Subsequently, the utility of the enecarbamate products was further developed in their roles as ketone, ester, and acid β-homoenolate synthetic equivalents

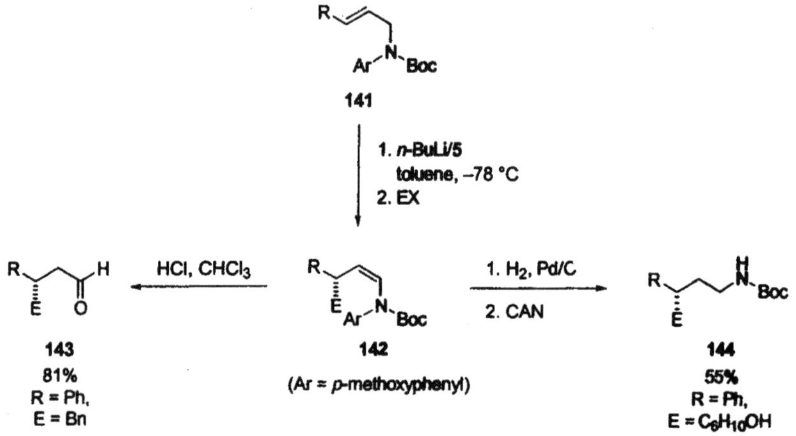

R	E	Yield 142 (%)	er	Config.
Ph	Me	73	97:3	*S*
Ph	Bn	70	98:2	*S*
Ph	COH(CH$_2$)$_5$	77	99:1	*R*
Cy	CH$_2$CH=CH$_2$	43	92:8	*S*

Scheme 44.

(Scheme 45) [101]. An additional α-lithiation/alkylation of the enecarbamate double bond allows access to ketone products **148** upon hydrolysis. Hydrolysis of the unsubstituted enecarbamate to the aldehyde **143**, followed by oxidation and esterification provided the carboxylic acids and esters **146**.

The enantiodetermining step in the reactions at –78°C has been determined to be an asymmetric deprotonation of the allylic amine to provide a configurationally stable organolithium intermediate, which reacts with either retention or

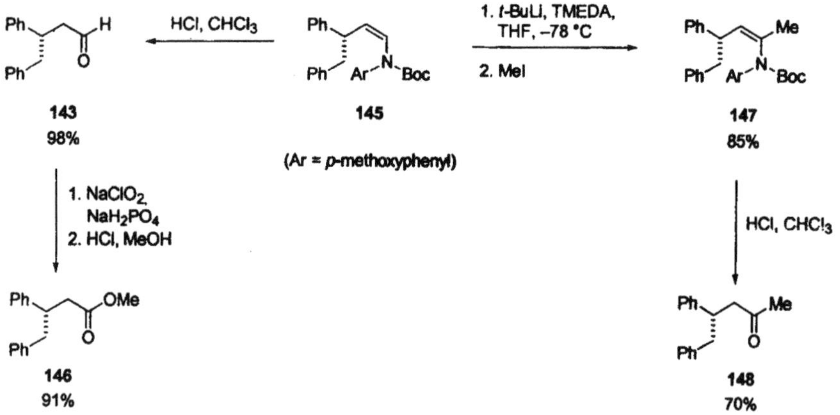

Scheme 45.

inversion in the substitution [4]. The stereochemistry of the reaction has been rationalized using the solid-state structure of the lithiated cinnamylamine, determined by X-ray crystallographic analysis [102, 103]. Both dynamic kinetic and dynamic thermodynamic resolutions have been observed in asymmetric substitution mechanisms when the lithiated intermediate was warmed.

Because of the difficulty of accessing (+)-sparteine, an alternative route to the enantiomeric enecarbamate products has been developed [100]. Synthesis of stannanes **149** and **150** by traditional deprotonation-substitution, followed by transmetallation and substitution with electrophiles provided products of the opposite configuration *ent*-**151** (Scheme 46). Accessing this stereoisomer is possible because of the substitution of lithiated **141** with Me_3SnCl occurs with inversion and the subsequent transmetallation with retention generates the epimer of lithiated **141**.

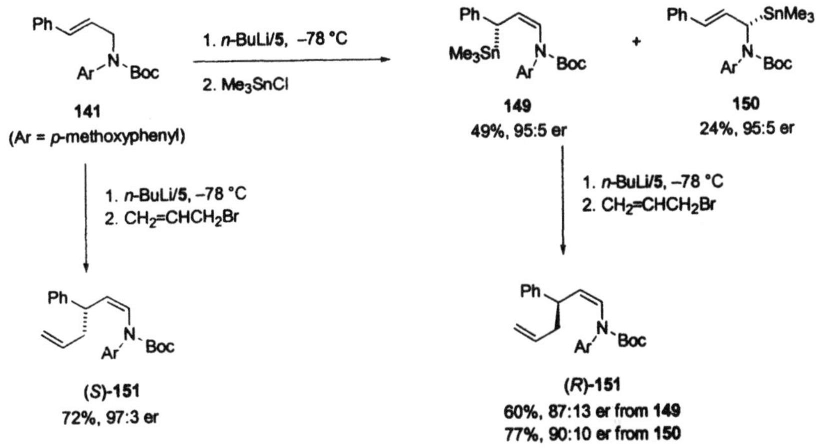

Scheme 46.

In addition to accessing enantiomers by tin/lithium exchange, stereochemically different substitution pathways have been observed with different electrophiles. Kim and coworkers demonstrated that substitution of lithiated **141** with carbon dioxide occurs with inversion of configuration, while substitution with methyl chloroformate proceeds with retention [85]. Subsequent transformations of the enantioenriched products provides either enantiomer of 3-phenyl pyrrolidinone (Scheme 47).

To expand of the scope of electrophiles in the lithiation-substitution reaction, aldehydes were employed and a highly diastereoselective homoaldol methodology was developed [104]. After generation of lithiated **141** under the standard conditions, transmetallation with Et_2AlCl or $TiCl(Oi\text{-}Pr)_3$, and substitution with various aldehydes provided the homoaldol products with both high diastereo- and enantioselectivities (Scheme 48).

The *anti*-addition products were observed, with the absolute and double bond configuration dependent on the transmetallating reagent. The stereochemical course was rationalized by transmetallation to the titanium or alumi-

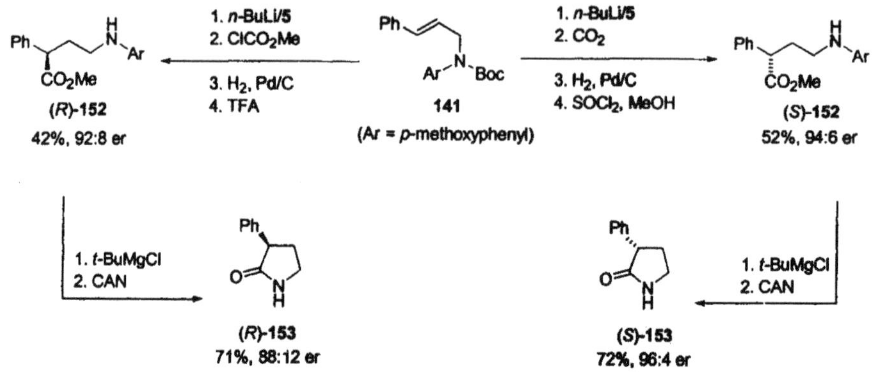

Scheme 47.

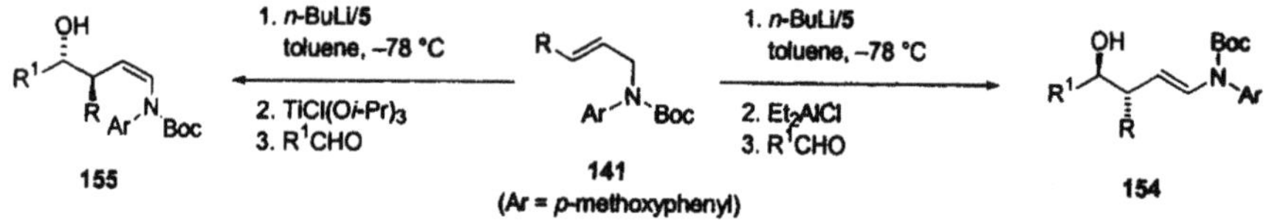

R	R^1	Yield 155 (%)[a]	er	E:Z
Ph	Ph	64	>99:1	2:98
Cy	Ph	38	94:6	2:98

[a]99:1 dr in both cases

R	R^1	Yield 154 (%)[a]	er	E:Z
Ph	Ph	85	97:3	90:10
Ph	Me	66	92:8	95:5
Cy	Ph	82	94:6	90:10
Cy	Me	72	93:7	97:3

[a]In all cases, 99:1 dr

Scheme 48.

num species with inversion of configuration, followed by substitution with retention through a six-membered transition state.

In addition to allylic organolithiums reacting with conventional electrophiles such as akyl halides and carbonyl compounds, activated olefins have also been employed to provide Michael addition products [86, 87, 105]. Both mono- and diactivated olefins including α,β-unsaturated ketones, esters, malonates, malonitriles and nitro compounds are compatible (Scheme 49). Addition of the organolithium to cyclic α,β-unsaturated carbonyl compounds occurred with retention of configuration, whereas addition to other activated olefins proceeds with inversion.

The construction of two contiguous stereogenic centers, as well as a unique 1,5-relationship between the two functional groups, renders the addition products synthetically useful chiral building blocks. Fused bicyclic compounds [106] can be accessed after further functionalization of the Michael adducts (Scheme 50).

In addition, a general route to substituted piperidines, including the antidepressant (–)-paroxetine, has been developed utilizing the additions to nitroalkenes (Scheme 51) [105].

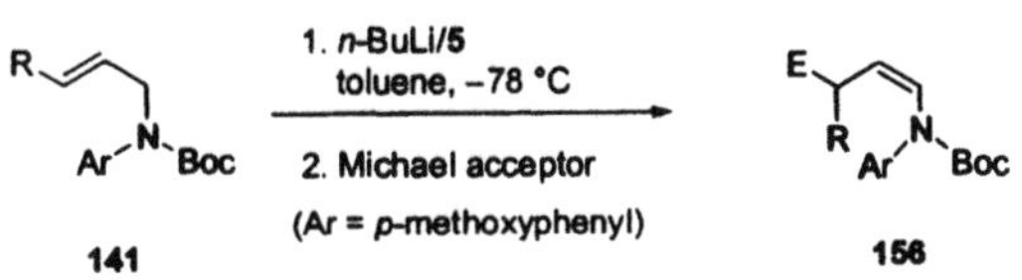

141 → 156: 1. *n*-BuLi/5, toluene, −78 °C; 2. Michael acceptor (Ar = *p*-methoxyphenyl)

R	Michael acceptor	Product	Yield (%)	er(minor)	dr
Ph	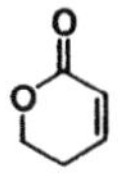 (a)	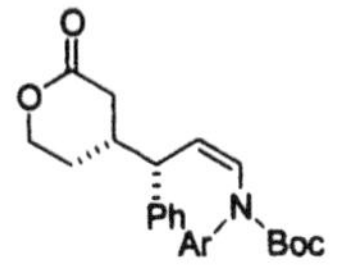	80	96:4	89:11
Ph	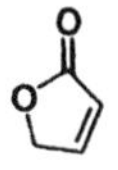 (a)	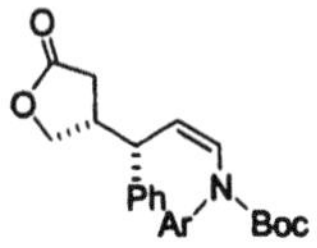	71	97:3	95:5
Me	 (a)	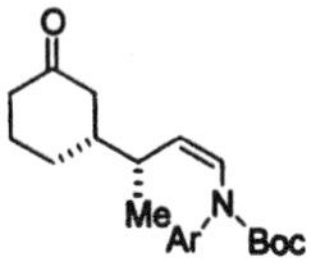	62	98:2 (97:3)	80:20
Ph	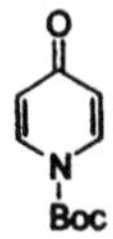 (a)	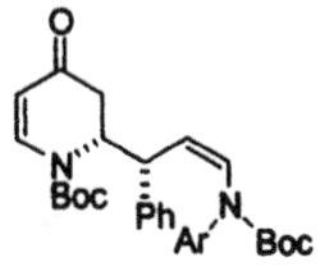	80	93:7	90:10
2-Furyl	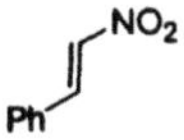	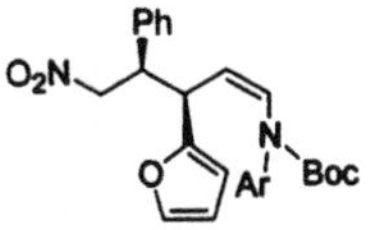	90	>97:3	93:7
Ph	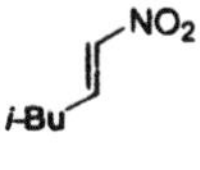	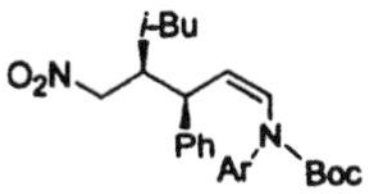	73	>97:3	98:2
Ph	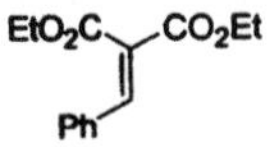	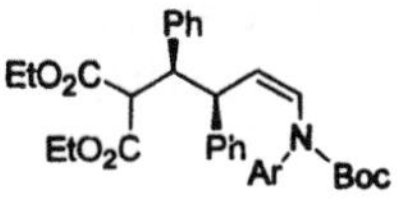	86	98:2	94:6
Ph	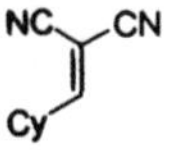	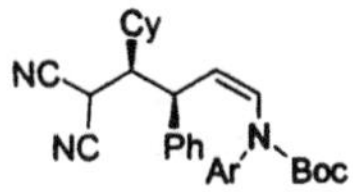	80	94:6 (95:5)	63:44

[a] with TMSCl added

Scheme 49.

While intermolecular lithiation-substitutions of allylic amines are most commonly studied, (−)-sparteine mediated intramolecular substitution reactions of lithiated *N*-Boc-allylamines have also been developed. Treatment of allylic

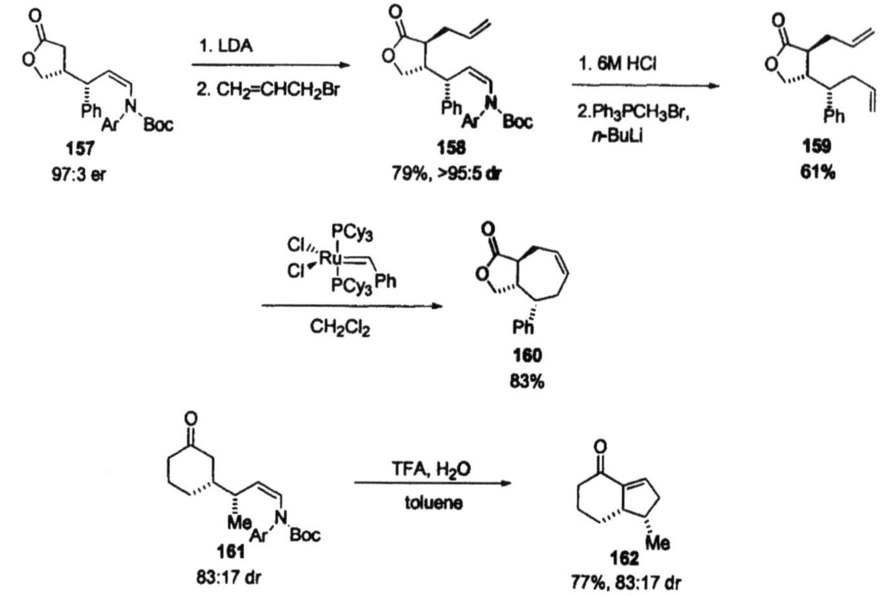

Scheme 50.

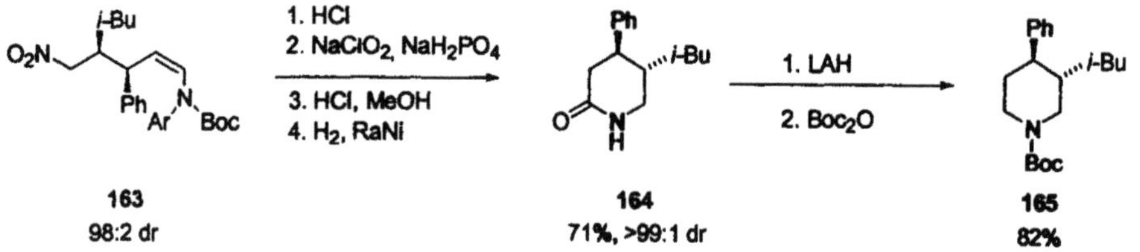

Scheme 51.

amines 166 with *n*-BuLi/(–)-sparteine at –78 °C provided 2-substituted pyrrolidines and piperidines 167 (Scheme 52) [107].

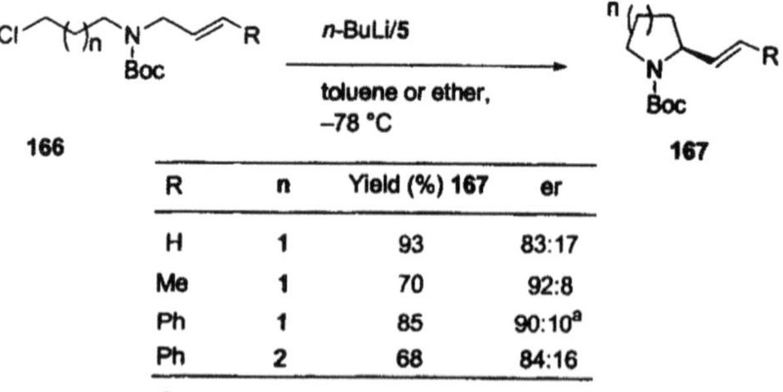

R	n	Yield (%) 167	er
H	1	93	83:17
Me	1	70	92:8
Ph	1	85	90:10[a]
Ph	2	68	84:16

[a]97:3 after recrystallization

Scheme 52.

The enantiodetermining step in the reaction was found to be an asymmetric substitution by transmetallation-substitution of the 2-trimethylstannyl derivative. The vinyl heterocycles serve as useful precursors to fused bicyclic lactams through Boc deprotection, subsequent acylation and ring closing metathesis (Scheme 53) [108].

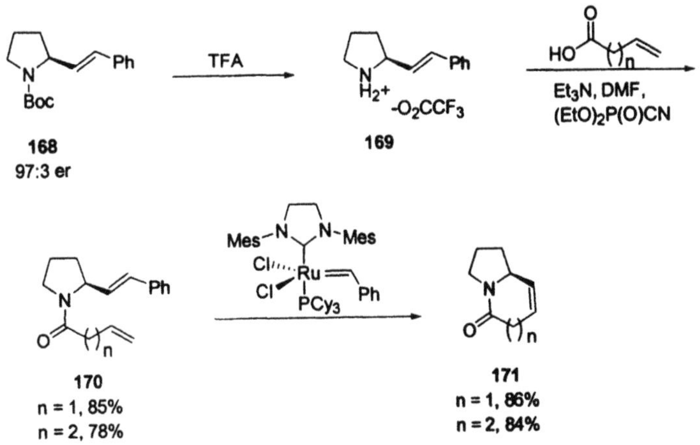

Scheme 53.

5
Conclusions

Methodology for asymmetric synthesis of amine and amine derivatives by lithiation-substitution sequences adjacent to nitrogen is at an emerging state of development. Applications of amine elaboration include alkaloid and enamide synthesis, as well as extensions to functional derivatives of these compounds. Novel conjugate addition of benzylic and allylic organolithium intermediates, which provide products with high diastereoselectives and enantioselectivities, are noteworthy. The pathways of asymmetric deprotonation, which require a diastereoselective deprotonation, a configurationally stable organolithium intermediate and a highly stereoselective reaction with an electrophile have received the most attention to date. The pathways of asymmetric substitution, whether under control by dynamic kinetic resolution or dynamic thermodynamic resolution require stereocontrol only in the reaction with the electrophile, and can be anticipated to be of future interest. A notable feature of the latter is the opportunities offered for rational stereocontrol, by control of the reaction conditions.

Note added in proof. Enecarbamates (e.g., **163**, Scheme 51), arising from (–)-sparteine-mediated lithiations of *N*-Boc allylic and benzylic amines and subsequent conjugate additions to nitroalkenes, have been elaborated to enantioenriched substituted cyclopentanones and aminocyclopentanes [109], as well as piperidines, pyrrolidines, and pyrimidinones [110]. Kinetic resolution of racemic α,β-unsaturated lactones by the organolithium species produced from asymmetric lithiation on *N*-Boc-*N*-(*p*-methoxyphenil)cinnamylamine provides conjugate addition products with three contiguous stereogenic centers in yields of 62–77%with diastereomeric ratios from 75:25 to >99:1 and enantiometric ratios for the major diastereomers from 94:6 to 98:2 [111]. Coldham and coworkers have reported the first highly enantioselective substitution of nonactivated organolithium species at ambient temperaure, via dynamic thermodynamic resolution: chiral ligands, such as (–)-sparteine or, preferably, litiated [*S*-(R^*,R^*)]-1-[(1-methyl-2-pyrrolidinyl)methyl]-2-pyrrolidinemethanol, were addded to initially racemic α-amino organolithium species [prepared by transmetallation of racemic *N*-alkyl (2-tributylstannyl)pyrrolidines] forming diastereomeric complexes which were then allowed to reach thermodynamic quilibrium, prior to reaction with a range of electrophiles to give products with greater than 90% ee [112]. Beak and Whisler have reported further synthetic applications of asymmetric homoenolate equivalents which arise from lithiation of doubly protected allylic amines **141** (Scheme 44) in the presence of (–)- sparteine [113].

References

1. Gawley RE, Hassner A (eds) (1998) Advances in asymmetric synthesis, vol 3. JAI press, Greenwich
2. Beak P, Basu A, Gallagher DJ, Park Y-S, Thayumanavan S (1996) Acc Chem Res 29:552; Basu A, Thayumanavan S (2002) Angew Chem Int Ed 41:717
3. Anderson DR, Faibish NC, Beak P (1999) J Am Chem Soc 121:7553
4. Pippel DJ, Weisenburger GA, Faibish NC, Beak P (2001) J Am Chem Soc 123:4919
5. Wiberg KB, Bailey WF (2001) J Am Chem Soc 123:8231
6. Wiberg KB, Bailey WF (2000) Angew Chem Int Ed 39:2127
7. Beak P, Anderson DR, Curtis MD, Laumer JM, Pippel DJ, Weisenburger GA (2000) Acc Chem Res 33:715
8. Faibish NC, Park Y-S, Lee S, Beak P (1997) J Am Chem Soc 119:11561
9. Basu A, Beak P (1996) J Am Chem Soc 118:1575
10. Basu A, Gallagher DJ, Beak P (1996) J Am Chem Soc 61:5718
11. Laumer JM, Kim DD, Beak P (2002) J Org Chem 67:6797
12. Ward RS (1995) Tetrahedron: Asymmetry 6:1475
13. Gawley RE, Zhang Q, McPhail AT (2000) Tetrahedron: Asymmetry 11:2093
14. Gaul C, Schärer K, Seebach D (2001) J Org Chem 66:3059
15. Gaul C, Seebach D (2000) Org Lett 2:1501
16. Gaul C, Arvidsson PI, Bauer W, Gawley RE, Seebach D (2001) Chem Eur J 7:4117
17. Gawley RE (1997) Curr Org Chem 1:71
18. Aggarwal VK (1994) Angew Chem Int Ed Engl 33:175
19. Pearson WH, Lindbeck AC, Kampf JW (1993) J Am Chem Soc 115:2622
20. Pearson WH, Lindbeck AC (1991) J Am Chem Soc 113:8546
21. Jeanjean F, Fournet G, Le Bars D, Goré J (2000) Eur J Org Chem 1297
22. Tomoyasu T, Tomooka K, Nakai T (1998) Synlett 1147
23. Tomoyasu T, Tomooka K, Nakai T (2000) Tetrahedron Lett 41:345
24. Ashweek NJ, Coldham I, Snowden DJ, Vennall GP (2002) Chem Eur J 8:195
25. Coldham I, Vennall GP (2000) Chem Commun 1569
26. Coldham I, Huften R, Price KN, Rathnell RE, Snowden DJ (2001) Synthesis 1523
27. Gawley RE, Zhang Q (1995) J Org Chem 60:5763
28. Chong JM, Park SB (1992) J Org Chem 57:2220
29. Burchat AF, Chong JM, Park SB (1993) Tetrahedron Lett 34:51
30. Beak P, Kerrick ST, Wu S, Chu J (1994) J Am Chem Soc 116:3231
31. Kerrick ST, Beak P (1991) J Am Chem Soc 113:9708
32. Gallagher DJ, Kerrick ST, Beak P (1992) J Am Chem Soc 114:5872
33. Gallagher DJ, Beak P (1995) J Org Chem 60:7092
34. Gallagher DJ, Wu S, Nikolic NA, Beak P (1995) J Org Chem 60:8148
35. Harrison JR, O'Brien P, Porter DW, Smith NM (2001) Chem Commun 1202
36. Bertini Gross KM, Beak P (2001) J Am Chem Soc 123:315
37. Dieter RK, Topping CM, Chandupatla KR, Lu K (2001) J Am Chem Soc 123:5132

38. Bertini Gross KM, Jun YM, Beak P (1997) J Org Chem 62:7679
39. Coldham I, Copley RCB, Haxell TFN, Howard S (2001) Org Lett 3:3799
40. Kise N, Urai T, Yoshida J-I (1998) Tetrahedron: Asymmetry 9:3125
41. Bailey WF, Beak P, Kerrick ST, Ma S, Wiberg KB (2002) J Am Chem Soc 124:1889
42. Elworthy TR, Meyers AI (1994) Tetrahedron 50:6089
43. Gawley RE, Zhang Q (1994) Tetrahedron 50:6077
44. Gawley RE, Zhang Q (1993) J Am Chem Soc 115:7515
45. Gawley RE, Hart GC, Bartolotti LJ (1989) J Org Chem 54:175
46. Gawley RE, Zhang Q, Campagna S (1995) J Am Chem Soc 117:11817
47. Wilkinson TJ, Stehle NW, Beak P (2000) Org Lett 2:155
48. Adamo MFA, Aggarwal VK, Sage MA (1999) Synth Commun 29:1747
49. Meyers AI, Fuentes LM (1983) J Am Chem Soc 105:117
50. Meyers AI, Fuentes LM, Kubota Y (1984) Tetrahedron 40:1361
51. Meyers AI, Guiles J (1989) Heterocycles 28:295
52. Meyers AI, Dickman DA, Boes M (1987) Tetrahedron 43:5095
53. Gottlieb L, Meyers AI (1990) J Org Chem 55:5659
54. Meyers AI, Bailey TR (1986) J Org Chem 51:872
55. Highsmith TK, Meyers AI, Pearson W (eds) (1992) Advances in heterocyclic natural product synthesis. JAI press, Greenwich
56. Gawley RE, Hart G, Goicoechea-Pappas M, Smith AL (1986) J Org Chem 51:3076
57. Gawley RE (1987) J Am Chem Soc 109:1265
58. Gawley RE, Smith GA (1988) Tetrahedron Lett 29:301
59. Rein KS, Gawley RE (1990) Tetrahedron Lett 31:3711
60. Rein KS, Gawley RE (1991) J Org Chem 56:1564
61. Meyers AI, Gonzalez MA, Struzka V, Akahane A, Guiles J, Warmus JS (1991) Tetrahedron Lett 32:5501
62. Loewe MF, Boes M, Meyers AI (1985) Tetrahedron Lett 26:3295
63. Meyers AI, Guiles J, Warmus JS, Gonzalez MA (1991) Tetrahedron Lett 32:5505
64. Meyers AI, Warmus JS, Gonzalez MA, Guiles J, Akahane A (1991) Tetrahedron Lett 32:5509
65. Meyers AI, Dickman DA (1987) J Am Chem Soc 109:1263
66. Huber IMP, Seebach D (1987) Helv Chim Acta 70:1944
67. Adam S, Pannecoucke X, Combret J-C, Quirion, J-C (2001) J Org Chem 66:8744
68. Meyers AI, Hutchings RH (1993) Tetrahedron 49:1807
69. Meyers AI, Hutchings RH (1996) Heterocycles 42:475
70. Meyers AI, Sohda T, Loewe MF (1986) J Org Chem 51:3108
71. Meyers AI, Miller DB, White FH (1988) J Am Chem Soc 110:4778
72. Beard RL, Meyers AI (1991) J Org Chem 56:2091
73. Meyers AI, Highsmith TK, Bounora PT (1991) J Org Chem 56:2960
74. Bragg RA, Clayden J, Bladon M, Ichihara O (2001) Tetrahedron Lett 42:3411
75. Gawley RE, Rein K, Chemburkar S (1989) J Org Chem 54:3002
76. Pippel, DJ, Curtis MD, Du H, Beak, P (1998) J Org Chem 63:2
77. Burton AJ, Graham JP, Simpkins NS (2000) Synlett 1640
78. Beak P, Du H (1993) J Am Chem Soc 115:2516
79. Lutz GP, Du H, Gallagher DJ, Beak P (1996) J Org Chem 61:4542
80. Schlosser M, Limat D (1995) J Am Chem Soc 117:12342
81. Voyer N, Roby J (1995) Tetrahedron Lett 36:6627
82. Park YS, Boys ML, Beak P (1996) J Am Chem Soc 118:3757
83. Park YS, Beak P (1997) J Org Chem 62:1574
84. Kim BJ, Park YS, Beak P (1999) J Org Chem 64:1705
85. Park YS, Weisenburger GA, Beak P (1997) J Am Chem Soc 119:10537
86. Curtis MD, Beak P (1999) J Org Chem 64:2996
87. Wu S, Lee S, Beak P (1996) J Am Chem Soc 118:715
88. Blake AJ, Ebden MR, Fox DN, Li WE, Simpkins NS (1998) Synlett 189
89. Ariffin A, Blake AJ, Ebden MR, Li WS, Simpkins NS, Fox DNA (1999) J Chem Soc Perkin Trans 1 2439

90. Clayden J, Menet CJ, Mansfield DJ (2002) Chem Commun 38; Clayden J, Menet CJ, Tchabanenko K (2002) Tetrahedron 58:4727
91. Ahlbrecht H, Beyer U (1999) Synthesis 365
92. Ahlbrecht H, Bonnet G, Enders D, Zimmerman G (1980) Tetrahedron Lett 21:3175
93. Ahlbrecht H, Enders D, Santowski L, Zimmerman G (1989) Chem Ber 122:1995
94. Ahlbrecht H, Sommer H (1990) Chem Ber 123:829
95. Ahlbrecht H, Kramer A (1996) Chem Ber 129:1161
96. Roder H, Helmchen G, Peters EM, Peters K, Von Schnering HG (1984) Angew Chem Int Ed Engl 23:898
97. Meyers AI, Dickman DA, Bailey TR (1985) J Am Chem Soc 107:7974
98. Meyers AI, Dupre B (1987) Heterocycles 25:113
99. Rein K, Goichoechea-Pappas M, Anklekar TV, Hart GC, Smith GA Gawley RE (1989) J Am Chem Soc 111:2211
100. Weisenburger GA, Beak P (1996) J Am Chem Soc 118:12218
101. Whisler MC, Soli ED, Beak P (2000) Tetrahedron Lett 41:9527
102. Pippel DJ, Weisenburger GA, Wilson SR, Beak P (1998) Angew Chem Int Ed 37:2522
103. Weisenburger GA, Faibish NC, Pippel DJ, Beak P (1999) J Am Chem Soc 121:9522
104. Whisler MC, Vaillancourt L, Beak P (2000) Org Lett 2:2655
105. Johnson TA, Curtis MD, Beak P (2001) J Am Chem Soc 123:1004
106. Lim SH, Curtis MD, Beak P (2001) Org Lett 3:711
107. Serino C, Stehle N, Park YS, Florio S, Beak P (1999) J Org Chem 64:1160
108. Lim SH, Ma S, Beak P (2001) J Org Chem 66:9056
109. Johnson TA, Curtis MD, Beak P (2002) Org Lett 4:2747
110. Johnson TA, Jang DO, Slafer BW, Curtis MD, Beak P (2002) J AM Chem Soc 124:11689
111. Lim SH, Beak P (2002) Org Lett 4:2657
112. Coldham I, Dufour S, Haxell TFN, Howard S, Vennall GP (2002) Angew Chem Int Ed 41:3887
113. Whisler MC, Beak P (2003) J Org Chem 68:1207

Topics Organomet Chem (2003) 5: 177–216
DOI 10.1007/b10336

Enantioselective Synthesis by Lithiation Adjacent to Sulfur, Selenium or Phosphorus, or without an Adjacent Activating Heteroatom

Takeshi Toru, Shuichi Nakamura

Department of Applied Chemistry, Nagoya Institute of Technology, Gokiso, Showa-ku, Nagoya 466-8555, Japan. *E-mail: toru@ach.nitech.ac.jp*

The enantioselective reactions of carbanions α to S, Se, P or halogens and those of benzyllithium compounds are reviewed, focusing on the enantiodetermining pathways in relation to the configurational stability of the lithium carbanions and the lithium carbanion-chiral ligand complexes.

Keywords. Configurational stability, α-Hetero carbanion, Enantioselective reaction, Asymmetric deprotonation, Asymmetric substitution

1 Introduction

Asymmetric reactions of prochiral compounds having a sufficiently acidic C-H bond are of great importance to access enantiomerically pure compounds. In this chapter, the enantioselective reaction of carbanions α to sulfur, selenium, phosphorus, and halogen as well as those of benzyllithium compounds are reviewed. These compounds serve as substrates for enantioselective reactions or as agents for the enantioselective introduction of functional groups. They are deprotonated by organolithium reagents in the presence of chiral tertiary amines or by enantiopure lithium amides, and the resulting complexes react with electrophiles to give enantioenriched products. Asymmetric induction occurs (Fig. 1)

a) on the prochiral methylene carbon,
b) on the methine carbon,
c) on the symmetrical carbon adjacent to a heteroatom (X-Y), or
d) on the electron-deficient carbon by facial selection with a carbanionic species.

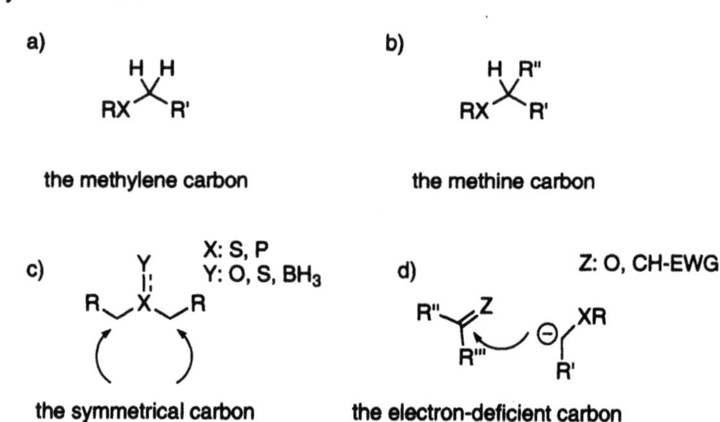

Fig. 1 a–d. Types of enantioselective reactions of α-hetero carbanions

1.1 Enantiodetermining Pathways

Asymmetric induction on prochiral methylene carbons has been extensively studied in the reactions of carbanions α to sulfur, selenium, and phosphorus as well as α-oxy and α-amino carbanions. These reactions consist of two consecutive reactions: a prochiral proton is replaced by lithium to give the organolithium species (a), which subsequently reacts with an electrophile to give the product (b) [Eq. (1)].

$$\mathrm{RXCH_2R'} \xrightarrow{(a)} \mathrm{RXCH(Li)R'} \xrightarrow{(b)} \mathrm{RXCH(E)R'} \qquad (1)$$

Asymmetric induction in these reactions occurs either in the first deprotonation step [Eq. (1a)] or in a postdeprotonation step [1] [(Eq. (1b)]. When the enantioselection occurs in the deprotonation step, a proton is stereoselectively removed by a chiral base from a prochiral substrate to provide a configurationally stable enantioenriched carbanion, which reacts with an electrophile giving an enantioenriched product. This enantiodetermining pathway is termed "asymmetric deprotonation". In fact, reactions of α-oxy and α-amino carbanions are often controlled through an asymmetric deprotonation pathway (Fig. 2) [1,2].

Enantioselectivity through an asymmetric deprotonation pathway is largely dependent on the configurational stability of the carbanion and the interaction between a substrate and a chiral base. On the other hand, when the enantioenriched product is formed in the reaction with an electrophile, even though the deprotonation gives the racemic carbanion, this pathway is termed "asymmetric substitution" [1]. Enantioselection through an asymmetric substitution path-

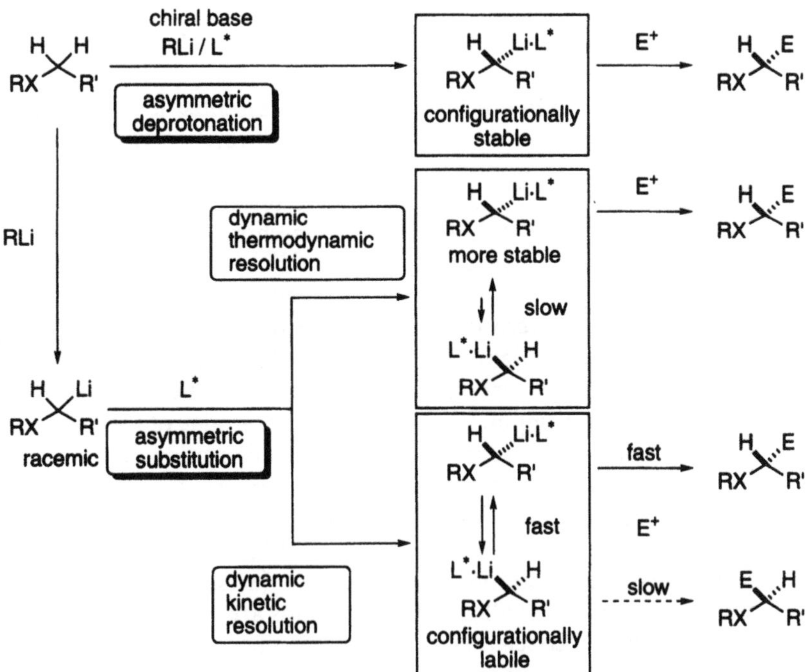

Fig. 2. Reaction pathways in enantioselective reactions of α-hetero carbanions

way is strongly related to the stability of the carbanion-chiral ligand complexes and the activation energy in the transition state of the reaction with an electrophile. The asymmetric substitution pathway is divided into two limiting pathways, dynamic thermodynamic resolution [3] and dynamic kinetic resolution pathways [4]. The reaction proceeds through a dynamic thermodynamic resolution pathway, when the complexes are configurationally stable enough to react with an electrophile, and the enantioselectivity is determined by the diastereomeric ratio of the lithium carbanion-chiral ligand complexes. Thus, the enantioselectivity is established before the reaction with an electrophile takes place. On the other hand, in dynamic kinetic resolution, following the Curtin-Hammett principle, the enantioselectivity is determined by the difference in the activation energy between the diastereomeric transition states in the reaction with an electrophile. In this case, the diastereomeric complexes are configurationally labile so as to allow the carbanionic center to undergo epimerization much faster than its reaction with an electrophile, and one of the complexes preferentially reacts with an electrophile to give the enantioenriched product.

The enantioselective reaction on a methine carbon is apparently different from that on a methylene carbon (Fig. 1b). Deprotonation of a racemic methine proton always affords the racemic carbanion which gives the racemic product when the carbanion is configurationally stable. The enantioenriched product would be formed only when the reaction proceeds either through an asymmetric substitution pathway, or through kinetic resolution in deprotonation.

1.2 Diagnostic Tests for Configurational Stability of Lithium Carbanions

High configurational stability of the carbanions is essential for high enantioselectivity through an asymmetric deprotonation pathway. In an asymmetric substitution pathway, the reaction proceeds through either pathway of dynamic thermodynamic resolution or dynamic kinetic resolution depending on the configurational stability of the lithium carbanion-chiral ligand complexes. Configurational stability of the α-hetero carbanion varies with the nature of the heteroatom X, solvent, and temperature. Several diagnostic reactions of the configurational stability of α-hetero carbanions have been developed [5]: Still and coworker estimated the configurational stability of lithium carbanions by comparison of the enantiomeric excesses of the products with those of the starting chiral stannyl or seleno compounds (Fig. 3). The close value of the enantiomeric excesses of these compounds corresponds to the configurationally stable lithium carbanion, and the different or often lower enantiomeric excesses to the labile carbanion. They first demonstrated that α-alkoxyalkyllithium compounds are configurationally stable for 15 min at –30 °C in THF [6].

This test showed that α-amino- [7] and α-oxy- [8] alkyllithium compounds except for their benzyl- or allyllithium compounds are configurationally highly stable, and α-thio-[9], α-selenoalkyllithium compounds [10], and lithiated phosphine oxides [11] are labile. Thus, this test proves that lithium carbanions are either configurationally stable or labile and affords reliable evidence to de-

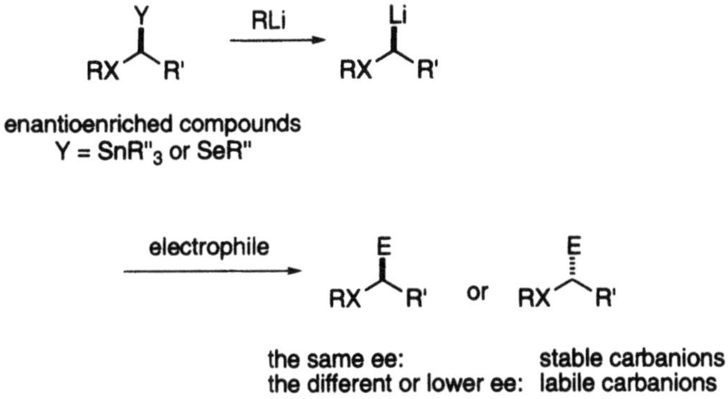

RLi

enantioenriched compounds
Y = SnR''_3 or SeR''

electrophile

or

the same ee: stable carbanions
the different or lower ee: labile carbanions

Fig. 3. Still's test

termine the enantiodetermining pathway, especially to confirm asymmetric deprotonation via configurationally stable carbanions. However, it is necessary to prepare optically pure organotin or organoselenium compounds in order to perform this test. Instead, Hoffmann and coworker have developed a test using racemic organolithium compounds together with racemic and chiral aldehydes [12]. This test will be discussed in detail later in this section. On the other hand, Beak and coworkers have reported tests for the determination of configurational stability of lithiated intermediates in the enantioselective reaction: one is the reaction of organolithium compounds with a substoichiometric amount of an electrophile in the presence of a chiral ligand [13, 14], and the other is a warm-cool procedure (Fig. 4) [14]. When the enantioselectivity of the product obtained in the reaction with a substoichiometric amount of an electrophile shows a different value from that obtained in the reaction with more than 1 equivalent of an electrophile, the lithium carbanion-chiral ligand complexes are configura-

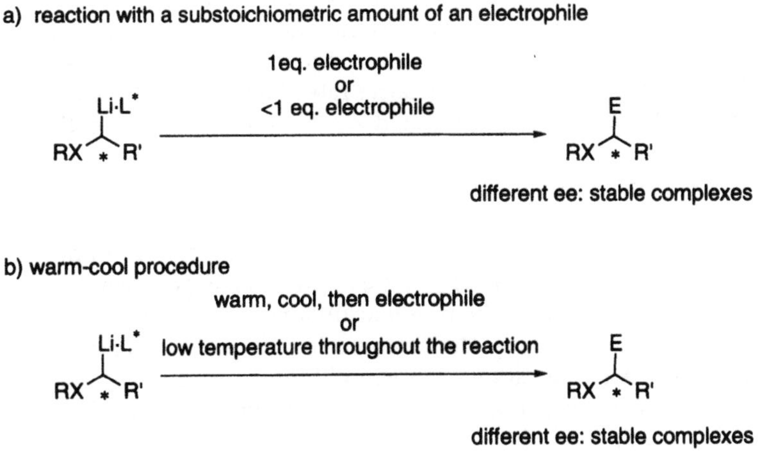

a) reaction with a substoichiometric amount of an electrophile

1eq. electrophile
or
<1 eq. electrophile

different ee: stable complexes

b) warm-cool procedure

warm, cool, then electrophile
or
low temperature throughout the reaction

different ee: stable complexes

Fig. 4 a,b. Beak's test

tionally stable and they do not undergo interconversion within the reaction time scale. The difference in enantiomeric excess obtained in the above reactions can be ascribed to the reactivity of the diastereomeric complexes with the electrophile and the thermodynamic stability of the complexes.

In the warm-cool procedure, the lithium carbanion-chiral ligand complexes, often prepared at low temperature, are warmed so as to be in equilibrium and then cooled to be reacted with an electrophile. When the product has a certain value of enantiomeric excess different from that obtained in the reaction at low temperature, the lithium carbanion-chiral ligand complexes are configurationally stable.

Another test for the configurational stability involves the enantioselective reaction of deuterated substrates [13, 15, 16] [Eq. (2)]. It can be predicted whether the asymmetric reaction proceeds through asymmetric deprotonation or not by the following results: the reaction pathway would be asymmetric deprotonation when the reaction of the enantioenriched deuterated substrate without a chiral ligand gives the product with the same enantiomeric ratio as that of the deuterated substrate, or the reaction of the racemic deuterated substrate in the presence of a chiral ligand gives the racemic, 100% deuterated product. If the latter reaction gives the enantioenriched product, the reaction pathway would be determined to be either asymmetric deprotonation or asymmetric substitution by the estimation based on the yield as well as on the enantiomeric excesses and the deuterium contents of the substrate and the product.

D
RX * R'
enantioenriched or
racemic compounds

1) base / (chiral ligand)
2) electrophile

E D
RX * R' (2)

Hoffmann and coworkers proposed an excellent method for evaluating the configurational stability of the lithium carbanion within the time scale of the rate of the reaction with an electrophile (Hoffmann test) [12]. The Hoffmann test is composed of two experiments. In the first experiment, a racemic organolithium compound is reacted with a racemic aldehyde such as 2-(*N*,*N*-dibenzylamino)-3-phenylpropanal, the relative rate of formation of the diastereomeric products (k_{SS}:k_{RS}) being estimated by an *anti*/*syn* ratio of the products (Fig. 5).

The racemic aldehyde for the first experiment should be chosen so that the k_{SS}:k_{RS} or k_{RS}:k_{SS} ratio is in the range of 1.5–3. In the second experiment, a racemic organolithium compound is reacted with the chiral aminophenylpropanal corresponding to the aldehyde used in the first experiment (Fig. 6). When the diastereomeric ratios of the products obtained in both experiments are not the same and, in addition, the diastereomeric ratio in the second experiment is around 50:50 (each enantiomeric organolithium compound should be transformed to the corresponding diastereomeric product in an equal ratio), the organolithium compound should be configurationally stable under the reaction conditions employed. Hence, the enantioselective reaction of the organolithium

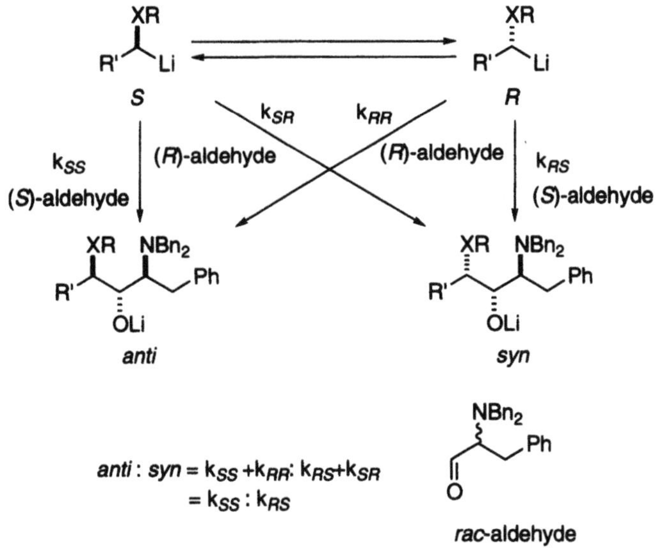

$$anti : syn = k_{SS} + k_{RR} : k_{RS} + k_{SR} = k_{SS} : k_{RS}$$

rac-aldehyde

Fig. 5. The first experiment of the Hoffmann test: reaction of a racemic compound with a racemic aldehyde

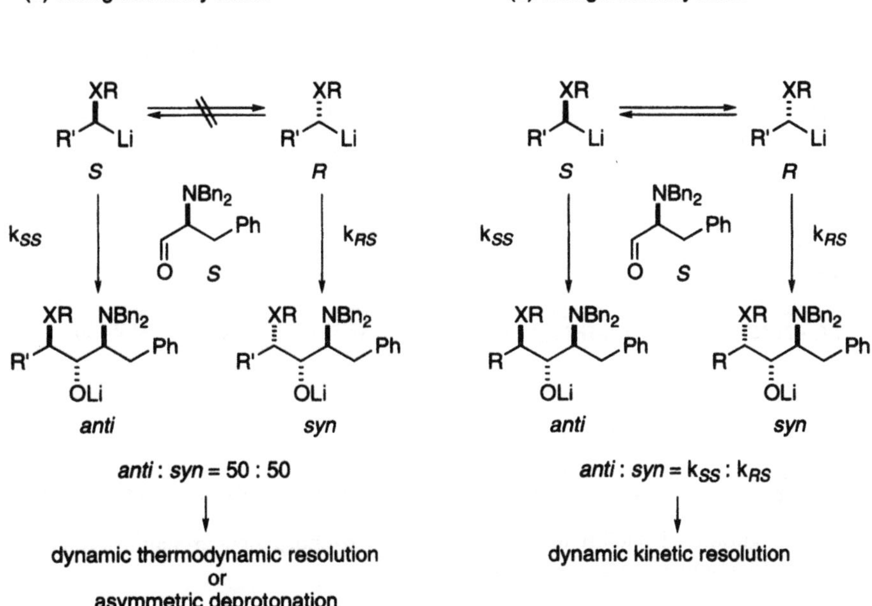

Fig. 6 a,b. The second experiment of the Hoffmann test: reaction of a racemic organolithium compound with a chiral aldehyde

compound used for the test in the presence of a chiral ligand is supposed to proceed through either pathway of dynamic thermodynamic resolution or asymmetric deprotonation; in order to obtain more information, a further Hoffmann test should be performed in the presence of a chiral ligand. When the *anti*/*syn* ratio of the product in the second experiment is in accord with the kinetic ratio (k_{SS}:k_{RS}) obtained in the first experiment, the organolithium compound is configurationally labile (Fig. 6b). The enantioselective reaction of such an organolithium compound would proceed through a dynamic kinetic resolution pathway.

Taking an example of a sulfur-stabilized benzyllithium compound, Hoffmann and coworkers performed two reactions of the racemic α-thio carbanion with a racemic or a chiral aldehyde [Eq. (3)] [17–19], and found that both reactions afford an identical diastereomeric ratio of the product. From these results, they have concluded that the sulfur-stabilized benzyllithium compound epimerizes faster than it reacts with the aldehyde, i.e., it is configurationally labile in THF at –78 °C.

PhCH(SPh)Li	+ OHC–CH(NBn_2)–CH_2Ph	→		PhCH(SPh)–CH(OH)–CH(NBn_2)–CH_2Ph	:	PhCH(SPh)–CH(OH)–CH(NBn_2)–CH_2Ph	(3)
racemic	racemic		90%	39	:	61	
racemic	(*S*)		90%	40	:	60	

On the other hand, the α-thio carbanion of butyl phenyl sulfide was estimated to be stable in 2-methyl-THF at –120 °C [17, 18]. The α-selenobenzyllithium [19] and the lithiated phosphine oxide [20, 21] are labile in THF at –78 °C (Fig. 7), whereas the α-selenobutyllithium and the α-sulfonylbenzyllithium are stable in Et_2O/THF at –105 °C [22] and in THF at –78 °C, respectively [19]. The α-silylbenzyllithium is configurationally labile at 0 °C in THF [19], the α-bromopentyllithium being stable in Trapp-solvent at –110 °C [17].

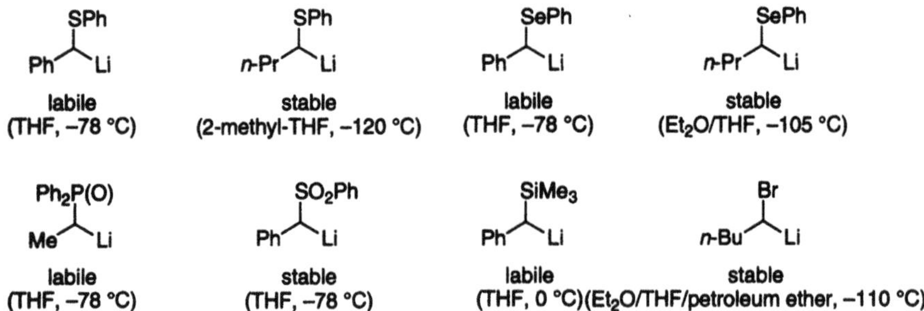

Fig. 7. Configurational stability of α-thio-, α-seleno-, α-phosphono-, α-sulfonyl, α-silyl and α-bromo carbanions estimated by the Hoffmann test

1.3 Racemization Mechanism of α-Hetero Carbanions

Carbanions α to the heteroatoms such as sulfur and selenium are configurationally rather labile in comparison with α-oxy carbanions. The racemization mechanism of α-thio- and α-seleno carbanions is essentially different from that of α-oxy carbanions. Sulfur and selenium being less electronegative than oxygen, make α-thio- and α-seleno carbanions have higher s character than α-oxy carbanions [23, 24] and the negative charge of α-thio- and α-seleno carbanions is partially stabilized by n-$\sigma_{X\text{-}C}^*$ negative hyperconjugation [25, 26]. Therefore, α-thio and α-seleno carbanions may have more sp^2-like orbitals, being closer to the transition states of inversion of the carbanionic centers in comparison with α-oxy carbanions. Thus, inversion of α-thio and α-seleno carbanions occurs more easily than that of α-oxy carbanions. However, the activation energy of racemization of α-thio and α-seleno carbanions does not correspond to that of the transition state of the inversion process. Hoffmann and coworkers studied in detail the mechanism of racemization of α-lithiated sulfides, α-lithiated selenides, and α-lithiated tellurides [27, 28]. These carbanions adopt an antiperiplanar arrangement of the C-Li and the X-R bonds in order to maximize delocalization by the n-$\sigma_{X\text{-}C}^*$ negative hyperconjugation. Their racemization process is composed of several distinct steps: 1) dissociation of a lithium cation from the carbanionic center to form a separated or a contact ion pair, 2) inversion of configuration of the carbanion, 3) subsequent rotation about the C-X bond, or vice versa, 2) first, rotation and 3) secondly, inversion, and 4) recombination of the carbanion with the Li cation (Fig. 8).

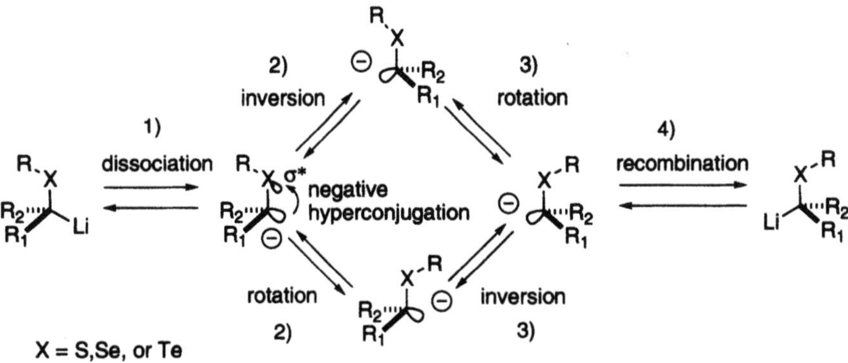

Fig. 8. Racemization mechanism of α-hetero carbanions

Hoffmann and coworkers have observed a marked steric effect on the racemization rate of α-lithiated sulfides, selenides and tellurides; branched, bulky substituents on the heteroatom increase the configurational stability (Fig. 9) [27, 29]. These results imply that the rotation about the C-X bond (Fig. 8) should be the rate-determining step of racemization [27]. Reich and coworker have also demonstrated a similar conclusion by ^{1}H-NMR analyses and kinetic studies [30]. Furthermore, they confirmed that the racemization rate of α-thio carban-

X = S, R = H, $\Delta G^{\ddagger}$ = 11.3 kcal/mol
X = Se, R = H, $\Delta G^{\ddagger}$ = 12.4 kcal/mol
X = Te, R = H, $\Delta G^{\ddagger}$ = 11.6 kcal/mol
X = Se, R = Me, $\Delta G^{\ddagger}$ = >12.9 kcal/mol

X = S: $\Delta G^{\ddagger}$ = >13.9 kcal/mol
X = Se: $\Delta G^{\ddagger}$ = >14.5 kcal/mol
X = Te $\Delta G^{\ddagger}$ = 14.1 kcal/mol

Fig. 9. Configurational stability of α-hetero carbanions

ions decreases as the ion pair is more separated. This is due apparently to enhanced negative hyperconjugation and hence increased barrier to rotation about the C-S bond [30, 31].

Hoppe and coworkers showed that the lithiated (*S*)-1-phenylethyl thiocarbamate (Fig. 10a) and the lithiated (*S*)-cyclohex-2-enyl thiocarbamate (Fig. 10b) are configurationally stable in THF at 0°C and at −70°C, respectively [32, 33]. They assumed that this is so because these resonance-stabilized tertiary carbanions have pronounced tendency to form solvent-separated ion pairs in THF. A silyl substituent makes the α-thio carbanion configurationally more stable than the parent thiocarbamate without the trimethylsilyl group (Fig. 10c) [34].

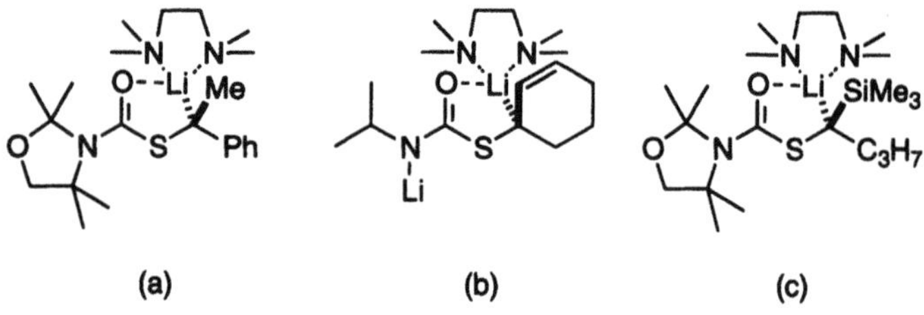

Fig. 10 a–c. Configurationally stable α-thio carbanions

Racemization of α-sulfonyl carbanions is also related to the n-$\sigma_{S\text{-}C}^*$ negative hyperconjugation [35, 36]. As the substituent (R) attached to the sulfonyl group is more electron-withdrawing, negative hyperconjugation becomes more effective and, therefore, the rate of racemization decreases (Fig. 11). For example, the lithiated trifluoromethyl sulfones are configurationally more stable than the lithiated *t*-butyl and the phenyl sulfones [35].

R = ^{t}Bu, $\Delta G^{\ddagger}$ = 13.0 kcal/mol (in THF/DMPU)
R = CF_3, $\Delta G^{\ddagger}$ = 17.3 kcal/mol (in THF/DMPU)

R = Ph, $\Delta G^{\ddagger}$ = 9.6 kcal/mol (inTHF)
R = CF_3, $\Delta G^{\ddagger}$ = 16.0 kcal/mol (in THF)

Fig. 11. Configurational stability of α-sulfonyl carbanions

In phosphorus-stabilized carbanions, interaction of the n orbital on the carbanionic center with the σ* orbital may increase the barrier to rotation about the C_{α}-P bond [37], but this interaction has been demonstrated to block rotation insufficiently in lithiated thiophosphonamides, even at low temperature (Fig. 12). It was also observed that HMPA stabilizes the carbanion in lithiated thiophosphonamides by dissociation of the lithium ion, decreasing the rate of rotation about the C-P bond [38].

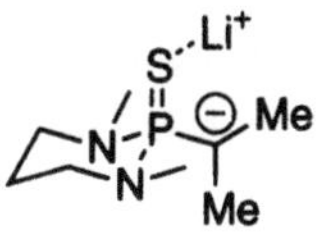

$\Delta G^{\ddagger}$ = 9.2 kcal/mol (in toluene/THF =1.6:1)
$\Delta G^{\ddagger}$ = 9.8 kcal/mol (in THF)
$\Delta G^{\ddagger}$ = 11.4 kcal/mol (in THF/HMPA)

Fig. 12. Configurational stability of α-lithiated thiophosphonamides

2
Enantioselective Reactions of Sulfur Compounds

2.1
Enantioselective Reactions of α-Thio Carbanions

Most α-thio carbanions are configurationally labile and racemize rapidly even at low temperature as shown in the preceding section. This may be the reason why the asymmetric reaction of α-lithiated thiocarbamates reported by Hoppe and coworkers does not show such high enantioselectivity [34] as attained in the reaction of similar dipole-stabilized α-oxy organolithium compounds [Eq. (4)] [2, 39]. Note that the stereochemistry of the products obtained from α-lithiated thiocarbamates is different from that obtained in the α-lithiated carbamate [Eq. (5)].

1) s-BuLi / (–)-sparteine
2) CO_2
ether, –78 °C

(S)

(–)-sparteine

R = ^{n}Pr 91%, 47% ee
R = Me 89%, 40% ee
R = ^{i}Pr 77%, 60% ee

(4)

1) s-BuLi / (–)-sparteine
2) CO_2
ether, -78 °C

(R)

52%, >95% ee

(5)

Takei and coworkers have reported the enantioselective reaction of arylthio-stabilized amide-homoenolates with aldehydes in the presence of (–)-sparteine, through a dianionic cyclic intermediate which give products with good diastereoselectivity but with low enantioselectivity [36% ee, Eq. (6)] [40]. Since the reaction at –100 °C showed even lower enantioselectivity (30% ee), they concluded that the dianionic intermediate kinetically formed racemizes at –78 °C.

n-BuLi / (–)-sparteine
THF, –78 °C

PhCHO

anti syn

85%, anti:syn = 87:13
36% ee (anti)

(6)

Toru and coworkers have reported a highly enantioselective reaction of α-lithiated benzyl phenyl sulfide with various electrophiles [41, 42], in which the most efficient chiral ligand was a bis(oxazoline). High enantioselectivities were obtained in the reaction with various carbonyl compounds, although the reaction with alkyl halides and carbon dioxide showed somewhat lower enantioselectivity (Table 1). Cumene was the best solvent giving the highest ee, whereas toluene slightly lowered the enantioselectivity.

Higher enantioselectivity was obtained when the reactions with ketones were performed at lower temperature (>99% ee). In addition, the Hoffman test using

Table 1. Enantioselective reaction of α-lithiated benzyl phenyl sulfide

Ph–S–CH(Ph)–$SnBu_3$ → [1) n-BuLi; 2) bis(oxazoline) (^{i}Pr), cumene, –78 °C] → Ph–S–*CH(Ph)–Li·Ln* → [electrophile] → Ph–S–CH(Ph)–E

Electrophile	yield (%)		syn:anti	ee (%)
Ph_2CO	79			99
acetone	71			>99
cyclohexanone	100	(–95 °C)		98
PhCHO	100	(–95 °C)	60:40	87 (*syn*), 96 (*anti*)
EtCHO	51		38:62	97 (*syn*), 96 (*anti*)
^{i}PrCHO	61		38:62	94 (*syn*), 95 (*anti*)
MeOTf	86			81
Me_3SiOTf	76			77
CH_2=$CHCH_2Br$	25			76
CO_2	87			74

2-(*N*,*N*-dibenzylamino)-3-phenylpropanal, TMEDA or the enantiopure bis(oxazoline), confirmed a dynamic kinetic resolution pathway for this reaction. Ab initio calculations showed that the C-Li bond prefers the conformation antiperiplanar to the S-C_{ips} bond [42] due to the n-σ* negative hyperconjugation of α-lithiated benzyl phenyl sulfide, as described in Sect. 1.3 (Fig. 13). Thus, this enantioselective reaction was assumed to proceed through a four-membered cyclic transition state stabilized by the negative hyperconjugation.

Interestingly, the stereochemical outcome in the reaction of α-lithiated benzyl phenyl sulfide is different from that in the reaction of α-lithiated benzyl methyl ether which proceeds through a dynamic thermodynamic resolution pathway [Eq. (7)] [15, 43]. Taking account of Wiberg's results, namely, that the α-oxy carbanion of dimethyl ether places the lone pair of carbanion *syn* to the methyl group by the electron repulsive effect [25], the difference in the stereochemical outcome between α-thio and α-oxy carbanions would be ascribed to their different preferred conformations. A similar change in the stereochemical outcome has been observed in reactions of dipole-stabilized α-thio and α-oxy organolithium compound, as shown in Eqs. (4) and (5).

Me–O–CH_2Ph → [1) n-BuLi, bis(oxazoline) (^{i}Pr), –78 °C; 2) CO_2, –110 °C, hexane,] → Me–O–CH(Ph)–CO_2H (7)

>95%, 95% ee

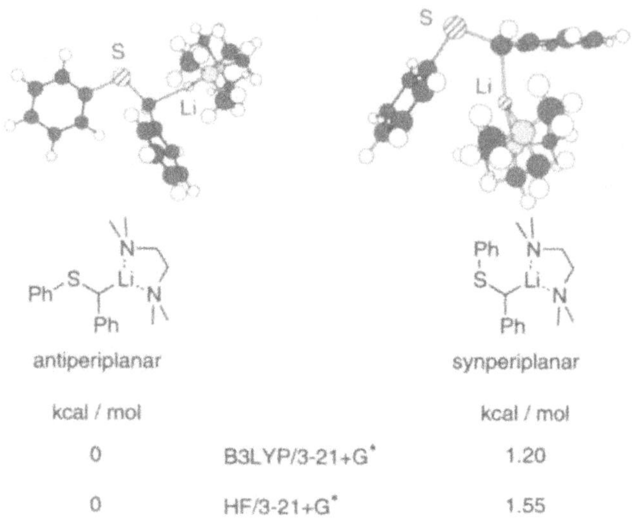

Fig. 13. Calculation of relative stability of conformers of α-lithiated benzyl phenyl sulfide-TMEDA complexes

When the substituent of the benzyl sulfide was changed from the phenyl to a pyridyl group, a drastic change in reaction features was observed. *n*-BuLi abstracts the proton without a ligand. Furthermore, the enantioselective reaction of α-lithiated benzyl 2-pyridyl sulfide with electrophiles gives the products with reversed stereochemistry to that obtained in the reaction of the α-lithiated benzyl phenyl sulfide (Table 2).

The reaction of α-lithiated benzyl 2-pyridyl sulfide was proved to proceed through a dynamic thermodynamic resolution pathway by the experiments us-

Table 2. Enantioselective reaction of α-lithiated benzyl 2-pyridyl sulfide

1) *n*-BuLi
2) bis(oxazoline) (R)
cumene, −78 → T^1 °C
Li·Ln*
electrophile
T^2 °C
R, E, Ph

electrophile	R	T^1 (°C)	T^2 (°C)	yield (%)	ee(%)
Ph_2CO	^{i}Pr	−50	−50	91	78
Ph_2CO	^{i}Pr	−50	−78	77	70
Ph_2CO	^{t}Bu	−78	−78	86	90
CO_2	^{i}Pr	−78	−78	60	70
MeI	^{t}Bu	−78	−78	72	89
Me_3SiOTf	^{t}Bu	−78	−78	91	93

ing a substoichiometric amount of an electrophile (0.2 equiv. of benzophenone at –78 °C, 50% ee) as well as by the warm-cool experiments as shown in Table 2 [13, 14]. Thus, α-lithiated benzyl 2-pyridyl sulfide reacts with an electrophile before interconversion between the diastereomeric lithium carbanion-chiral ligand complexes. The (*R*)-diastereomeric complex was shown to be more stable than the (*S*)-complex by ab initio and semiempirical calculations (Fig. 14). In this intermediate, the pyridyl nitrogen plays an important role in determining the direction of approach of an electrophile. An electrophile attacks the carbanion without coordination to the cationic lithium, which is fully coordinated with three nitrogens, two of the bis(oxazoline) and one of the pyridine. Thus, the reaction proceeds with inversion of the carbanionic center.

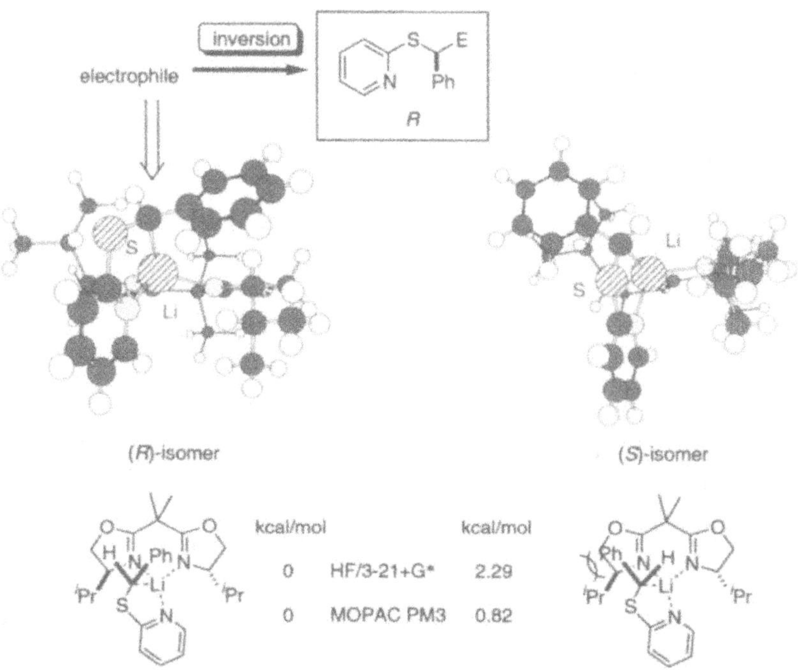

Fig. 14. Stability of α-lithiated benzyl 2-pyridyl sulfide-bis(oxazoline) complexes

Inversion of configuration in the above reaction is in contrast to retention of stereochemistry generally observed in the enantioselective reaction with carbonyl compounds. There are a few reports on the S_E2 reaction with a carbonyl compound proceeding with inversion [32, 44]. Recently, Hoppe and coworkers have observed that the configurationally stable chiral α-thio carbanion derived from chiral thiocarbamates reacted with carbonyl compounds to create the quaternary center with complete inversion of the carbanionic center [Eq. (8)] [32], whereas the reaction of the corresponding α-oxy carbanion with electrophiles (including carbonyl compounds) proceeds with retention of configuration [45].

(8)

>99% ee

The S_E2 reaction on the sp^2-like hybridized carbanion tends to proceed with inversion of the stereochemistry [45]. It is reasonable that α-thio carbanions, having higher s character than α-oxy carbanions, more frequently invert the stereochemistry on the anionic carbon in the reaction with electrophiles. As in the case of α-lithiated benzyl 2-pyridyl sulfide, the reaction of α-lithiated benzyl 2-quinolyl sulfide also proceeds through a dynamic thermodynamic resolution pathway. The quinolyl group is an excellent protecting group of thiols [46] and can be removed to yield the corresponding chiral thiols without racemization (Table 3) [47].

Table 3. Enantioselective reaction of α-lithiated benzyl 2-quinolyl sulfide and subsequent conversion into chiral thiols

electrophile	sulfide		thiol
	yield (%)	ee (%)	ee (%)
Ph_2CO	95	89	87
acetone	40	80	80
cyclohexanone	46	81	80
benzaldehyde	55 (*syn* / *anti* = 45 / 55)	*syn* : 95 *anti* : 93	
CO_2	54	83	82
MeOTf	85	78	78

Asymmetric reaction of the sulfides bearing a tricarbonyl(η^6-arene)chromium complex was shown to be successful by Gibson and Simpkins [48, 49]. The benzylic methylene groups in tricarbonyl(η^6-phenylmethyl alkyl sulfide)chromium(0) and tricarbonyl(η^6-1,3-dihydroisobenzothiophene)chromium(0) were highly asymmetrically functionalized by deprotonation with a chiral bis-lithium amide and subsequent electrophilic reactions (Tables 4 and 5).

Table 4. Enantioselective reaction of tricarbonyl(η^6-phenylmethyl alkyl sulfide)chromium(0)

1) chiral bis-lithium amide (Ph, Ph, N, Li Li, N, Ph, Ph), LiCl; 2) R^2X; THF, –78 °C

R^1	R^2X	yield (%)	ee (%)
Me	$PhCH_2Br$	91	88
Et	Me_3SiCl	72	82
CH_2Ph	MeI	84	91

Table 5. Enantioselective reaction of tricarbonyl(η^6-1,3-dihydroisobenzothiophene)chromium(0)

1) chiral bis-lithium amide (Ph, Ph, N, Li Li, N, Ph, Ph), LiCl; 2) electrophile; THF, –100 °C

electrophile	yield (%)	ee (%)
MeI	95	94
Ph_2CO	88	95
Me_3SiCl	95	89

Gibson and coworkers have pointed out an interesting feature of these reactions: the reaction of the α-lithiated sulfides, derived from the chromium complexes of benzyl sulfides, gives products having a configuration opposite to that obtained in the reaction of the corresponding ethers [Table 4 and Eq. (9)] [50], although this type of stereochemical reversibility has not been observed in enantioselective reactions of chromium complexes of 1,3-dihydroisobenzothiophene and the 1,3-dihydroisobenzofuran [51].

1) Ph Ph N N Li Li Ph Ph , LiCl; 2) electrophile; THF, –78 °C

R = Me, Bn

electrophile: MeI, PhSSPh
>97% ee

(9)

The reaction of the chromium complexes of benzyl ethers has been assumed to proceed through a selective deprotonation giving the configurationally stable intermediates, which react with electrophiles from their *exo* face [Eq. (10)] [52]. Provided that the reaction of benzyl sulfides proceeds through similar intermediates, the different stereochemical results between benzyl sulfides and ethers would be rationalized by the difference in the deprotonation pattern.

selective deprotonation; exo face; R'X

(10)

Uemura and Katsuki have independently reported the enantioselective [2,3]-sigmatropic rearrangement of allylic sulfur ylides using a catalytic amount of chiral rhodium(II) and cobalt(III) complexes, respectively. The carbenoids derived from diazoacetate were enantioselectively added to the sulfur atom of cinnamyl phenyl sulfide to afford the product with low to good enantioselectivity [53, 54] [Eq. (11)]. McMillen and coworkers have also reported a similar enantioselective [2,3]-sigmatropic rearrangement of allylic sulfur ylides using a Cu(I)-bis(oxazoline) catalyst [55].

chiral catalyst

[2,3]

R	Catalyst / conditions	Result
R = ^{t}Bu	Rh_2(5*S*-MEPY)$_4$ (1 mol%) $CHCl_3$, 0 °C	81%, *anti* : *syn* = 60 : 40 18% ee, 17% ee
R = Et	Co-salen(III) (0.05 mol%) CH_2Cl_2, rt	81%, *anti* : *syn* = 85 : 15 64% ee (anti)

(11)

Rh_2(5*S*-MEPY)$_4$

Co-salen(III)

2.2 Enantioselective Reactions of α-Sulfonyl Carbanions

Carbanions α to the sulfonyl group can be easily prepared due to their high acidity in comparison with α-thio carbanions. Although α-sulfonyl carbanions have a more sp^2-hybridized character than α-thio carbanions [24, 56], α-sulfonyl carbanions have been shown by Gais and coworker to have a somewhat higher inversion barrier to racemization than α-thio carbanions [35]. For example, when an enantiomerically pure trifluoromethyl sulfone was first deprotonated with *n*-BuLi and subsequently protonated with trifluoroacetic acid, the sulfone was recovered with 90% ee [Eq. (12)].

1) *n*-BuLi (3 min) 2) CF_3CO_2H THF, –105 °C

(*S*) >99% ee → (*S*) 90% ee

(12)

Silylation of α-sulfonyl carbanions may proceed through an asymmetric deprotonation pathway [Eqs. (13) and (14)] [57]. Thus, Simpkins has shown that silylated products are obtained with certain enantioselectivities when TMSCl is present in the reaction mixture during the deprotonation reaction with a camphor-derived lithium amide (internal quench condition), whereas the external quench of the generated carbanions with TMSCl gives racemic products.

Li N Ph CH3 or Li N Ph
Me3SiCl
71%, 67% ee (13)

Li N Ph CH3 or Li N Ph
Me3SiCl
48%, 21% ee (14)

Enantioselective reaction of the lithium carbanion derived from an allyl sulfone in the presence of a chiral diamine has been reported by Mukaiyama and coworkers to afford a product with low enantioselectivity (15% ee) in comparison with the organomagnesium compound [60% ee, Eq. (15)] [58]. Since the deprotonation is carried out at high temperature, probably this reaction does not proceed through an asymmetric deprotonation.

1) base / chiral ligand, 0 °C
2) acetone , –78 °C
THF
(15)

chiral ligand

Base	yield (%)	ee (%)
n-BuLi	91	15
EtMgBr	47	60

2.3
Asymmetric Reactions of α-Sulfinyl Carbanions

The enantioselective reaction of α-sulfinyl carbanions involves both desymmetrization of the prochiral alkyl groups and diastereoselection. Simpkins and coworkers have reported the asymmetric induction of a *trans*-thiane oxide with a camphor-derived chiral lithium amide giving products with complete diastereoselectivity and with good enantioselectivity (Table 6) [59, 60].

Table 6. Enantioselective deprotonation of *trans*-thiane oxide using a chiral lithium amide

1) chiral lithium amide; 2) electrophile, THF, –78 °C

electrophile	yield (%)	ee (%)
Me_3SiCl	91	60
MeI	81	64

More recently, they have reported an enantioselective reaction via an α-sulfinyl carbanion involving rearrangement to a vinyl sulfoxide [Eq. (16)] [61]. Treatment of an episulfoxide with a chiral lithium amide followed by methylation yielded a methysulfinylcyclopentene with high enantioselectivity. In this reaction, the enantioselection occurs in the first deprotonation step.

1) chiral dilithium amide; 2) MeI, THF, –78 °C → 85%, 82% ee (16)

Kinetic resolution was shown by Kunieda and coworker to be another enantioselective process of α-sulfinyl carbanions. Slight asymmetric induction was achieved in the reaction of racemic α-lithiomethyl *p*-tolyl sulfoxide with ethyl carboxylates in the presence of (–)-sparteine [Eq. (17)] [62].

$PhCO_2Et$ + Tol–S(O)–CH_2Li (2.0 eq.), (–)-sparteine, THF, 0 °C → 70%, 15% ee (17)

2.4 Enantioselective Reactions of α-Lithiated Dithioacetals

Metalated dithioacetals are known to be useful as acyl anion equivalents. Thus, successful diastereoselective reactions have led to many applications of dithioacetals [63]. However, only a few enantioselective reactions of symmetric dithioacetals have been reported. One example is the enantioface-selective conjugate addition of a lithiated dithiane to α,β-unsaturated esters, in the presence of a chiral ligand derived from L-phenylalanine [Eq. (18)] [64].

1) n-BuLi
Me$_2$N O MeO
2) ^{i}Pr CO$_2$Et
toluene, –78 °C
^{i}Pr CO$_2$Et Ph
32%, 67% ee

(18)

It is known that the use of (–)-isosparteine as a ligand sometimes gives much better enantioselectivity than (–)-sparteine [65]. The following reaction is an example. Enantioface-selective reaction of 2-lithio-1,3-dithiane with benzaldehyde in the presence of (–)-α-isosparteine gave the addition product with 70% ee, whereas the reaction with (–)-sparteine showed almost no selectivity [66]. The assumed transition state for this reaction is illustrated in Eq. (19). However, the reaction failed to achieve high enantioface-selection with aliphatic aldehydes.

1) n-BuLi / chiral ligand
2) PhCHO
Et$_2$O, –78 °C
OH Ph
(–)-α-isosparteine 73%, 70% ee
(–)-sparteine 84%, 3% ee
(–)-α-isosparteine

(19)

3
Enantioselective Reactions of Selenium Compounds

Only a few enantioselective reactions of α-seleno carbanions have been reported. First, Hoffmann and coworkers observed, using ^{77}Se-NMR spectral analyses, large equilibrium constants for diastereomeric complexes derived from α-seleno carbanions and various chiral diamines [67]. They studied the enantioselective reaction of an α-seleno carbanion generated from a diselenoacetal, with benzaldehyde in the presence of chiral diamines and found that a chiral cyclopentanediamine ligand showed good enantioselectivity [Eq. (20)] [68]. The reaction of the α-seleno carbanion with racemic or chiral aldehydes in a Hoffmann test showed that it proceeds through a dynamic thermodynamic resolution pathway. Since the α-seleno carbanion-chiral diamine complex was found to react with benzaldehyde more rapidly than the uncomplexed organolithium species, it was hoped to realize this reaction with a catalytic amount of the chiral ligand, but a strong association of the lithium alkoxide formed in the reaction with the chiral diamine required the use of a stoichiometric amount of the diamine for this reaction.

1) *t*-BuLi
2) (chiral cyclopentane-1,2-bis(NMe$_2$))
3) PhCHO, ether, –80 °C

SePh, SePh → SePh, Ph, OH + SePh, Ph, OH (20)

syn 74% ee *anti* 86% ee

83%, *syn*:*anti* = 68:32

As in the case of the sulfur ylide [Eq. (11)], the corresponding selenium ylide also undergoes enantioselective rearrangement. Diazoacetate was reacted with the allylic selenide in the presence of a chiral rhodium(II) catalyst to give the product via [2,3]-sigmatropic rearrangement, and somewhat higher enantioselectivity was achieved [Eq. (21)] [53] than that in the reaction of the corresponding sulfide.

Ph SePh + N=N=CH–CO$_2$Et $\xrightarrow[\text{CHCl}_3,\ 0\ °\text{C}]{\text{Rh}_2(5S\text{-MEPY})_4\ (1\text{mol\%})}$ [Ph SePh$^{\oplus}$ $^{\ominus}$CH–CO$_2$Et]

$\xrightarrow{[2,3]}$ Ph * * SePh, CO$_2$Et (21)

dr 58:42
25% ee, 41% ee

4 Enantioselective Reactions of Phosphorus Compounds

The enantioselective reactions of carbanions α to phosphorus, such as α-lithiated carbanions of activated phosphines, phosphonates, and phosphine oxides, have been studied extensively because of their broad synthetic utility. In fact, enantioselective reactions of α-lithiated phosphonates with carbonyl compounds give olefins having axial chirality. Furthermore, the enantioselective reaction of α-carbanions of phosphine-boranes affords a method for the preparation of the C_2-symmetric chiral ligands and P-chiral ligands. In this section, enantioselective reactions of phosphorus compounds are described, including their reaction mechanisms and synthetic applications.

4.1 Enantioselective Horner-Wadsworth-Emmons Reactions

The Horner-Wadsworth-Emmons (HWE) reaction of a carbonyl compound with an α-lithiated phosphonate is an extremely useful method for introducing a double bond. In order to create a chiral compound having a double bond, 4-substituted cyclohexanones, 3-substituted cyclohexanones, 2,2-disubstituted cyclohexane-1,3-diones, or ketenes would be potential carbonyl substrates for the enantioselective HWE reaction, as Rein and coworker proposed [Fig. (15)] [69]. Only 4-substituted cyclohexanones have so far been applied to the enantioselective HWE reaction.

The HWE reaction is known to proceed in a stepwise manner [Eq. (22)]: first, reversible aldolate formation, and second, irreversible double bond formation with liberation of the phosphate. Enantioselection occurs according to either of the following pathways:

1) the enantioselectivity is dependent on the rate of formation of the diastereomeric aldolates, when the selection of one of the prochiral methylene protons of the phosphonate and the selection of one of the prochiral carbonyl faces are effected by a chiral additive, and elimination of the phosphate is faster than the reversible reaction, or
2) the enantioselectivity is dependent on the rate of elimination of the phosphate from the diastereomeric aldolate-chiral ligand complexes.

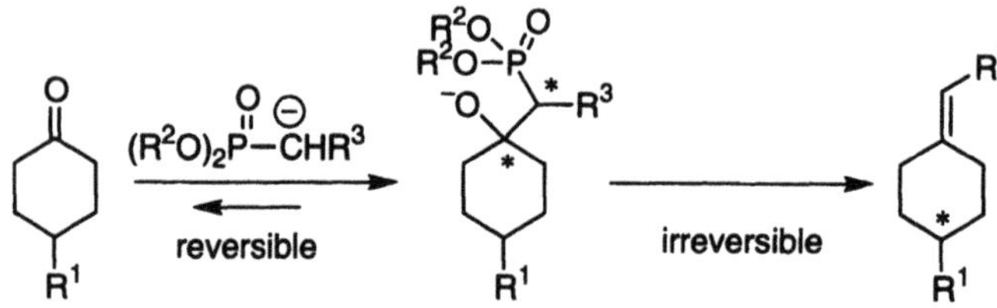

(22)

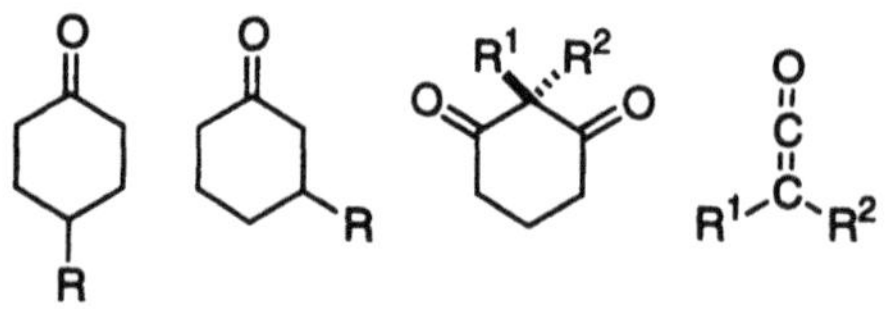

Fig. 15. The carbonyl substrates proposed for enantioselective HWE reactions

In contrast to the number of studies on asymmetric HWE reactions using chiral phosphonates, only a few enantioselective HWE reactions using a combination of achiral phosphonates and chiral ligands are known [69, 70]. Koga and coworkers reported the first enantioselective HWE reaction of diethylphosphonoacetonitrile and 4-*tert*-butylcyclohexanone using a stoichiometric amount of lithium 2-aminoalkoxides as a chiral base [Eq. (23)] [71]. The α,β-unsaturated nitrile was obtained in 92% yield with 52% ee. When the racemic aldolate intermediate was treated with a chiral diamine, a similar result was obtained. These results show that dissociation of the lithium aldolate to the α-lithiated phosphonoacetonitrile and recombination to the aldolate reversibly occurs during the reaction, and the enantioselectivity is controlled by the rate of the elimination reaction of the phosphate.

O
+ (EtO)$_2$P(O)–CH$_2$CN
Me Ph
^{t}BuCH$_2$–NH OLi
toluene, –78 °C
3 h
NC
R
^{t}Bu
92%, 52% ee
(23)

Tomioka and coworkers have reported the enantioselective reaction of 4-*tert*-butylcyclohexanone with lithium phosphonates in the presence of a chiral dimethoxyethane. They succeeded in the isolation of the diastereomeric aldolates, one of which was formed with high enantioselectivity. Thermal treatment of this aldolate in propionic acid/NaOAc gave benzylidene-4-*tert*-butylcyclohexane without any change in enantioselectivity [Eq. (24)] [72]. These results indicate that the enantioselectivity of the reaction is controlled not only in the first deprotonation step but also by the discrimination of the carbonyl face, although the enantiodetermining step in this reaction is not in accord with Koga's results mentioned above.

O
+ (EtO)$_2$P(O)–CH$_2$Ph
Ph Ph
MeO OMe
n-BuLi
toluene, –78 °C
0.5 h
^{t}Bu
Ph
PO(OEt)$_2$
OH
89%, 82% ee
+
^{t}Bu OH
Ph PO(OEt)$_2$
5%, 12% ee
Ph
S
^{t}Bu
85%, 84% ee
(24)

The fluoroolefin moiety is known to be useful as an amide bond mimic of bioactive compounds [73]. Nagao and coworkers have reported the enantioselective formation of fluoroalkylidenes by HWE reactions. Reaction of 2-fluoro-2-diethylphosphonoacetates with $Sn(OTf)_2$, *N*-ethylpiperidine, and a chiral pyrrolidine base afforded the desired α-fluoro-α,β-unsaturated esters in up to 80% ee [Eq. (25)] [74].

1) $Sn(OTf)_2$/ N-Et, 0 °C, 30 min
2) , 0 °C, 1 h
Me
3) ^{t}Bu , –60 °C, 40 min
CH_2Cl_2
$CO_2{}^iPr$
F
R
^{t}Bu
79%, 80% ee

(25)

The reactions described above require a stoichiometric amount of a chiral ligand to obtain the products in optically active form. Shioiri and coworkers have reported a catalytic version of the enantioselective HWE reaction using phosphonoacetates. Various chiral quaternary ammonium salts were examined as phase-transfer catalysts, generating enantioselectivity in chiral α,β-unsaturated esters which ranged from 37% ee to 55% ee. The best result (55% ee) was obtained by using a quaternary ammonium salt derived from cinchonine and RbOH as a base [Eq. (26)] [75].

+ $(EtO)_2P(O)-CH_2CO_2Et$
1) PTC (20 mol%) benzene, RbOH rt, 192 h
2) HCl / EtOH, 60 °C
EtO_2C
R
^{t}Bu
75%, 55% ee

(26)

HO
N^+
Br^-
N
^{t}Bu
PTC

Toda and coworker have disclosed that the solid-state enantioselective Wittig reaction of (carbethoxymethylene)triphenylphosphorane using a chiral 1,4-dioxaspiro[5,4]decane as a chiral host compound gave 4-*tert*-butylcarbethoxymethylidenecyclohexane with 57% ee [Eq. (27)] [76], although the reaction using various chiral carboxylic acids, as host compounds (such as mandelic acid and *N*-phthaloylleucine), gave the product with less than 10% ee [77].

(27)

$Ph_3P{=}CHCO_2Et$, 70 °C, 4 h → 58%, 57% ee

4.2
Enantioselective Reactions of α-Lithiated Phosphine Oxides

Simpkins and coworker have reported desymmetrization of a *meso*-phospholane oxide involving discrimination of prochiral protons by a chiral lithium amide (Table 7) [78]. The addition of LiCl was effective to enhance the selectivity, probably due to lesser aggregation of the lithium salt [79]. In certain cases the enantiomeric excesses of the products could be increased up to 97% ee by recrystallization. The obtained phospholane oxides can be readily reduced to optically active phosphines, which are known to be useful as chiral ligands.

Table 7. Enantioselective reaction of a *meso*-phospholane oxide

1) chiral lithium amide, LiCl; 2) electrophile, THF, –100 °C

electrophile	yield (%)	ee (%)
MeI	87	85
acetone	85	87
PhCHO	82	92

Deprotonation-protonation is an alternative choice for asymmetric induction. Deprotonation of the *meso*-phospholane oxide with *s*-BuLi-(–)-sparteine and subsequent protonation with acetic acid afforded *trans*-phospholane oxide with 45% ee together with the recovered *meso*-phospholane oxide (Table 8) [80].

Vedejs and coworker succeeded in a kinetically controlled enantioselective protonation reaction, in which α-lithiated methylbenzylphosphine oxide was treated with *n*-BuLi at –78°C and, subsequently, with a chiral camphor-derived amine to achieve protonation [Eq. (28)] [81]. Thereafter, Warren and coworkers

Table 8. Enantioselective protonation-deprotonation of a *meso*-phospholane oxide

1) *s*-BuLi / (–)-sparteine
2) protonating agent
THF, –78 °C

(*trans,cis*) meso (*trans,trans*)

protonating agent	ratio (*trans,cis*):(*trans,trans*)	(*trans,cis*) ee (%)
tert-BuOH	98 : 2	2
CH_3CO_2H	33 : 66	45
Ph_3CH	5 : 95	58

proved by the Hoffmann test that carbanions α to phosphine oxides are configurationally labile [21]. However, treatment of isobutylphosphine oxide with *n*-BuLi-(–)-sparteine at –78 °C and subsequently with TMSCl resulted in low enantioselectivity [Eq. (29)] [82].

1) *n*-BuLi
2) NHPh
toluene, -78 °C
81% ee (28)

1) *n*-BuLi / (–)-sparteine
2) Me_3SiCl
ether / THF, –78 °C
48%, 10% ee (29)

4.3 Enantioselective Reactions of α-Lithiated Phosphine Derivatives

There are no reports on the enantioselective reaction of the carbanion α to non-activated phosphines due to the low acidity of their α-protons. However, formation of phosphine-borane complexes enables deprotonation at the α position [83]. Evans and coworkers have found that enantiotopic methyl groups of phosphine-boranes can be efficiently discriminated by *s*-BuLi-(–)-sparteine [Eq. (30)] [84]. This result is in accord with many other reports that the chiral complex composed of *s*-BuLi and (–)-sparteine is generally the most efficient enantioselective deprotonating combination [1, 2, 3, 85]. However, in the reaction of phosphine sulfides this is not the case. The *n*-BuLi-(–)-sparteine combination shows better enantioselectivity (79% ee) than when using *s*-BuLi-(–)-sparteine [Eq. (31), 55% ee]. The oxidative coupling by copper(II) pivalate afforded C_2-symmetric P-chiral diphosphine-boranes with excellent enantioselectivity via the enantiomerically enriched α-lithiated phosphine-boranes formed by the action of *s*-BuLi-(–)-sparteine. Formation of the *meso*-isomer reduces the amount of the antipode [Eq. (32)]. This reaction was applied by Imamoto and coworkers to the synthesis of C_2-symmetric P-chiral diphosphine ligands and enantioenriched secondary phosphine-boranes [86].

1) *s*-BuLi / (–)-sparteine
2) Ph_2CO
Et_2O, –78 °C
84%, 87% ee (30)

1) *n*-BuLi / (–)-sparteine
2) Ph_2CO
Et_2O, –78 °C
79%, 79% ee (31)

1) *s*-BuLi / (–)-sparteine
2) $Cu(OPiv)_2$
Et_2O, –78 °C
+ *meso*
67%, 99% ee
(*S,S*) : *meso* = 88:12 (32)

Kobayashi and coworkers have demonstrated that optically active phospholane-2-carboxylic acid [n=1, Eq. (33)] and phosphorinane-2-carboxylic acid (n=2) can be prepared by asymmetric carboxylation of the phosphine-borane complexes. High *trans* selectivity and excellent enantiomeric excesses of the *trans* isomers were achieved on treatment of 1-phenylphospholane-borane and 1-phenylphosphorinane-borane complexes with *s*-BuLi-(–)-sparteine and subsequently with CO_2. Notably, enantioselectivity was improved up to 92% ee when 1-phenylphospholane-borane was treated with *s*-BuLi-(–)-sparteine at 25 °C, compared with 83% ee at –78 °C [87].

1) *s*-BuLi / (–)-sparteine, T °C
2) CO_2, –78 °C
Et_2O

n = 1, –78 °C, 24%
trans : *cis* = 3 : 1, 83% ee (*trans*)

25 °C, 55%,
trans : *cis* = 4 : 1, 92% ee (*trans*)

n = 2, –78 °C, 40%,
trans : *cis* = 18 : 1, 90% ee (*trans*) (33)

Livinghouse and coworkers have reported highly enantioselective alkylation of a P-lithiophosphine-borane in the presence of (–)-sparteine [88] [Eq. (34)]. The enantiomeric excess in the reaction was dependent on the reaction time and temperature. The lithium phosphine-borane was stirred for 1 h at 25 °C in the presence of (–)-sparteine prior to alkylation with alkyl halides at –78 °C, giving products with high enantioselectivity (up to 93% ee), whereas warming the reaction mixture to 0 °C prior to alkylation resulted in lower enantioselectivity (35% ee). From these results, they have concluded that the reaction of lithium phosphine-borane in the presence of (–)-sparteine proceeds through a dynamic thermodynamic resolution pathway. Furthermore, they also proved that the P-lithio-*tert*-butylphenylphosphine-borane-(–)-sparteine complexes are configurationally stable at –78 °C by the fact that deprotonation of a chiral phosphine-borane and subsequent reaction with alkyl halides proceed with retention of configuration at phosphorus without changing the value of the enantiomeric excess [Eq. (35)] [89]. They observed products with higher ees in the reaction of the P-lithiophosphine-borane with alkyl dihalides [Eq. (36)], as in the reaction shown in Eq. (32).

BH_3 / Ph–P(H)–^{t}Bu —[*n*-BuLi / (–)-sparteine, Et_2O]→ BH_3 / Ph–P*(Li·Ln)–^{t}Bu —[MeI, –78 °C]→ BH_3 / Ph–P(Me)–^{t}Bu (34)

–78 → 25 °C: 88%, 93% ee
–78 → 0 °C: 35% ee

BH_3 / Ph–P(H)–Me (>99% ee) —[1) *n*-BuLi; 2) 2-(chloromethyl)benzothiophene (CH_2Cl), THF, –78 °C]→ BH_3 / Ph–P(Me)–CH₂-benzothienyl (94%, >99% ee) (35)

BH_3 / Ph–P(H)–^{t}Bu —[1) *n*-BuLi / (–)-sparteine, –78 → 25 °C; 2) α,α′-dibromo-o-xylene (Br, Br), –78 °C, Et_2O]→ (*R*,*R*) + meso, (*R*,*S*) (36)

21.7 : 1
68%, >99% ee

5
Enantioselective Reactions of Halo Compounds

In general, organomagnesium compounds are configurationally more stable than organolithium compounds [90]. An α-bromomagnesium compound was demonstrated by Hoffmann and coworkers to have high configurational stability at –78 °C in a reaction starting with a chiral α-bromo sulfoxide. This reaction proceeds with retention of configuration [Eq. (37)] [91].

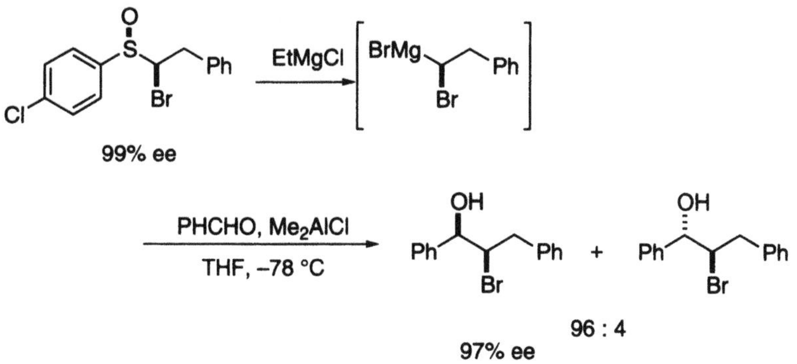

(37)

They also succeeded in discrimination of prochiral iodine atoms. In the reaction of 1,1-diiodo-2-phenylethane with a chiral Grignard reagent, one of the iodides preferentially underwent iodine-magnesium exchange; subsequent aldol reaction with benzaldehyde in the presence of Me_2AlCl resulted in the formation of an iodohydrin with 53% ee [Eq. (38)] [92].

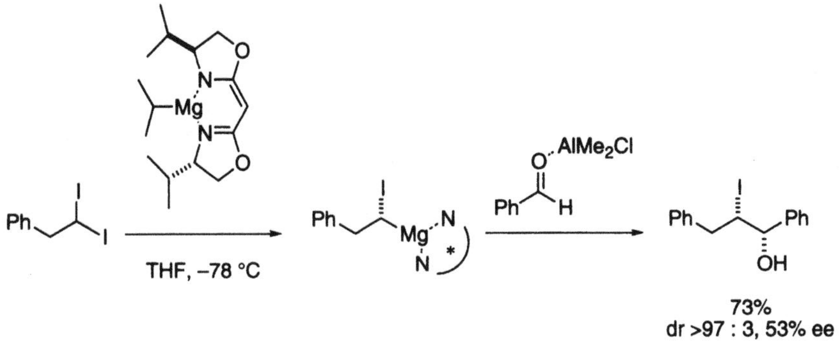

(38)

6 Enantioselective Reactions of Benzyllithium Compounds without an Adjacent Heteroatom

Since benzyllithium compounds having no adjacent heteroatom are configurationally labile [93], their enantioselective reactions may proceed through an asymmetric substitution pathway. In 1971, Nozaki and coworkers commented in the first enantioselective reaction of benzyllithium compounds that the enantiomeric ratio of organolithium compounds might be influenced by a chiral additive [94]. Deprotonation of ethylbenzene was performed with *n*-BuLi-(–)-sparteine at 70°C, followed by carboxylation at –65°C [Eq. (39)].

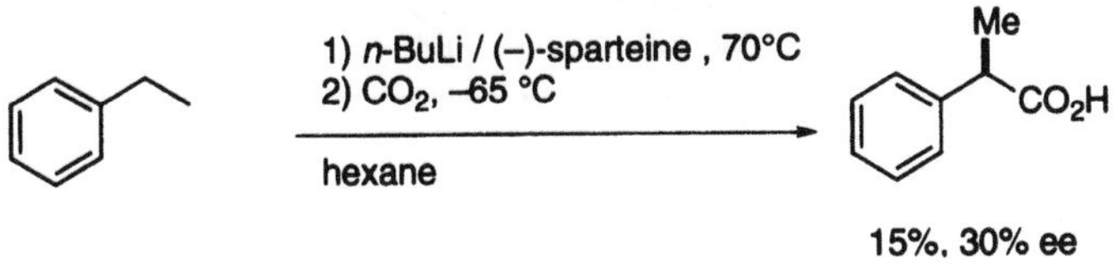

(39)

Taylor and Wei have reported an excellent alternative way to generate the benzyllithiums, starting with 2-substituted styrenes. Addition of *n*-BuLi to 2-substituted styrenes in the presence of (–)-sparteine successfully afforded benzyllithiums which were trapped with CO_2 to give products with high enantioselectivity, the absolute stereochemistry of the products not being determined [Eq. (40)] [95].

1) *n*-BuLi / (–)-sparteine
2) CO_2
cumene, –95 °C

58%, 72% ee (40)

Recently, Beak's group has disclosed a number of highly enantioselective reactions of benzyllithum compounds having various activating groups in a remote position. They have proved these reactions proceed through either a dynamic thermodynamic resolution pathway, or a dynamic kinetic resolution pathway. When 2 equiv. of the base are used, β-lithiation of *N*-methyl-3-phenylpropionamide is favored over α-lithiation. The lithium carbanion at the β-position is stabilized by the coordination of the amide nitrogen to lithium, this effect being termed the complex-induced proximity effect [96]. This was confirmed by the ^{6}Li- and ^{13}C-NMR spectral analyses of the β-lithiated amide. High enantioselectivity was achieved in subsequent reactions with various electrophiles in the presence of (–)-sparteine (Table 9) [97].

Table 9. Enantioselective reaction of *N*-methyl 3-phenylpropionamide

s-BuLi / (–)-sparteine
–78 °C, 1:1 MTBE / THF

electrophile

electrophile	yield (%)	ee (%)
MeI	84	78
n-BuI	77	88
$PhCH_2Br$	78	80
Ph_2CO	84	73

This reaction showed high enantioselectivity even when (–)-sparteine was added after lithiation of the carboxamide. Also, the racemic lithium carbanion derived from the racemic tin precursor via metal exchange reaction, gave products with high enantioselectivity [Eq. (41)], whereas when the carbanion prepared from the corresponding chiral tin compound was reacted with an electrophile such as TMSCl without (–)-sparteine it yielded racemic product. These results indicate that the reaction of the lithiated 3-phenylpropionamide proceeds through an asymmetric substitution pathway. Furthermore, a warm-cool procedure and then reaction with a substoichiometric amount of an electrophile confirmed a dynamic thermodynamic resolution pathway for this reaction.

1) *s*-BuLi
2) (–)-sparteine
3) Me_3SiCl
–78 °C, THF

(41)

X = H, Y = H, 72%, 82% ee
X = $SnBu_3$, Y = H, 48%, 78% ee
X = D, Y = D, 75% (94% d_1), 90% ee

The enantioselective reaction of lithiated *N*-pivaloyl-*o*-ethylaniline with various electrophiles was also successful [98] (Table 10). The temperature of the reaction significantly influenced the stereoselectivity. The reaction of lithiated *N*-

Table 10. Enantioselective reaction of *o*-ethyl-*N*-pivanilide

PivNH
1) *s*-BuLi
2) (–)-sparteine
MTBE, –25 °C

PivNLi Li•sparteine
electrophile
–78 °C
PivNH E

electrophile	yield (%)	ee (%)
CH_2=$CHCH_2Br$	67	82
$PhCH_2Br$	56	80
PhCHO	67	82

pivaloyl-*o*-ethylaniline with TMSCl in the presence of (–)-sparteine at -78°C gave the silylated product with low enantioselectivity [Eq. (42)]. On the other hand, aging of the mixture of (–)-sparteine and benzyllithium for 45 min at -25°C prior to the addition of TMSCl at -78°C, remarkably increased the enantioselectivity. Furthermore, when 0.1 equiv. of TMSCl was used, the silylated product was obtained with much higher enantioselectivity than in the reaction with a stoichiometric amount of the electrophile. From these results the reaction of lithiated *N*-pivaloyl-*o*-ethylaniline was shown to proceed through a dynamic thermodynamic resolution pathway.

PivNH → (*s*-BuLi, Et_2O) → PivNLi Li → 1) (–)-sparteine (conditions A,B, or C) 2) Me_3SiCl → PivNH $SiMe_3$

(42)

A: (–)-sparteine, –78 °C (75 min)
52%, 21% ee
B: (–)-sparteine, –25 °C (45 min), then –78 °C (30 min)
72%, 90% ee
C: (–)-sparteine, –78 °C (15 min)
0.1 equiv of Me_3SiCl, 82% ee

The chiral benzyllithium compound generated by the reaction of the enantioenriched stannyl precursor with *s*-BuLi-(–)-sparteine was reacted with cyclohexanone to give the product with good enantioselectivity. Interestingly, this reaction proceeds with inversion of the configuration of the stannylated carbon [Eq. (43)].

PivNH $SnMe_3$ → 1) *s*-BuLi / (–)-sparteine 2) cyclohexanone; MeO^tBu, –78 °C → HO PivNH

(43)

66% ee 68%, 62% ee

High levels of enantioinduction were observed in the reaction of *N,N*-diisopropyl-*o*-ethylbenzamide with *s*-BuLi-(–)-sparteine and subsequently with alkyl halides or a stannyl chloride (Table 11) [99]. Since lithiated *N,N*-diisopropyl-*o*-ethylbenzamide was confirmed to be configurationally labile by a Hoffmann test, the reaction should proceed through a dynamic kinetic resolution pathway. Note that alkylation with alkyl tosylates gives the products having the reverse stereochemistry to that obtained with alkyl halides. The alkyl halide (a non-complexing, reactive electrophile) approaches the carbanion from the sterically less encumbered direction to give the inversely substituted product, whereas the reaction with the alkyl tosylate (a coordinative, low reactive electrophile) proceeds with retention of the stereochemistry at the carbanionic center. Highly diastereo- and enantioselective reactions of the lithiated ethylbenzamide with imines have been reported by Snieckus and Derdau [Eq. (44)] [100].

Table 11. Enantioselective reaction of *N,N*-diisopropyl-*o*-ethylbenzamide

$(^iPr)_2N$ ... *s*-BuLi / (–)-sparteine, pentane/^{t}BuOMe, –78 °C → Li•sparteine; RCl; ROTs

RCl	yield (%)	ee (%)
CH_2=$CHCH_2Cl$	89	92
n-BuCl	95	80
Bu_3SnCl	78	87

ROTs	yield (%)	ee (%)
CH_2=$CHCH_2OTs$	46	88
n-BuOTs	52	97
$PhCH_2OTs$	26	77

1) *s*-BuLi / (–)-sparteine
2) PhCH=NPh
toluene, –78 °C → rt

Me, NHPh, Ph

41%, 95% de, 98% ee (44)

Hoppe and coworkers succeeded in crystallization of lithiated 1-methyl- or 1-butyl-3*H*-indene and (–)-sparteine complexes (Table 12) [101]. The crystallization preferentially gave one of the stereoisomers which was reacted with acid chlorides to afford acylated indenes with high enantioselectivity. In a comparison of the absolute configuration of the acylated indenes with that of the crystallized complexes determined by the X-ray analysis, it was found that the reaction with acid chlorides proceeds with retention of configuration at the lithiated carbanion.

It has been reported by Levacher and coworkers that the enantioselective methylation of the lithiated diarylmethane shows good enantioselectivity [102]. In addition, they studied the protonation reaction of lithiated 1-phenyl-1-(2-pyridyl)ethane with various proton sources and found that protonation by the use of 2,6-di-*tert*-butyl-4-methylphenol in the presence of (–)-sparteine shows good enantioselectivity. Even better selectivity can be obtained in the protonation with chiral *N,N*-dimethylephedrine [Eqs. (45) and (46)]. Furthermore, as another example of the enantioselective protonation, they have disclosed highly enan-

Table 12. Enantioselective reaction of 1-methyl- and 1-butyl-3*H*-indenes

a: R = Me
b: R = *n*-Bu

n-BuLi / (–)-sparteine, Et_2O / hexane → Li•sparteine → Electrophile; crystallization

R	electrophile	yield (%)	ee (%)
Me	MeOCOCl	64	>95
Me	MeCOCl	63	>95
Me	PhCOCl	74	>95
n-Bu	PhCOCl	79	>95

tioselective deuteration of lithiated 4-phenyl tetrahydroisoquinoline [Eq. (47)] [103].

1) *s*-BuLi / (–)-sparteine 2) MeI; hexane, –78 °C — 44%, 52% ee (45)

Method A or B; ether, –78 °C

Method A
1) *s*-BuLi, (–)-sparteine, 2) [phenol: ${}^tBu$, Me, HO, ${}^tBu$] 65% ee (46)

Method B
1) *s*-BuLi, 2) [HO, Ph, NMe_2, Me] 84% ee

1) *n*-BuLi, (–)-sparteine 2) MeOD; Et_2O, –45 °C — 88% ee (47)

7
Conclusion

In this chapter, we have presented the enantioselective reactions of α-hetero carbanions and benzylic carbanions and have demonstrated the relation between the enantiodetermining step and the configurational stability of these carbanions. The mechanism of enantioselective reactions and the mechanism of racemization of α-hetero carbanions with adjacent S, Se, and P atoms have been shown to be often different from those of α-oxy and α-amino carbanions. Generally, chiral diamine ligands such as sparteine, isosparteine, bis(oxazoline)s and cycloalkanediamines, and chiral lithium amides and alkoxides as deprotonating agents have been effectively used, although the enantioselectivities of these reactions have not always been very high. One of the attractive challenges for future workers will be the development of catalytic versions of the asymmetric reactions of the α-hetero carbanions described.

Note added in proof. DFT calculations have demonstrated how THF solvation of the monomer of a chiral α-thioallyllithium compound (similkar to Fig. 10b) impedes racemization, while upon dimerization the inversion process accelerates [104]; these findings agree with earlier experiments performed by Hoppe and coworkers. More recently, (*S*)-S-(2-cyclohexenyl) *N,N*-diisopropylmonothiocarbamate has been deprotonated by s-BuLi/TMEDA to form a configurationally stable organolithium {*ent* Fig. 10b [i-PrNLi = $(i\text{–}Pr)_2N$]}, which does not exhibit solvent dependent configurational stability, and which is regioselectively α-alkylated by alkyl halides with complete stereoinversion to form monothiocarbamates which afford highly enantioenriched tertiary 2-cyclohexene-1-thiols on reductive cleavage [105]. Diphosphines (*S,S*)-1,2-bis(mesitylmethylphosphino)ethane, and (*S,S*)-1,2-bis(9-anthrylmethylphosphino)ethane, which contain 2,6-disubstituted aryl P-substituents, were prepared by Evans-sparteine-assisted enantioelective deprotonation – Cu(II) oxidative coupling [Eq. (31)] of $P(Ar)(Me)_2BH_3$ (Ar = mesityl or 9-anthryl), but the enantioselectivity did not exceed 37% ee [106]. The X-ray crystal structure of 1,3-diphenylallithium, complexed with (–)-sparteine, has been determined. The crystallized complex is stereohomogeneus; however, its subsequent electrophilic substitution reactions proceed with low enantioselectivity (uo to 22% ee) [107].

References

1. (a) Beak P, Basu A, Gallagher DJ, Park YS, Thayumanavan S (1996) Acc Chem Res 29:552; (b) Basu A, Thayumanavan S (2002) Angew Chem Int Ed 41:716
2. Hoppe D, Hense T (1997) Angew Chem Int Ed Engl 36: 2282
3. Beak P, Anderson DR, Curtis MD, Laumer JM, Pippel DJ, Weisenburger GA (2000) Acc Chem Res 33:715
4. (a) Noyori R, Tokunaga M, Kitamura M (1995) Bull Chem Soc Jpn 68:36; (b) Ward RS (1995) Tetrahedron: Asymmetry 6:1475; (c) Caddick S, Jenkins K (1996) Chem Soc Rev 447
5. Aggarwal VK (1994) Angew Chem Int Ed Engl 33:175
6. Still WC, Sreekumar C (1980) J Am Chem Soc 102:1201
7. (a) Vedejs E, Moss WO (1993) J Am Chem Soc 115:1607; (b) Burchat AF, Chong JM, Park SB (1993) Tetrahedron Lett 34:51; (c) Gawley RE, Zhang Q (1994) Tetrahedron 50:6077
8. (a) Lesimple P, Beau JM, Sinay P (1985) J Chem Soc Chem Commun 894; (b) Hutchinson DK, Fuchs PL (1987) J Am Chem Soc 109:4930; (c) Lindermann RJ, Ghannam A (1990) J Am Chem Soc 112:2392; (d) Rychovsky SD, Plzak K, Pickering D (1994) Tetrahedron Lett 35: 6799; (e) Carstens A, Hoppe D (1994) Tetrahedron 50:6097

9. (a) McDougal PG, Condon BD, Laffosse, Jr. MD, Lauro AM, VanDerveer D (1988) Tetrahedron Lett 29:2547; (b) McDougal PG, Condon BD (1989) Tetrahedron Lett 30:789; (c) Lutz GP, Wallin AP, Kerrick ST, Beak P (1991) J Org Chem 56:4938; (d) Brickmann K, Brückner R (1993) Chem Ber 126:1227; (e) Brickmann K, Hamblock F, Spolaore E, Brückner R (1994) Chem Ber 127:1949; (f) Wang F, Tang J, Labaudiniere L, Marek I, Normant JF (1995) Synlett 723
10. (a) Krief A, Evrard G, Badaoui E, De Beys V, Dieden R (1989) Tetrahedron Lett 30:5635; (b) Hoffmann RW, Bewersdorf M (1990) Tetrahedron Lett 31:67; (c) Reich HJ, Bowe MD (1990) J Am Chem Soc 112:8994; (d) Hoffmann RW, Bewersdorf M (1992) Liebigs Ann Chem 643; (e) Krief A, Badaoui E, Dumont W (1993) Tetrahedron Lett 34:8517
11. (a) O'Brien P, Warren S (1995) Tetrahedron Lett 36:8473; (b) O'Brien P, Warren S (1996) J Chem Soc Perkin Trans 1 2567
12. (a) Hoffmann RW, Lanz J, Metternich R, Tarara G, Hoppe D (1987) Angew Chem Int Ed Engl 26:1145; (b) Hirsch R, Hoffmann RW (1992) Chem Ber 125:975; (c) Hoffmann RW, Rühl T, Chemla F, Zahneisen T (1992) Liebigs Ann Chem 719
13. Faibish NC, Park YS, Lee S, Beak P (1997) J Am Chem Soc 119:11561
14. (a) Gallagher DJ, Du H, Long SA, Beak P (1996) J Am Chem Soc 118:11391; (b) Thayumanavan S, Basu A, Beak P (1997) J Am Chem Soc 119:8209; (c) Weisenburger GA, Faibish NC, Pippel DJ, Beak P (1999) J Am Chem Soc 121:9522
15. Komine N, Wang LF, Tomooka K, Nakai T (1999) Tetrahedron Lett 40:6809
16. (a) Hoppe D, Paetow M, Hintze F (1993) Angew Chem Int Ed Engl 32:394; (b) Anderson DR, Faibish NC, Beak P. (1999) J Am Chem Soc 121:7553; (c) Clayden J, Helliwell M, Pink JH, Westlund N (2001) J Am Chem Soc 123:12449
17. Hoffmann RW, Julius M, Chemla F, Ruhland T, Frenzen G (1994) Tetrahedron 50:6049
18. Hoffmann RW (1993) In: Enders D, Gais HJ, Keim W (eds), Organic synthesis via organometallics (OSM4). Vieweg & Sohn Verlagsges, Braunschweig, p 79
19. Hoffmann RW, Rühl T, Harbach J (1992) Liebigs Ann Chem 725
20. O'Brien P, Powell HR, Raithby PR, Warren S (1997) J Chem Soc Perkin Trans 1 1031
21. O'Brien P, Warren S (1996) Tetrahedron Lett 37:4271
22. Hoffmann RW, Julius M, Pltmann K (1990) Tetrahedron Lett 31:7419
23. Bent HA (1961) Chem Rev 61:275
24. Boche G (1989) Angew Chem Int Ed Engl 28:277
25. Wiberg KB, Castejon H (1994) J Am Chem Soc 116:10489
26. Lehn JM, Wipff G, Demuynck J (1977) Helv Chim Acta 60:1239
27. (a) Ruhland T, Dress R, Hoffmann RW (1993) Angew Chem Int Ed Engl 32:1467; (b) Hoffmann RW, Dress RK, Ruhland T, Wenzel A (1995) Chem Ber 128:861; (c) Dress RK, Rölle T, Hoffmann RW (1995) Chem Ber 128:673
28. Ahlbrecht H, Harbach J, Hoffmann RW, Ruhland T (1995) Liebigs Ann 211
29. (a) Hoffmann RW, Koberstein R (1999) Chem Commun 33; (b) Hoffmann RW, Koberstein R (2000) J Chem Soc Perkin Trans 2 595
30. Reich HJ, Dykstra RR (1993) Angew Chem Int Ed Engl 32:1469
31. Ruhland T, Hoffmann RW, Schade S, Boche G (1995) Chem Ber 128:551
32. (a) Hoppe D, Kaiser B, Stratmann O, Fröhlich R (1997) Angew Chem Int Ed Engl 36:2784; (b) Stratmann O, Kaiser B, Fröhlich R, Meyer O, Hoppe D (2001) Chem Eur J 7:423
33. Marr F, Fröhlich R, Hoppe D (1999) Org Lett 1:2081; Marr F, Fröhlich R, Wibbeling B, Diedrich C, Hoppe D (2002) Eur J Org Chem 2970
34. Kaiser B, Hoppe D (1995) Angew Chem Int Ed Engl 34:323
35. (a) Gais HJ, Hellmann G, Günther H, Lopez F, Lindner HJ, Braun S (1989) Angew Chem Int Ed Engl 28:1025; (b) Gais HJ, Hellmann G (1992) J Am Chem Soc 114:4439
36. (a) Bors DA, Streitwieser Jr A (1986) J Am Chem Soc 108:1397; (b) Koch R, Anders E (1994) J Org Chem 59:4529; (c) Raabe G, Gais HJ Fleischhauser J (1996) J Am Chem Soc 118 4622
37. (a) Kranz M, Denmark SE, Swiss KA, Wilson SR (1996) J Org Chem 61:8551; (b) Kranz M, Denmark SE (1995) J Org Chem 60:5867; (c) Cramer CJ, Denmark SE, Miller PC, Dorow, RL, Swiss KA, Wilson SR (1994) J Am Chem Soc 116:2437; (d) Zarges W, Marsch M, Harms K, Haller F, Frenking G, Boche G (1991) Chem Ber 124:861; (e) Koch R, Anders E (1995) J Org Chem 60:5861

38. Denmark SE, Swiss KA, Wilson SR (1993) J Am Chem Soc 115:3826
39. Hoppe D, Hintze F, Tebben P (1990) Angew Chem Int Ed Engl 29:1422
40. Shinozuka T, Kikori Y, Asaoka M, Takei H (1995) J Chem Soc Perkin Trans 1 119
41. Nakamura S, Nakagawa R, Watanabe Y, Toru T (2000) Angew Chem Int Ed 39:353
42. Nakamura S, Nakagawa R, Watanabe Y, Toru T (2000) J Am Chem Soc 122:11340
43. Tomooka K, Wang LF, Komine N, Nakai T (1999) Tetrahedron Lett 40:6813
44. Basu A, Beak P (1996) J Am Chem Soc 118:1575
45. Derwing C, Frank H, Hoppe D (1999) Eur J Org Chem 3519
46. Zhang J, Matteucci MD (1999) Tetrahedron Lett 40:1467
47. Nakamura S, Furutani A, Toru T (2002) Eur J Org Chem 1690
48. Gibson (née Thomas) SE, Ham P, Jefferson GR, Smith MH (1997) J Chem Soc Perkin Trans 1 2161
49. (a) Ariffin A, Blake AJ, Ewin RA, Simpkins NS (1998) Tetrahedron: Asymmetry 9:2563; (b) Gibson (née Thomas) SE, Reddington EG (2000) Chem Commun 989
50. (a) Cowton ELM, Gibson (née Thomas) SE, Schneider MJ, Smith MH (1996) Chem Commun 839; (b) Ariffin A, Blake AJ, Ewin RA, Li W, Simpkins NS (1999) J Chem Soc Perkin Trans 1 3177
51. Ewin RA, MacLeod AM, Price DA, Simpkins NS, Watt AP (1997) J Chem Soc Perkin Trans 1 401
52. Gibson SE, Potter PCV, Smith MH (1996) Chem Commun 2757
53. Nishibayashi Y, Ohe K, Uemura S (1995) J Chem Soc Chem Commun 1245
54. (a) Fukuda T, Katsuki T (1997) Tetrahedron Lett 38:3435; (b) Fukuda T, Irie R, Katsuki T (1999) Tetrahedron 55:649
55. McMillen DW, Varga N, Reed BA, King C (2000) J Org Chem 65:2532
56. (a) Chassaing G, Marquet A (1978) Tetrahedron 34:1399; (b) Boche G, Marsch H, Harms K, Sheldrick GM (1985) Angew Chem Int Ed Engl 24:573; (c) Koch R, Anders E (1994) J Org Chem 59:4529
57. (a) Simpkins NS (1988) Chem Ind 387; (b) Cox PJ, Simpkins NS (1991) Tetrahedron: Asymmetry 2:1
58. Akiyama T, Shimizu M, Mukaiyama T (1984) Chem Lett 611
59. Cox PJ, Persad A, Simpkins NS (1992) Synlett 194
60. Armer R, Begley MJ, Cox PJ, Persad A, Simpkins NS (1993) J Chem Soc Perkin Trans 1 3099
61. Blake AJ, Westaway SM, Simpkins NS (1997) Synlett 919
62. Kunieda N, Kinoshita M (1981) Phosphorus and Sulfur 10:383
63. (a) Colombo L, Gennari C, Scolastico C, Guanti G, Narisano E (1981) J Chem Soc Perkin Trans 1 1278; (b) Colombo L, Gennari C, Resnati G, Scolastico C (1981) J Chem Soc Perkin Trans 1 1284; (c) Solladié G, Colobert F, Ruiz P, Hamdouchi C, Carreno MC, Ruano GJL (1991) Tetrahedron Lett 32:3695; (d) Aggarwal VK, Schade S, Adams H (1997) J Org Chem 62:1139
64. Tomioka K, Sudani M, Shinmi Y, Koga K (1985) Chem Lett 329
65. (a) Kang J, Cho WO, Cho HG, Oh HJ (1994) Bull Korean Chem Soc 15:732; (b) Kang J, Cho WO, Cho HG (1994) Tetrahedron: Asymmetry 5:1347; (c) Hodgson DM, Lee GP (1997) Tetrahedron: Asymmetry 8:2303
66. Kang J, Kim JI, Lee JH (1994) Bull Korean Chem Soc 15:865
67. Hoffmann RW, Klute W, Dress RK, Wenzel A (1995) J Chem Soc Perkin Trans 2 1721
68. (a) Klute W, Dress R, Hoffman RW (1993) J Chem Soc Perkin Trans 2 1409; (b) Hoffmann RW, Klute W (1996) Chem Eur J 2:694
69. (a) Rein T, Reiser O (1996) Acta Chem Scand 50:369; (b) For a review on asymmetric Wittig-type reactions see; Rein T, Pedersen TM (2002) Synthesis 579
70. Tanaka K, Fuji K, (1998) J Synth Org Chem Jpn 56:521
71. Kumamoto T, Koga K (1997) Chem Pharm Bull 45:753
72. Mizuno M, Fujii K, Tomioka K (1998) Angew Chem Int Ed 37:515
73. Allmendinger T, Furet P, Hungerbühlere (1990) Tetrahedron Lett 31:7297
74. Sano S, Yokokawa K, Teranishi R, Shiro M, Nagao Y (2002) Tetrahedron Lett 43:281
75. Arai S, Hamaguchi S, Shioiri T (1998) Tetrahedron Lett 39:2997
76. Toda F, Akai H (1990) J Org Chem 55:3446

77. Bestmann HJ, Lienert J (1970) Chem Ztg 94:487
78. Hume SC, Simpkins NS (1998) J Org Chem 63:912
79. (a) Hall PL, Gilchrist JH, Harrison AT, Fuller DJ, Collum DB (1991) J Am Chem Soc 113:9575; (b) Bunn BJ, Cox PJ, Simpkins NS (1993) Tetrahedron 1993:207
80. Guillen F, Moinet C, Fiaud J (1997) Bull Soc Chim Fr 134:371
81. Vedejs E, Garcia-Rivas JA (1994) J Org Chem 59:6517
82. O'Brien P, Warren S (1996) Synlett 579
83. (a) Schmidbaur H (1980) J Organomet Chem 200:287; (b) Imamoto T, Oshiki T, Onozawa T, Kusumoto T, Sato K (1990) 112:5244
84. Muci AR, Campos KR, Evans DA (1995) J Am Chem Soc 117:9075
85. (a) Wiberg KB, Bailey WF (2000) Angew Chem Int Ed 39:2127; (b) Wiberg KB, Bailey WF (2001) J Am Chem Soc 123:8231
86. (a) Nagata K, Matsukawa S, Imamoto T (2000) J Org Chem 65:4185; (b) Sugama H, Saito H, Danjo H, Imamoto T (2001) Synthesis 2348; (c) Ohashi A, Kikuchi S, Yasutaka M, Imamoto T (2002) Eur J Org Chem 2535
87. Kobayashi S, Shiraishi N, Lam WWL, Manabe K (2001) Tetrahedron Lett 42:7303
88. Wolfe B, Livinghouse T (1998) J Am Chem Soc 120:5116
89. Heath H, Wolfe B, Livinghouse T, Bae SK (2001) Synthesis 2341
90. (a) Hoffmann RW, Kusche A (1994) Chem Ber 127:1311; (b) Hoffmann RW, Julius M, Chemla F, Ruhland T, Frenzen G (1994) Tetrahedron 50:6049; (c) Klute W, Krüger M, Hoffmann RW (1996) Chem Ber 129:633
91. (a) Hoffmann RW, Nell PG (1999) Angew Chem Int Ed 38:338; (b) Hoffmann RW, Nell PG, Leo R, Harms K (2000) Chem Eur J 6:3359
92. Schulze V, Hoffmann RW (1999) Chem Eur J 5:337
93. Hoffmann RW, Rühl T, Chemla F, Zahneisen T (1992) Liebigs Ann Chem 719
94. Nozaki H, Aratani T, Toraya T, Noyori R (1971) Tetrahedron 27:905
95. Wei X, Taylor RJK (1997) Tetrahedron: Asymmetry 8:665
96. (a) Beak P, Hunter JE, Jun YM, Wallin AP (1987) J Am Chem Soc 109:5403; (b) Beak P, Meyers AI (1986) Acc Chem Res 19:356
97. (a) Beak P, Du H (1993) J Am Chem Soc 115:2516; (b) Lutz GP, Du H, Gallagher DJ, Beak P (1996) J Org Chem 61:4542; (c) Gallagher DJ, Du H Long SA, Beak P (1996) J Am Chem Soc 118:11391
98. (a) Basu A, Beak P (1996) J Am Chem Soc 118:1575; (b) Basu A, Gallagher DJ, Beak P (1996) J Org Chem 61:5718; (c) Thayumanavan S, Basu A, Beak P (1997) J Am Chem Soc 119:8209
99. Thayumanavan S, Lee S, Liu C, Beak P (1994) J Am Chem Soc 116:9755
100. Derdau V, Snieckus V (2001) J Org Chem 66:1992
101. Hoppe I, Marsch M, Harms K, Boche G, Hoppe D (1995) Angew Chem Int Ed Engl 34:2158
102. Prat L, Dupas G, Duflos J, Queguiner G, Bourguignon J, Levacher V (2001) Tetrahedron Lett 42:4515
103. Prat L, Mojovic L, Levacher V, Dupas G, Queguiner G, Bourguignon J, (1998) Tetrahedron: Asymmetry 9:2509
104. Brandt P, Haeffner F (2003) J Am Chem Soc 125:48
105. Marr F, Hoppe D (2002) Org Lett 4:4217
106. Maienza F, Spindler F, Thommen M, Pugin B, Malan C, Mezzetti A (2002) J Org Chem 67:5239
106. Marr F, Fröhlich R, Hoppe D (2002) Tetrahedron: Asymmetry 13:2587

Topics Organomet Chem (2003) 5: 217–250
DOI 10.1007/b10339

Enantioselective Synthesis by Lithiation Adjacent to Oxygen and Subsequent Rearrangement

David M. Hodgson[1], Katsuhiko Tomooka[2], Emmanuel Gras[1]

[1] Dyson Perrins Laboratory, Department of Chemistry, University of Oxford, South Parks Road, Oxford OX1 3QY, UK. *E-mail: david.hodgson@chem.ox.ac.uk*

[2] Department of Applied Chemistry, Tokyo Institute of Technology, Meguro-ku, Tokyo 152-8552, Japan. *E-mail: ktomooka@o.cc.titech.ac.jp*

Enantioselective carbanion rearrangements induced by lithiation α to oxygen have recently progressed dramatically. In this chapter, an overview of this field is presented. The first section concerns asymmetric Wittig rearrangements. Asymmetric retro-Brook rearrangements are then covered, followed by rearrangements of phosphate to phosphonate. The last section reviews asymmetric rearrangements of lithiated epoxides.

Keywords. Wittig rearrangement, Retro-Brook rearrangement, Oxiranyl anions, Carbene insertion, Electrophile trapping

1 Wittig Rearrangements

1.1 Introduction to Wittig Rearrangements

Allylic, propargylic, and benzylic protons adjacent to oxygen are sufficiently acidic to give good yields of carbanions upon treatment with strong bases such as alkyllithium reagents. In certain systems, the α-oxy carbanion acts as a migrating terminus in a rearrangement. These α-oxy carbanion rearrangements are synthetically valuable in terms of an O-Y bond to C-Y bond transformation (Table 1). In principle, the carbanion rearrangement with a chiral, enantioenriched base ("chiral base") could produce an optically active product.

In general, such an enantioselective rearrangement involves two possible enantio-determining steps, namely the lithiation step and the bond-rearrangement step. Also, a dynamic equilibrium may or may not exist between the two

Table 1. Rearrangement of α-oxy carbanions

transformation	Y-O-C⁻-Z → ⁻O-C(Y)-Z	
O-C → C-C	R-O-C⁻-R' → ⁻O-C(R)-R'	[1,2]-Wittig rearrangement
	O-C⁻-R' (allyl/propargyl-R) → ⁻O-C(R')-C(R)-alkene	[2,3]-Wittig rearrangement
	R-O-C⁻=C-R' ⇄ R-O-CH=CH-C⁻-R' → ⁻O-CH=CH-C(R)-R'	[1,4]-Wittig rearrangement
O-Si → C-Si	R_3Si-O-C⁻-R' → ⁻O-C(SiR_3)-R'	[1,2]-retro-Brook rearrangement
O-P → C-P	$(RO)_2P(=O)$-O-C⁻-R' → ⁻O-C($P(=O)(OR)_2$)-R'	phosphate-phosphonate rearrangement

enantiomeric organolithium intermediates involved. When a chiral base selectively abstracts one of the enantiotopic protons from the prochiral substrate **A**, either of the chiral organolithium intermediates **B** or *ent*-**B** can be formed preferentially. If the migrating terminus of the resulting carbanion **B** is configurationally stable and the rearrangement proceeds stereospecifically (retention or inversion), the rearrangement provides enantioenriched product **C** or *ent*-**C**, in which the initial lithiation step is stereo-determining (Scheme 1).

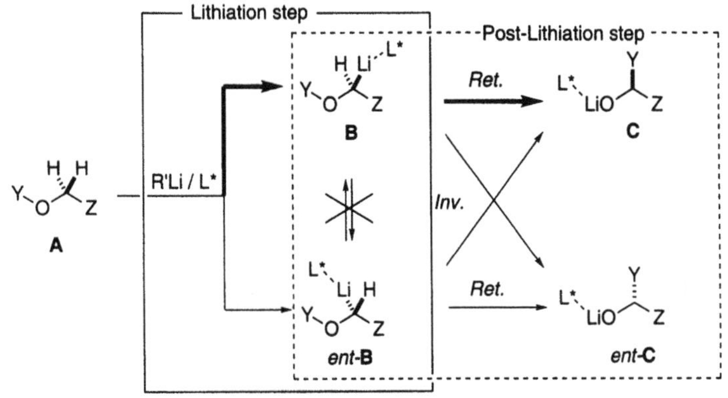

Scheme 1.

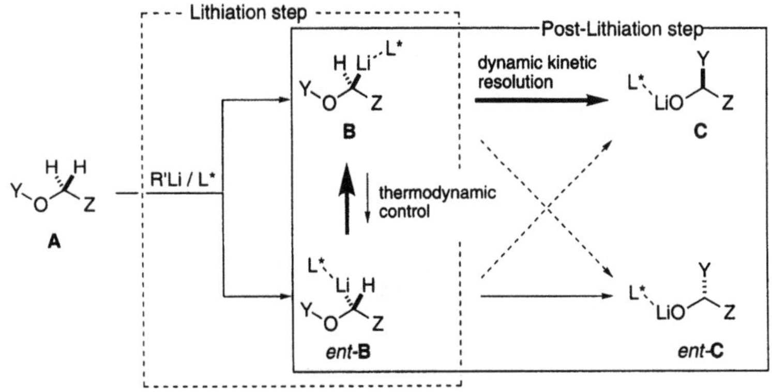

Scheme 2.

On the other hand, when equilibration exists between the two species **B** and *ent*-**B**, and one of the two stereoisomers exists preferentially over the other (in the presence of a chiral ligand) and then stereospecifically rearranges, the reaction provides the enantioenriched product **C** or *ent*-**C** in which the stereochemistry was determined at the post-lithiation step by a thermodynamic control mechanism (Scheme 2). Alternatively, if the equilibration occurs rapidly, and either lithio species rearranges much faster than the other species, either **C** or *ent*-

C is formed preferentially in which the stereochemistry was determined at the post-lithiation step by a dynamic kinetic resolution mechanism. Therefore, depending on the configurational stability of migrating terminus and on the mechanism of the reaction, the role of the chiral base varies (i.e., enantioselective lithiation agent and/or chiral coordinating agent in the rearrangement). A key to success is the judicious choice of chiral base. As a chiral base, several kinds of chiral lithium amides and alkyllithium-chiral coordinating agent complexes have been developed for the enantioselective rearrangement.

1.2 [1,2]-Wittig Rearrangement and Related Reactions

The [1,2]-Wittig rearrangement is a carbanion rearrangement of ethers or acetals which involves a 1,2-alkyl shift onto a α-oxy carbanion terminus [Eq. (1)] [1,2].

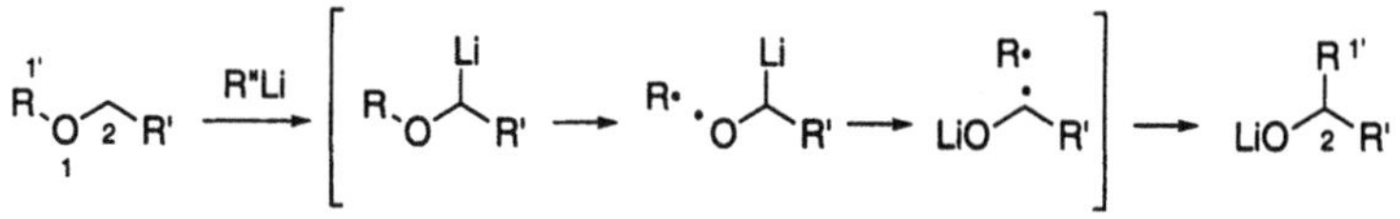

(1)

This type of carbanion rearrangement is recognised to proceed by means of a radical dissociation-recombination mechanism. In 1983, Mazaleyrat and Welvart reported the kinetic resolution of a racemic binaphthyl ether, which involves an enantioselective variant of the [1,2]-Wittig rearrangement [Eq. (2)] [3].

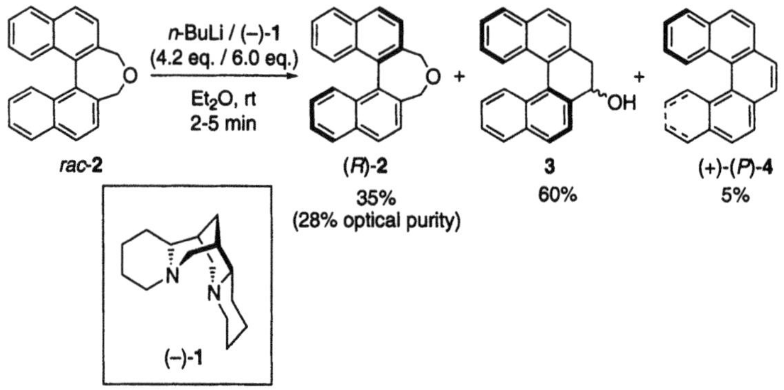

(2)

The reaction of binaphthyl ether *rac*-**2** with *n*-BuLi and a chiral coordinating agent, (–)-sparteine (**1**), gave the rearrangement product (+)-**3** [enriched in (*S*)-axial chirality] and pentahelicene (+)-**4**, along with the recovered ether (*R*)-**2** in an enantioenriched form (28% optical purity). This result reveals that (*S*)-**2** rearranges faster than (*R*)-**2** in the presence of (–)-**1**.

Recently, a more versatile enantioselective variant was accomplished by use of the chiral bis(oxazoline) **5** as a chiral coordinating agent [Eq. (3)] [4].

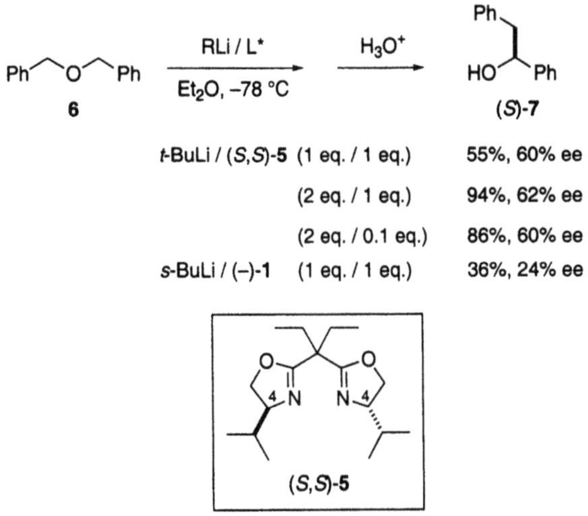

Ph⁀O⁀Ph **6** —RLi / L*, Et_2O, –78 °C→ —H_3O^+→ (*S*)-**7**

t-BuLi / (*S*,*S*)-**5**	(1 eq. / 1 eq.)	55%, 60% ee
	(2 eq. / 1 eq.)	94%, 62% ee
	(2 eq. / 0.1 eq.)	86%, 60% ee
s-BuLi / (–)-**1**	(1 eq. / 1 eq.)	36%, 24% ee

(*S*,*S*)-**5**

(3)

The rearrangement of dibenzyl ether **6** using the premixed complex *t*-BuLi/(*S*,*S*)-**5** (1.0 equiv. each) in ether at –78 °C, afforded a 55% yield of alcohol (*S*)-**7** in 60% ee, along with 43% recovery of **6**. When *s*-BuLi/(–)-**1** was used instead of *t*-BuLi/(*S*,*S*)-**5**, (*S*)-**7** was obtained in only 24% ee. Significantly, when *t*-BuLi (2.0 equiv.)/**5** (1.0 equiv.) complex was initially used, the yield of alcohol **7** increased to 94% while maintaining the same level of ee (62% ee). This indicates that two equivalents of *t*-BuLi are required for the reaction to complete. Based on this result, an "asymmetric catalytic version" of the present rearrangement has been developed. The rearrangement of **6** using 10 mol % of **5** and 2 equiv. of *t*-BuLi under the same conditions provided the alcohol (*S*)-**7** in equally high chemical yield and ee (86%, 60% ee). The amount of the coordinating agent **5** can be reduced to 5 mol% without any significant change in the results (81%, 54% ee). Scheme 3 depicts a plausible asymmetric catalytic cycle in which the enantioselectivity might be determined at the radical recombination step involving the chiral ligand-bound anion radical.

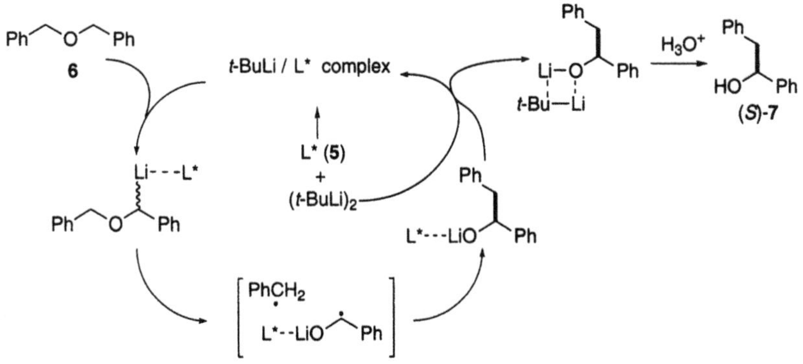

Scheme 3.

The reaction of a racemic α-deuterated ether **6**-d_3 in this particular rearrangement provided valuable information on the mechanism [Eq. (4)]. Optically active (*S*)-**7**-d_3 (>90% D-content) was obtained in 87% yield, with essentially the same ee as previously observed for **6**, indicating that the stereochemistry of the initially generated α-oxy carbanion has nothing to do with the product stereochemistry. In other words, the stereochemistry of the rearrangement product **7** was determined at a post-lithiation step.

rac-**6**-d_3 → *t*-BuLi / (*S,S*)-**5**, Et_2O, −78 °C → H_3O^+ → (*S*)-**7**-d_3 (87%) (>90% *d*, 59% ee) (4)

The enantioselective [1,2]-Wittig rearrangement of various ethers using this protocol [*t*-BuLi (2.0 equiv.)/**5** (2.0 equiv.)] is summarised in Table 2.

Table 2. Enantioselective [1,2]-Wittig rearrangement

R-O-CH2-R' → 1. *t*-BuLi / (*S,S*)-**5** (2.0 eq. / 2.0 eq.); 2. H_3O^+ → HO-CH(R)-R'

substrate		yield	% ee
Ph-CH2-O-CH2-Ph	−78 °C	84%	61% ee
	−110°C	31%	71% ee
bis(3,5-dimethylbenzyl) ether	−78 °C	65%	56% ee
Ph-C(Me)2-O-CH2-C≡C-X, X=TMS		52%	40% ee
X=TBDPS		49%	43% ee
Ph-CH2-O-CH(Ph)-C≡C-X (racemic), X=TMS		55%	54% ee
X=TIPS		65%	65% ee

The 1,2-shift of an sp^2-carbon such as CH=CHR′, C(=NOR)H, and C(=O)NR′$_2$ onto the α-oxy carbanion which is formally related to the [1,2]-Wittig rearrangement, has been reported [5–7]. The enantioselective version of carbamoyl migration was accomplished by use of *s*-BuLi/(−)-**1** complex as a chiral base [8]. The reaction of carbamate **8** with pre-mixed *s*-BuLi/(–)-**1** (1.2 equiv. each) in ether at –78 °C to room temperature gave the (*R*)-α-hydroxy amide **9** in 46% yield with 96% ee, along with the olefin **10** (29%) [Eq. (5)]. This stereochemical outcome reveals that the 1,2-carbamoyl migration occurs with complete retention of configuration at the α-oxy carbanion. The steric course of the 1,2-carbamoyl migration is in the opposite sense to that of the [1,2]-Wittig rearrangement (predominantly inversion [2d, 9]), and the different stereochemical outcome was explained by an addition-elimination pathway in which intramolecular addition of the α-oxy carbanion to the carbamoyl-carbonyl occurs in completely stereoretentive fashion.

(i-Pr)$_2$N–C(=O)–O–CH$_2$–R (**8**, R = C_2H_4Ph) $\xrightarrow[\text{Et}_2\text{O, -78 °C}]{s\text{-BuLi / (–)-1}}$ [R'$_2$N–C(=O)–O–CH(Li·L*)–R] $\xrightarrow[\text{2. } H_3O^+]{\text{1. -78 °C} \rightarrow \text{rt}}$ (*R*)-**9** 46% (96% ee) + **10** 29% (5)

1.3
[2,3]-Wittig Rearrangement

The [2,3]-Wittig rearrangement is a special class of [2,3]-sigmatropic rearrangement which involves an α-oxy carbanion as the migrating terminus to afford a variety of homoallylic alcohols or allenic alcohols [Eq. (6)] [2b, 10].

(6)

The regiospecific carbon-carbon bond formations with allylic transposition of the oxygen function, as well as the stereoselective formation of vicinal chiral centres via the [2,3]-Wittig rearrangement enjoy widespread application in many facets of stereoselective synthesis. The enantioselective version of the [2,3]-Wittig rearrangement should involve an enantioselective generation of the chiral migrating terminus from an achiral substrate with a chiral base. In this section, the enantioselective [2,3]-Wittig rearrangement is categorised into two classes, according to the type of chiral base employed.

1.3.1
Chiral Lithium Amide Protocol

The pioneering work on the enantioselective [2,3]-Wittig rearrangement was carried out by Marshall's group, in the ring-contracting rearrangement of a 13-membered cyclic ether using lithium bis(1-phenylethyl)amide **11** as a chiral base [Eq. (7)] [11]. Upon treatment with (*S,S*)-**11** (3 equiv.) in THF at –70 to –15 °C, ether **12** afforded the enantioenriched alcohol **13** in 82% yield with 69% op. The reaction was applied in the synthesis of (+)-aristolactone (**14**).

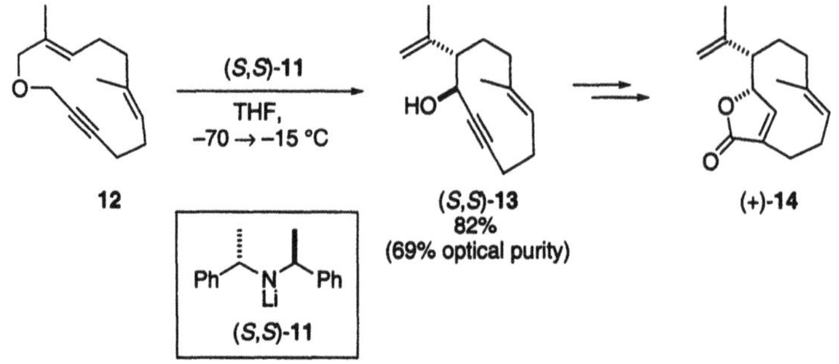

12 → (S,S)-11, THF, –70 → –15 °C → (S,S)-13 82% (69% optical purity) ⇌ (+)-14 (7)

(S,S)-11

However, a similar rearrangement of related 9- and 17-membered cyclic substrates **15** and **17** with amide **11** provided lower ees (25 and 29% ee, respectively) (Scheme 4) [11, 12]. Accordingly, the level of enantioselection appears to depend critically upon the substrate ring size. In fact, no appreciable level of enantioselectivity was observed in the rearrangement of acyclic substrate **19** with amide **11**. For the 9-membered ring contraction of **15**, exclusive formation of *cis*-substituted cyclohexenol **16** is due to steric constraints which disfavour a highly strained *trans*-bicyclo[4.3.0] transition-state arrangement leading to the *trans*-isomer. Also, even though ether **15** possesses two different sets of abstractable protons (α and α′), rearrangement can only occur with the observed regiochemistry owing to the highly unlikely generation of a cyclohexene containing a *trans*-double bond from the alternative deprotonation (α′)-rearrangement.

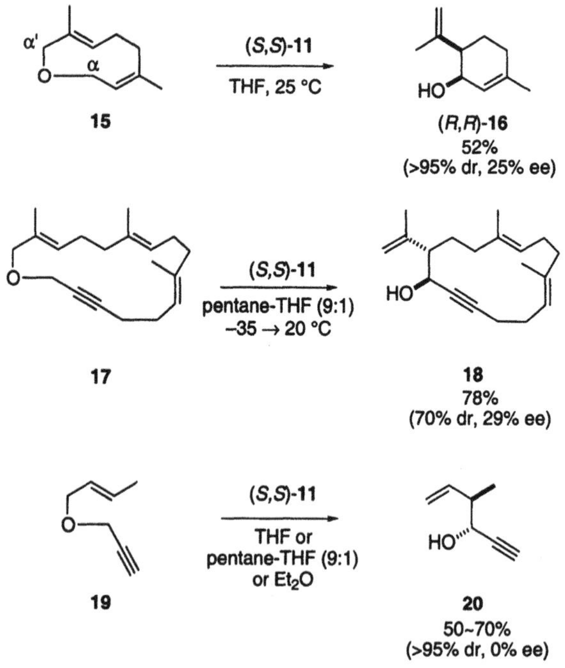

15 → (S,S)-11, THF, 25 °C → (R,R)-16 52% (>95% dr, 25% ee)

17 → (S,S)-11, pentane-THF (9:1), –35 → 20 °C → 18 78% (70% dr, 29% ee)

19 → (S,S)-11, THF or pentane-THF (9:1) or Et_2O → 20 50~70% (>95% dr, 0% ee)

Scheme 4.

A transition state model proposed by Marshall for the rearrangement of ether **12** is shown in Scheme 5. The chiral amide (*S*,*S*)-**11**, in a partial chair-like conformation, may abstract either of the enantiotopic propargylic protons (Scheme 5). Complex **A** should be of lower energy than complex **B**, as the former lacks the unfavourable phenyl/alkyne interaction present in complex **B**, thus the pro-*R* proton was removed predominantly. This rationale suggests that the enantioselectivity is predominantly determined at the lithiation step.

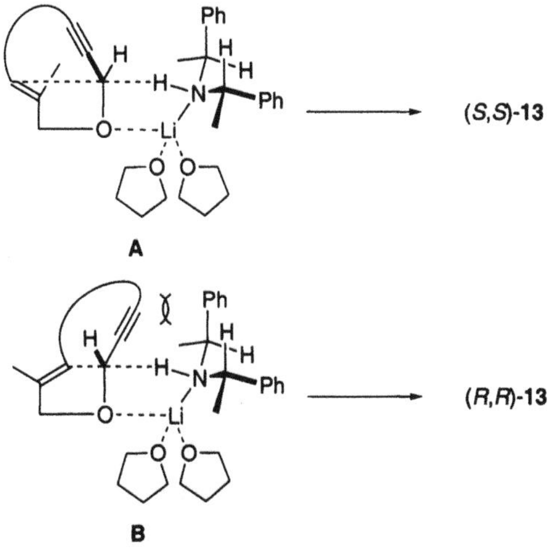

A → (*S*,*S*)-**13**

B → (*R*,*R*)-**13**

Scheme 5.

A high level of enantioselectivity in the rearrangement of an acyclic system has been reported with tricarbonylchromium(0) complexes of allyl benzyl ethers, using chiral lithium amide base **21** [13] [Eq. (8)]. Upon treatment with 1.1 equiv. of lithium amide **21** and 1 equiv. of LiCl at –78 to –50°C, ether **22** afforded the rearrangement product (*R*)-**23** in 80% yield with 96% ee. The effect of olefinic substituents on the chemical yields and enantioselectivity of this [2,3]-Wittig rearrangement was also studied [see Table 3, Sect. 1.3.3].

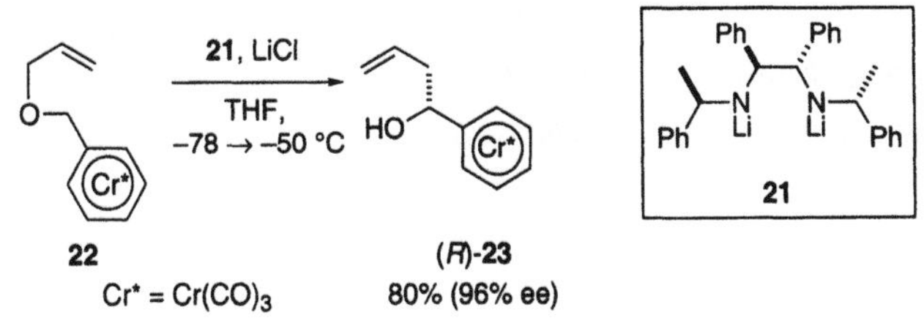

22 → (**21**, LiCl; THF, –78 → –50 °C) → (*R*)-**23** 80% (96% ee) (8)

Cr* = $Cr(CO)_3$

21

1.3.2
Alkyllithium-Chiral Coordinating Agent Complex Protocol

Recently, a number of successful results have been reported for the enantioselective [2,3]-Wittig rearrangement induced by the combined use of an achiral (or racemic) alkyllithium and a chiral non-racemic coordinating agent [e.g., sparteine, isosparteine, norpseudoephedrine, bis(oxazoline)]. The first example of an alkyllithium and (–)-sparteine (**1**) or (–)-α-isosparteine (**24**)-mediated rearrangement was reported by Kang and coworkers [14]. The rearrangement of (*Z*)-allyl propargyl ether **25** induced by *s*-BuLi/(–)-isosparteine (**24**) provides the [2,3]-product *syn*-**26** in 71% ee, whereas the use of (–)-sparteine (**1**) shows poor enantioselectivity (12% ee) [Eq. (9)].

25 → *s*-BuLi / L*, hexane, –78 °C → *syn*-**26** + *anti*-**26**; (–)-**24**

L*	dr	*syn*-26	*anti*-26
L* = (–)-**24**	dr 74:26	71% ee	46% ee
L* = (–)-**1**	dr 38:62	12% ee	2% ee

(9)

This protocol was applied to the rearrangement of *o*-allyloxymethylbenzamide **27** [Eq. (10)] [15], crotyl 2-furfuryl ether **29** [Eq. (11)] [16], allylbenzothiazolyl ether **31** [Eq. (12)] [17], and α-propargyloxyacetic acid **33** [Eq. (13)] [18]. Moreover, an "asymmetric catalytic version" of the rearrangement has been reported [15]. The reaction of **27** with 2.2 equiv. of *s*-BuLi in the presence of 0.2 equiv. of (–)-**1** gave the [2,3]-product **28** in 48% ee, whereas a rearrangement using 2.2 equiv. of (–)-**1** provided the alcohol **28** in 60% ee [Eq. (10)]. A similar rearrangement of a racemic α-deuterated ether **27**-*d* was found to afford 60% yield of **28**-*d* (>98% D-content) with significantly decreased enantioselectivity (35% ee) [Eq. (14)]. This result suggests that the enantioselectivity is predominantly determined at the lithiation step.

27 → *n*-BuLi / (–)-**1**, pentane, –95 °C → (*S*)-**28**

n-BuLi / (–)-1	(*S*)-28
(2.2 eq. / 2.2 eq.)	83% (60% ee)
(2.2 eq. / 0.2 eq.)	44% (48% ee)

(10)

29 → *n*-BuLi / (–)-**1** (1.3 eq. / 1.5 eq.), toluene, –78 °C → *syn*-(1*S*, 2*R*)-**30**, 37% (42% ee)

(11)

n-BuLi / (–)-**1** (1.1 eq. / 1.5 eq.)
toluene, –78 °C
31 → (+)-**32** 90% (46% ee) (12)

n-BuLi / (–)-**1** (2.2 eq. / 2.2 eq.) toluene, –78 °C; CH_2N_2, Et_2O
33 R = *n*-C_7H_{15} → (*S*)-**34** 35% (48% ee) (13)

n-BuLi / (–)-**1** pentane, –95 °C
rac-**27**-*d* → (*S*)-**28**-*d* 60%, >98% D (35% ee) (14)

The norpseudoephedrine-derived amino ether **35** was also used as a chiral coordinating agent for the enantioselective [2,3]-Wittig rearrangement [19]. The rearrangement of propargyl ether **36** induced by *n*-BuLi/**35** provided allenyl alcohol (*S*)-**37** in 62% ee [Eq. (15)]. In contrast, a similar reaction with (–)-**1** provided (*S*)-**37** in only 9% ee.

n-BuLi / **35** (1.05 eq. /1.10 eq.) toluene, –78 °C
36 R = c-hex → (*S*)-**37** 51% (62% ee) (15)
35

Chiral bis(oxazoline) **5** is an effective chiral coordinating agent for enantiocontrol in the [2,3]-Wittig rearrangement. The rearrangement of (*Z*)-crotyl benzyl ether **38** with *t*-BuLi/(*S*,*S*)-**5** (1.5 equiv. each) in hexane provided [2,3]-shift product (1*R*,2*S*)-**39** in 40% ee [Eq. (16)] [20]. The feasibility of the asymmetric catalytic version was also examined. In ether, the rearrangement with 20 mol % of **5** was found to provide the same level of enantioselectivity as when using 150 mol % (34% ee).

38 → syn-(1R, 2S)-39 (16)

t-BuLi / (S,S)-5, −78 °C

(1.5 eq. / 1.5 eq.), hexane 89% *syn* (40% ee)
(1.5 eq. / 1.5 eq.), ether 89% *syn* (34% ee)
(1.5 eq. / 0.2 eq.), ether 89% *syn* (34% ee)

A similar rearrangement of racemic α-deuterated ether **38**-*d* afforded 70% yield of **39**-*d* (98% D-content) with significantly decreased enantioselectivity (14% ee) [Eq. (17)]. This result suggests that the enantioselectivity may be determined substantially at the lithiation step, since the enantioselectivity should be the same as observed with the non-deuterated substrate, if determined at the post-lithiation step.

rac-38-d → syn-(1R, 2S)-39 (17)

t-BuLi / (S,S)-5, hexane, −78 °C

70%, 98% D
(86% *syn*, 14% ee)

It is noteworthy that the rearrangement of crotyl propargyl ether **40** with *t*-BuLi/(*S*,*S*)-**5** at −95 °C (carried out in pentane) provided [2,3]-shift product **41** in high enantiopurity (89% ee, 47% yield) with high (>95%) *anti*-selectivity [Eq. (18)].

40 → anti-(3S, 4S)-41 (18)

t-BuLi / (S,S)-5 (1.5 eq. / 1.5 eq.)

hexane, −78 °C 90% (93% *anti*, 75% ee)
pentane, −95 °C 47% (>95% *anti*, 89% ee)

1.3.3
Tabular Survey

The enantioselective [2,3]-Wittig rearrangements of acyclic substrates are grouped by substrate structure in Table 3.

1.4
[1,4]-Wittig Rearrangement

The Wittig rearrangement of allyl alkyl ethers has been shown to lead to a mixture of alcohols and aldehydes arising from 1,2- and 1,4-migration of alkyl group [Eq. (19)] [2b, 21]. The 1,4-migration, namely [1,4]-Wittig rearrangement, is formally a symmetry-allowed concerted processes. However, detailed mecha-

Table 3. Enantioselective [2,3]-Wittig rearrangements of acyclic substrate

migrating terminus	substrate	reagent, conditions	product(s) yield (%), stereopurity		ref.
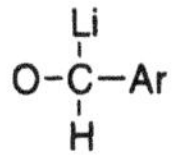 Li O-C-Ar H	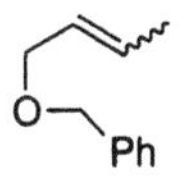 O–CH₂Ph (allyl)		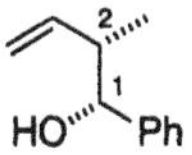 (1*R*, 2*S*)	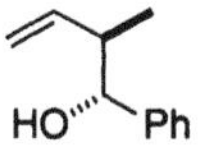 (1*R*, 2*R*)	
	(*E*)	*n*-BuLi / **35**, toluene, –78 °C	31 (16% ee)	17 (12% ee)	[19]
	(*Z*)	*n*-BuLi / **35**, toluene, –78 °C	45 (64% ee)	5 (80% ee)	[19]
	(*Z*)	*t*-BuLi / **5**, hexane, –78 °C	89 (40% ee)	11 (--)	[20]
	(*Z*) 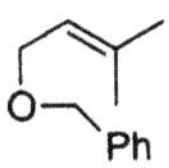	*t*-BuLi / **5** cat. ether, –78 °C	89 (34% ee) 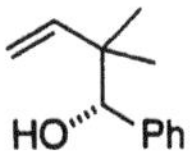	11 (--)	[20]
		n-BuLi / **35**, toluene, –78 °C	68 (40% ee)		[19]
		n-BuLi / **1**, toluene, –78 °C	70 (19% ee)		[18]
	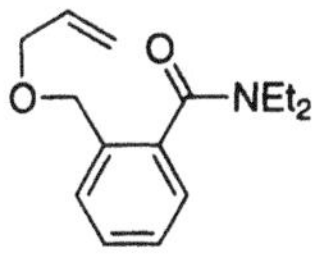 NEt₂		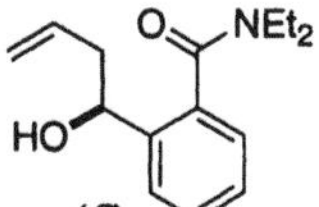 (*S*)		
		n-BuLi / **1**, pentane, –78 °C	65 (44% ee)		[15]
		n-BuLi / **1**, pentane, –95 °C	83 (60% ee)		[15]
		n-BuLi / **1** cat. pentane, –95 °C	44 (48% ee)		[15]
	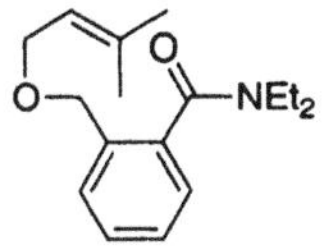 NEt₂		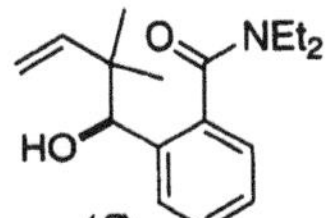 (*S*)		
		n-BuLi / **1**, pentane, –95 °C	71 (62% ee)		[15]
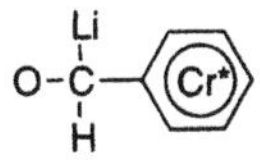 Li O-C-(Cr*) H Cr* = $Cr(CO)_3$	R^1, R^2, R^3		R^1, R^2, R^3	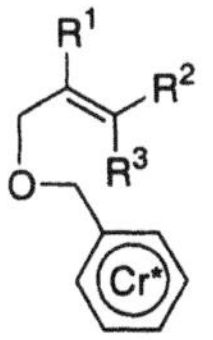	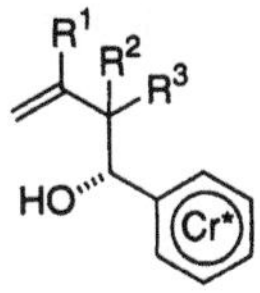
	R^1=R^2=R^3=H	**21**, LiCl, THF, –78 → 50 °C	80 (96% ee)		[13]
	R^1=H, R^2=R^3=Me		82 (84% ee)		[13]
	R^1=Me, R^2=R^3=H		33 (91% ee)		[13]
	R^1=R^3=H, R^2=Me		82 (95% syn, 96% ee)		[13]
	R^1=R^2=H, R^3=Me		24 (50% syn, 90% ee)		[13]

Table 3. (continued)

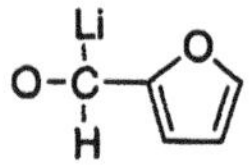

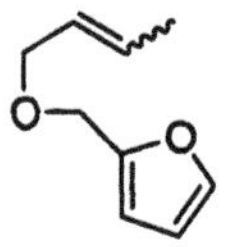

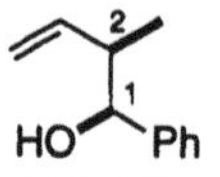

Carbanion	Substrate	Conditions	Product: yield (selectivity)	Ref.
O–C(Li)(H)–(2-furyl)	(*E*)	*n*-BuLi / **1**, toluene, –78 °C	(1*R*, 2*S*) 36 (90% syn, 28% ee)	[16]
	(*Z*)		37 (>99% syn, 42% ee)	[16]
O–C(Li)(H)–(2-thiazolyl)	R=H	*n*-BuLi / **1**, toluene, –78 °C	90 (46% ee)	[17]
	R=Ph		79 (50% ee)	[17]
	Ar = Ph, R = H	*n*-BuLi / **1**, toluene, –78 °C	90 (44% ee)	[17]
	Ar = 2-$MeOC_6H_4$, R = Me		83 (40% ee)	
O–C(Li)(H)–CO_2H(Li)	R=*n*-C_7H_{15}	*n*-BuLi / **1**, toluene, –78 °C	35 (48% ee)	[18]
	R=iso-Bu		36 (40% ee)	[18]
	R=iso-Pr		44 (42% ee)	[18]

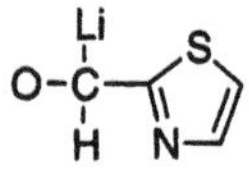

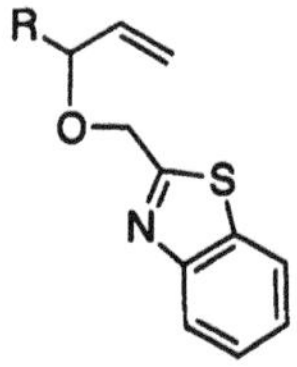

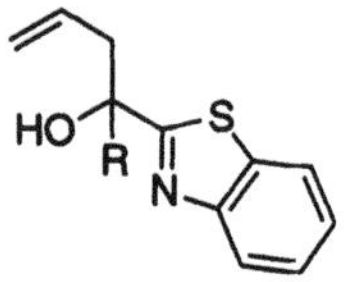

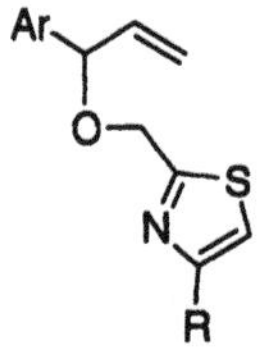

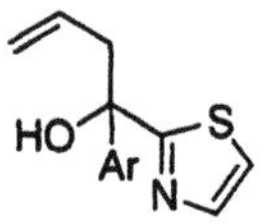

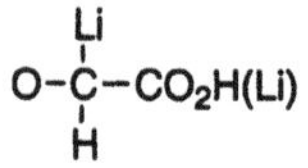

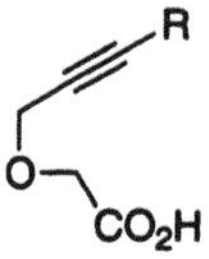

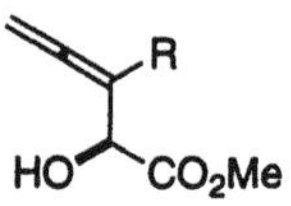

Table 3. (continued)

O–C(Li)(H)–C≡C–Y

(3*R*, 4*S*) (3*R*, 4*R*)

Substrate	Conditions	Product	Ref.
(*E*), Y=TMS	*s*-BuLi / **1**, hexane, –78 °C	dr = 82:18 [16% ee (3*R*, 4*S*)] [11% ee (3*R*, 4*R*)]	[14]
	s-BuLi / **24**, hexane, –78 °C	dr = 77:23 [29% ee (3*R*, 4*S*)] [31% ee (3*R*, 4*R*)]	[14]
	t-BuLi / **5**, hexane, –78 °C	dr = 80:20 [32% ee (3*S*, 4*R*)] [---]	[20]
(*Z*), Y=TMS	*s*-BuLi / **1**, hexane, –78 °C	dr =92:8 [24% ee (3*R*, 4*S*)] [---]	[14]
	s-BuLi / **24**, hexane, –78 °C	dr = 100:0 [42% ee (3*R*, 4*S*)] [---]	[14]
	t-BuLi / **5**, hexane, –78 °C	dr = 95:5 [45% ee (3*S*, 4*R*)] [---]	[20]
(*E*), Y=Me	*t*-BuLi / **5**, hexane, –78 °C	dr =7:93 [---] [75% ee (3*S*, 4*S*)]	[20]
	t-BuLi / **5**, hexane, –95 °C	dr = <5:>95 [---] [89% ee (3*S*, 4*S*)]	[20]
	s-BuLi / **1**, hexane, –78 °C	dr = 32:68 [---] [2% ee (3*R*, 4*S*)]	[20]
(*Z*), Y=Me	*t*-BuLi / **5**, hexane, –78 °C	dr =94:6 [39% ee (3*S*, 4*R*)] [---]	[20]

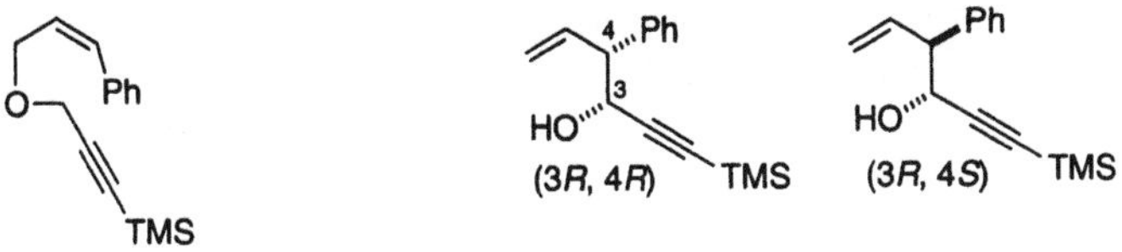

(3*R*, 4*R*) (3*R*, 4*S*)

Substrate	Conditions	Product	Ref.
(*Z*)-Ph, TMS	*s*-BuLi / **1**, hexane, –78 °C	dr = 38:62 [12% ee (3*R*, 4*R*)] [2% ee (3*R*, 4*S*)]	[14]
	s-BuLi / **24**, hexane, –78 °C	dr = 74:26 [71% ee (3*R*, 4*R*)] [2% ee (3*R*, 4*R*)]	[14]
	t-BuLi / **5**, hexane, –78 °C	dr = 45:55 [12% ee (3*S*, 4*S*)] [2% ee (3*S*, 4*R*)]	[20]

(*S*)

Substrate	Conditions	Product	Ref.
R = *c*-hex	*n*-BuLi / **35**, toluene, –78 °C	51 (62% ee)	[19]
	n-BuLi / **1**, toluene, –78 °C	23 (9% ee)	[19]
	t-BuLi / **5**, pentane, –78 °C	44 [67% ee (*R*)]	[20]

nistic studies suggest that this class of rearrangement proceeds via a non-concerted radical process [22].

[1,2]-Wittig

H_3O^+

[1,4]-Wittig

H_3O^+

(19)

Synthetic applications of the [1,4]-Wittig rearrangement have been rather limited, because of the low yields and restricted range of substrates. However, recently a novel and synthetically useful rearrangement system has been developed. The key to success is the choice of an ethynylvinylmethanol-derived migrating terminus [Eq. (20)] [23]. The first enantioselective version was also accomplished with this class of substrates. For example, the rearrangement of ether *rac*-**42** with *t*-BuLi-bis(oxazoline) **5** afforded ketone **43** in 48% ee [Eq. (21)] [24]. Furthermore, the sequential [1,4]-Wittig rearrangement-aldol reaction has been shown to provide moderate diastereo- and enantioselectivity for the product β-hydroxy ketone **44**.

1. alkyllithium
2. E^+

(20)

t-BuLi / (*S*,*S*)-**5**
(1.3 eq. / 1.6 eq.)
hexane, –78 °C

E^+

rac-**42**

E^+=H_3O^+ **43** (R = H)
67% (63% dr, 48% ee)

E^+=HCHO **44** (R = CH_2OH)
50% (69% dr, 50% ee)

(21)

2 Retro-Brook Rearrangement

The reverse process of a Brook rearrangement, the O- to C-trialkylsilyl group shift from a silyl ether to a carbanion terminus, is well-known and has been intensively studied [Eq. (22)] [25].

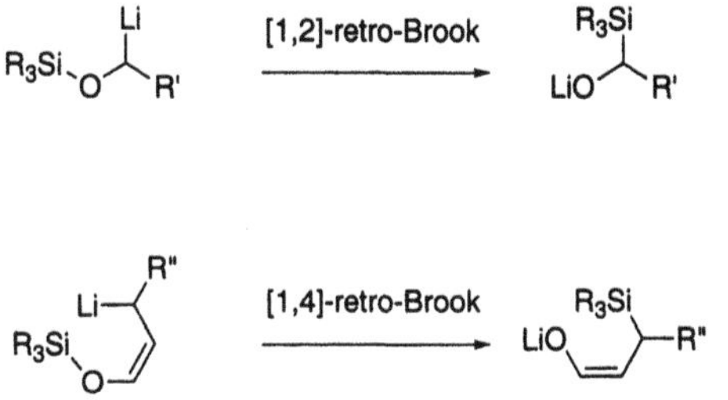

(22)

This rearrangement is sometimes referred to as the silyl-Wittig rearrangement. Mechanistic studies on the retro-Brook rearrangement have revealed that the process is highly stereospecific with regard to the configuration at the migrating terminus. Therefore, enantiocontrol in a retro-Brook rearrangement should be possible with the chiral coordinating agent protocol. In practice, the enantioselective version of a retro-Brook rearrangement was accomplished as the subsequent reaction of enantioselective cyclocarbolithiation [Eq. (23)] [26]. The cyclisation precursor **45** was treated with *s*-BuLi/(–)-**1** in ether at –78 °C to –40 °C, providing the cyclised and silyl-rearranged product (*R*,*R*,*R*)-**46** in 58% yield as a single stereoisomer along with the cyclisation/1,3-cycloelimination product **47** (9%).

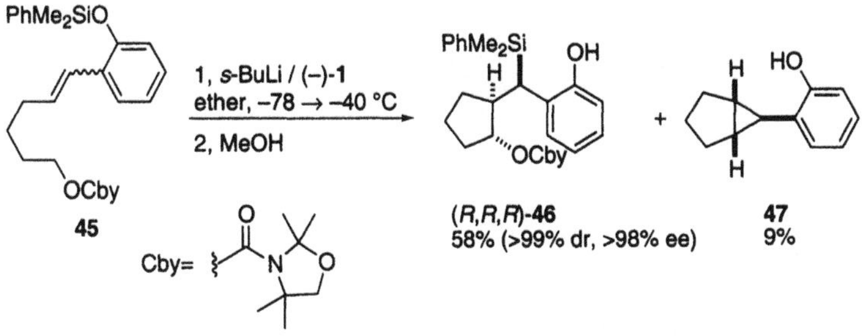

(23)

The stereoselective formation of **46** reveals that:

1) the chiral base abstracts the pro-*S*-proton in **45** to give the lithiated carbamate **48** (Scheme 6), which undergoes the 5-*exo-trig* ring closure with retention of configuration at the Li-bearing carbanion centre in a *syn*-fashion;

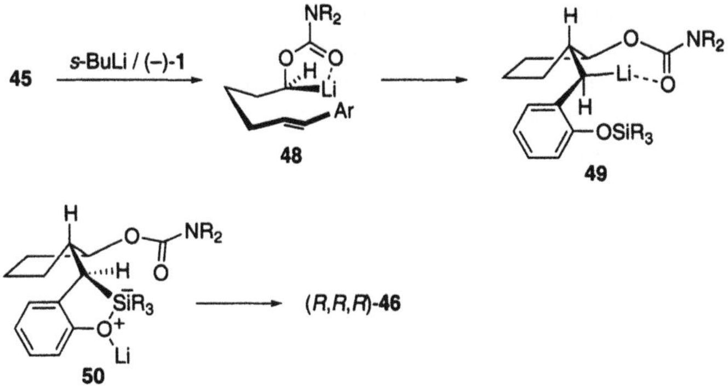

Scheme 6.

2) the resulting configurationally labile benzylic anions are epimerised to the thermodynamically stable stereoisomer **49**;
3) the intermediate **49** can easily form the complex **50** bearing a pentacoordinated Si-centre with retention of the configuration at the benzylic position;
4) the complex **50** then furnishes the lithium phenoxide which is subsequently converted to **46** by hydrolysis.

3
Phosphate-Phosphonate Rearrangement

Phosphate-derived α-oxy carbanions can rearrange into α-hydroxy phosphonates. This class of rearrangement is known to proceed with retention of configuration at the carbanion terminus [27]. The enantioselective version of this rearrangement has been developed using a chiral lithium amide as base [Eq. (24)] [28]. The reaction of benzyl dimethyl phosphate **51** with amide (*R*,*R*)-**11** in THF gave the hydroxy phosphonate (*R*)-**52** in 30% yield, and in enantioenriched form (52% optical purity). In contrast, when *s*-BuLi/(–)-**1** was used as a base in ether at –78 °C, the enantiopurity of the phosphonate **52** was very low (8% optical purity, 65% yield).

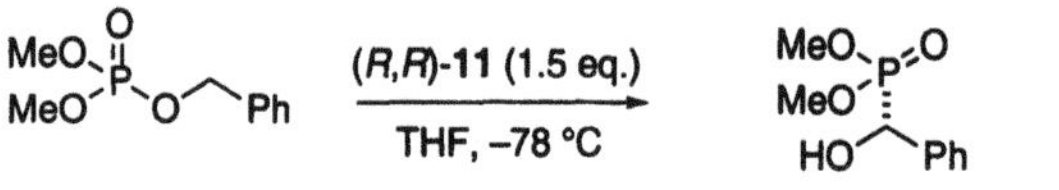

MeO(MeO)P(=O)OCH$_2$Ph **51** —[(*R*,*R*)-**11** (1.5 eq.), THF, –78 °C]→ (MeO)$_2$P(=O)CH(OH)Ph (*R*)-**52** 30% (52% optical purity) (24)

4
Lithiated Epoxide Rearrangements

4.1
Introduction to Lithiated Epoxide Rearrangements

In the second half of this chapter the enantioselective generation and reactivity of α-lithiated epoxides **53** (also named oxiranyl anions) is reviewed; such organolithiums are reactive intermediates in which the anion is carried by the oxirane, as shown in Fig. 1. After a short introduction the different reactivities that such intermediates can exhibit, which depend on the other substituents of the oxirane and, of course, on the reaction conditions, will be discussed [29, 30].

Epoxides are widely utilised as versatile synthetic intermediates, and the epoxide functional group is also present in a number of interesting natural prod-

53: α-lithiated epoxide

Fig. 1.

ucts [31]. In basic media epoxides exhibit a wide range of reactivity, undergoing nucleophilic ring openings to give functionalised alcohols **54** [32, 33], as well as α-deprotonation and β-eliminations (Scheme 7).

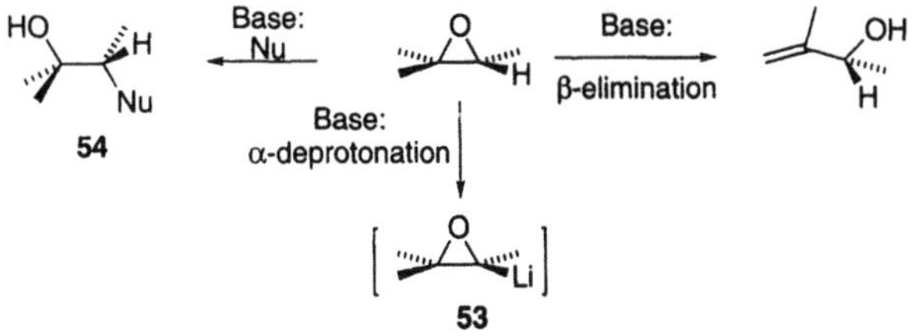

Scheme 7.

Reaction of a lithium amide with an epoxide bearing an available β-hydrogen often results in β-elimination giving an allylic alcohol [34]. This elimination is normally thought to occur via a *syn* pathway through formation of a 1 : 1 epoxide-base complex **55**, which results in a six-membered transition state (shown for cyclohexene oxide in Scheme 8). While this picture can be a useful aid in understanding the (enantio)selectivity of the elimination, in reality there are likely more complex aggregates from which reaction may also proceed. The participation of allylic alkoxides **56** generated during the reaction may further complicate the simple picture [35, 36]. The application of epoxide β-elimination to the desymmetrisation of *meso*-epoxides is well developed [37], and remains an active field of study [38–40]. However, this reactivity mode of epoxides, not proceeding via an organometallic, lies beyond the scope of the current chapter.

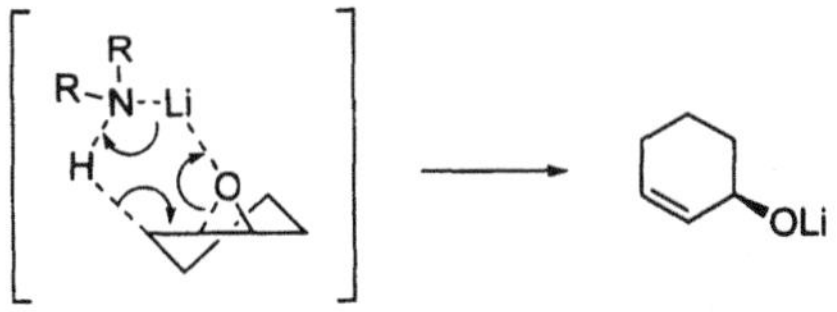

Scheme 8.

Although the β-elimination of epoxides is a well-described process, due to the electron-withdrawing effect of the oxygen and the acidifying effect of the three-membered ring, the deprotonation can also occur at the α-position. Over half a century ago Cope and Tiffany invoked an α-deprotonation pathway to explain the rearrangement of cyclooctatetraene oxide to 2,4,6-cyclooctatrien-1-one [41]. Subsequently, there have been numerous studies concerning α-lithiation of epoxides, and it is now known that such lithiated epoxides can undergo a range of transformations (as outlined below).

For an epoxide bearing an α-anion stabilising group (e.g., ester, phenyl), α-deprotonation (rather than β-elimination) is the preferred pathway when using lithium amides. In the absence of such a stabilising group, the lithium amide and the epoxide are in equilibrium with the lithiated epoxide and the amine [34, 38, 42]; this usually results in (irreversible) β-elimination, if it is possible, dominating the reaction profile. For the latter epoxides α-deprotonation is also the favoured (usually exclusive) reaction pathway in the presence of a strong organolithium base, such as an alkyllithium, but under these conditions the deprotonation is not reversible. Once an α-lithiated epoxide is generated, its stability is an issue. In the presence of an anion-stabilising group it can classically react as a nucleophile [43]. Otherwise it can exhibit carbenoid reactivity [44]; reactivity via an α-lithiooxy carbene **57**, generated by α-ring opening, can also be invoked (Fig. 2).

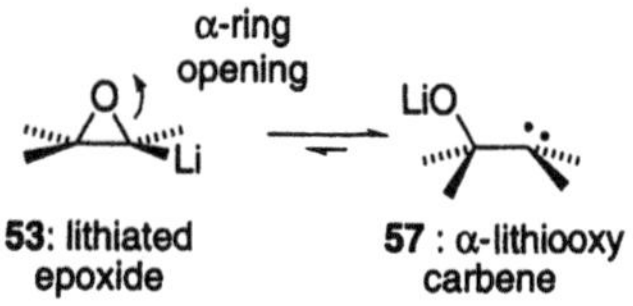

Fig. 2.

Due to their carbenoid properties lithiated epoxides can undergo various C-H insertions and 1,2-hydride shifts as well as cyclopropanations (Scheme 9, Path A); a free α-lithiooxy carbene may be involved in some of these transformations. Interestingly, the observation of an allylic alcohol as the outcome of a

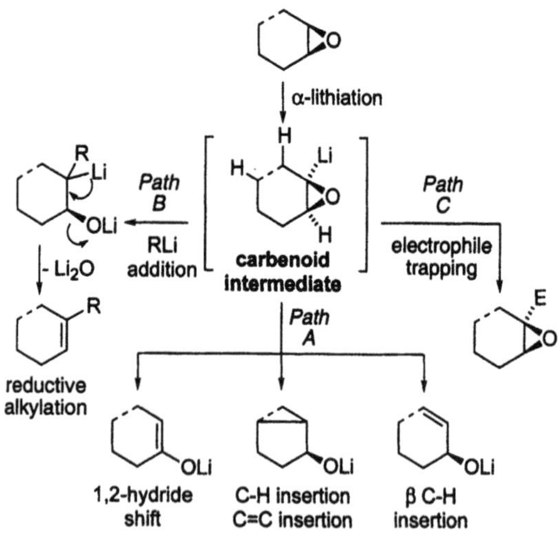

Scheme 9.

base-induced epoxide rearrangement may be due to an α-deprotonation and/or a β-elimination process. Morgan and Gronert have shown that the structural features of the epoxide can have a significant influence on which reaction pathway is followed [45] Collum and Ramirez suggested that the structures of the organolithium aggregates are also of major importance [46].

While they are negatively charged, the ring strain of the three-membered heterocycle is still an important feature of the lithiated epoxide. That, combined with the weakening of the bond between the carbon bearing the charge and the oxygen, due to its greater polarisation, makes lithiated epoxides very electrophilic species [44]. Their reactions with (usually strong) nucleophiles can then lead to alkenes via a "reductive alkylation" pathway (Scheme 9, Path B).

Finally, lithiated epoxides are potentially nucleophilic reagents, and they can therefore react with electrophiles, under specific conditions (Scheme 9, Path C).

The different reactivities outlined in Scheme 9 only hint at the very exciting chemistry of lithiated epoxides and suggest a potentially wide range of applications in synthesis.

4.2 The Carbenoid Chemistry of Lithiated Epoxides: C-H Insertion and 1,2-Shifts

Although they are sometimes utilised for their nucleophilic properties, as outlined above lithiated epoxides are carbenoid species, and they are well known to react through C-H and C=C insertion or 1,2-shift processes. Following his early studies on cyclooctatetraene oxide, Cope extended his work to stilbene derivatives **58** and **60** (Scheme 10) [47]. The ketone **59** and aldehyde **61** were derived from (formal) 1,2-shift of the group (either H or phenyl, respectively) *cis* to lithium on the lithiated epoxide.

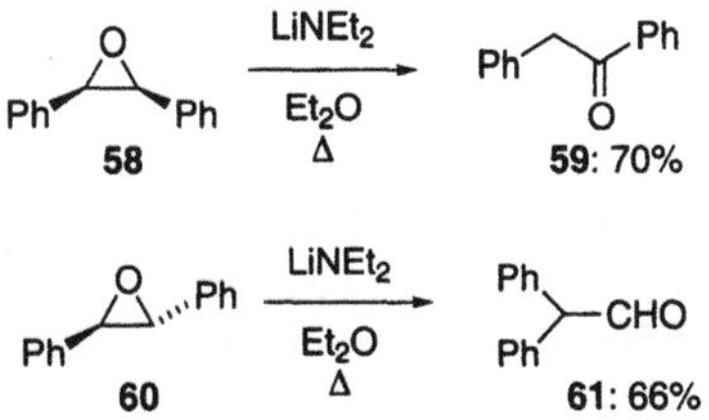

Ph 58 Ph — $LiNEt_2$, Et_2O, Δ → Ph 59: 70% Ph

Ph 60 Ph — $LiNEt_2$, Et_2O, Δ → Ph Ph CHO 61: 66%

Scheme 10.

Contemporaneously, Cope also reported the first examples of base-induced transannular rearrangements (C-H insertion) of medium-sized cycloalkene oxides **62** and **64** (Scheme 11) [48]. These transformations could proceed via a carbenoid species (the lithiated epoxide) and/or an α-alkoxy carbene (such as **57** in Fig. 2).

The most recent developments in the above area mainly concern enantioselective variants of these reactions (vide infra). The asymmetric version of transannular C-H insertions from enantioenriched lithiated epoxides has been extensively studied in the Hodgson laboratory over the last six years. In these

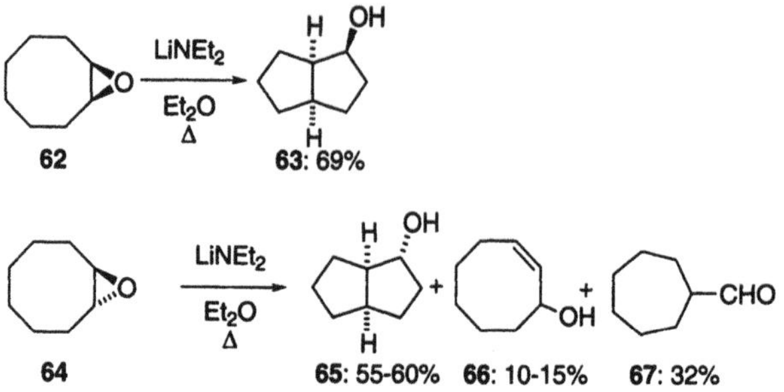

LiNEt$_2$, Et$_2$O, Δ: 62 → 63: 69%

LiNEt$_2$, Et$_2$O, Δ: 64 → 65: 55-60% + 66: 10-15% + 67: 32%

Scheme 11.

cases, asymmetric induction arises from enantioselectivity in the α-deprotonation, i.e., the enantioselective generation of an intermediate lithiated epoxide. The enantioselective α-deprotonation of the epoxide is likely kinetically controlled via diastereomeric transition states (Fig. 3, R* being chiral, non-racemic).

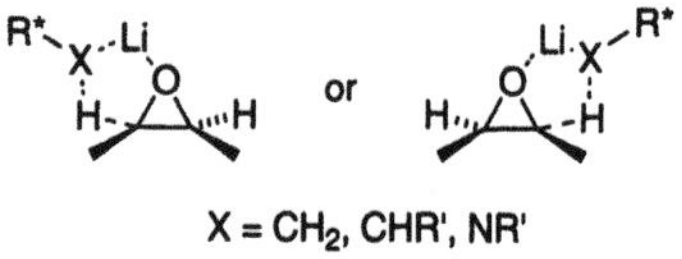

X = CH$_2$, CHR', NR'

Fig. 3.

4.2.1
Transannular C-H Insertion

cis-Cyclooctene oxide **62**, used by Cope et al. to carry out the lithium amide-induced rearrangement shown in Scheme 11, is a *meso*-epoxide; the α-deprotonation prior to rearrangement therefore breaks the symmetry. If this α-deprotonation is carried out with differentiation of the two enantiotopic protons, an enantioselective rearrangement of the epoxide will occur. However, the chiral lithium amide (*S,S*)-**11** [Eq. (7)] only induces low ee (10%), in the rearrangement of *cis*-cyclooctene oxide [49].

The same rearrangement (**62** to **63**, Scheme 11) was carried out with an alkyllithium in ether by Boeckmann in 1977 [50]; it was therefore of interest to study this rearrangement using a combination of an organolithium with a chiral diamine such as (–)-sparteine **1** [Eq. (2)] [51]. Secondary alkyllithiums together with **1** were found to be the most efficient organolithiums both in terms of yield and ee [52]. Using *s*-BuLi with cyclooctene oxide **62** at –98 °C allowed the isolation of the isomeric bicyclic alcohol **63** (Scheme 11) in 79% yield and 70% ee; using *i*-PrLi yielded 74% of 11 in 83% ee. As previously observed [53], the combination of *t*-BuLi and (–)-sparteine **1** proved inefficient in terms of asymmetric induction. Ether proved to be the preferred solvent. Finally, sub-stoichiometric amounts of (–)-sparteine still allowed the reaction to proceed, although not to completion and with lower ee. Nevertheless, this indicates the potential for asymmetric catalysis. The scope of the enantioselective rearrangement included other medium-sized cycloalkene oxides, such as cyclononene oxide **68** and cyclodecene oxide **70** (Scheme 12).

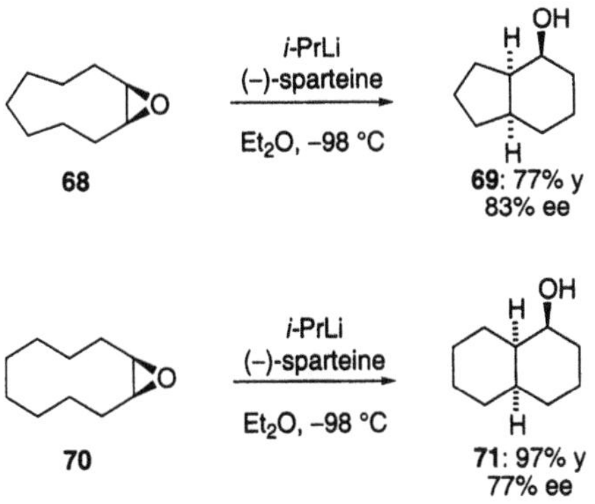

i-PrLi
(–)-sparteine
Et_2O, –98 °C
68
69: 77% y
83% ee

i-PrLi
(–)-sparteine
Et_2O, –98 °C
70
71: 97% y
77% ee

Scheme 12.

As only (–)-sparteine is readily available, extension of this asymmetric methodology to other chiral ligands was investigated [54]. Bisoxazoline ligands are particularly attractive, since both enantiomers are normally available. In the transannular desymmetrisation chemistry the valine-derived bisoxazoline possessing a diisobutyl-substituted bridge (*S,S*)-**5** [Eq. (3)] was found to be the most promising in this class of ligands. Moving from isopropyl to more sterically demanding substituents at C-4 resulted in a significant erosion of both yield and ee. The diethyl-substituted bridge C-4 isopropyl-bisoxazoline was slightly less effective at performing the transformation. In contrast to (–)-sparteine **1**, the best ee with ligand (*S,S*)-**5** was obtained with *s*-BuLi (66% *ee* of *ent*-**63** from cyclooctene oxide).

Similarly to the bisoxazolines, (–)-α-isosparteine **24**, a diastereomer of (–)-sparteine, possesses C_2 symmetry (Scheme 13). No general trend in the efficiency of organolithium-mediated transformations with (–)-α-isosparteine **24** compared with (–)-sparteine **1** can be discerned from the literature, (–)-α-isosparteine being sometimes less [55, 56], sometime more efficient [57, 58]. In the α-deprotonation of *meso*-epoxides, the efficiency of (–)-α-isosparteine proved to be substrate dependent. Whereas it gave a similar ee to sparteine for **63** in the reaction of cyclooctene oxide **62** (81%), in the rearrangement of cyclodecene oxide **70** the use of (–)-α-isosparteine proved less efficient [38% ee, vs. 51% ee for **71** with (–)-sparteine, under otherwise identical conditions]. Interestingly, with cyclooctene oxide the use of sub-stoichiometric amounts of (–)-α-isosparteine resulted in an increase of ee for **63**, the best results being obtained with 0.2 equiv. of this diamine (86% yield, 84% ee). This apparently surprising result may be

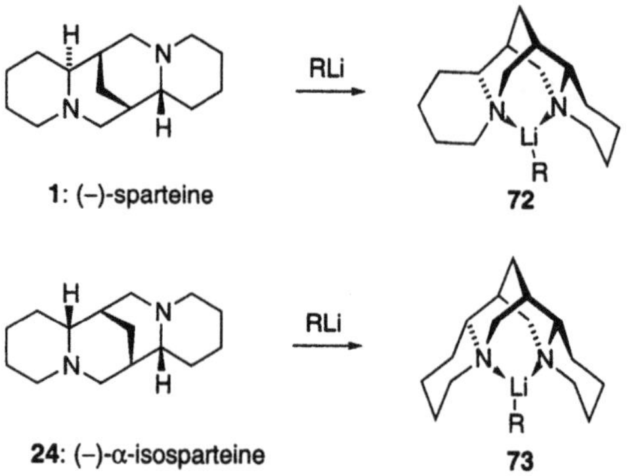

1: (–)-sparteine → RLi → 72

24: (–)-α-isosparteine → RLi → 73

Scheme 13.

due to a better homogeneity of the solution when using sub-stoichiometric (–)-α-isosparteine. Moreover, the more sterically demanding complex **73** (compared with **72**, Scheme 13) may aid dissociation of the lithium alkoxide initially arising from the rearrangement, thus promoting catalysis.

The effect of added $BF_3 \cdot Et_2O$ in the presence the base was recently investigated by Alexakis [59]. In comparison to the results above (from **62** to **63**), the addition of the Lewis acid had a tremendous accelerating effect, together with an erosion of the best ee obtained in the earlier work with *s*-BuLi (from 77% to 61%). Both the rate enhancement and the erosion of ee could be explained by the coordination of the Lewis acid to the oxygen of the epoxide, since this complexation could favour the deprotonation by weakening the C-O bonds and making the α-protons more acidic by increasing the electron-withdrawing effect of the oxygen. Therefore, both the chiral alkyllithium-diamine complex and ligand-free secondary alkyllithium could competitively perform the deprotonation, leading to an overall decrease in the enantiotopic proton selection. However, when compared with the same earlier work an improvement in both yield and ee was observed when *n*-BuLi was used (98% yield in 48% ee, compared with 74% yield in 31% ee). When phenyllithium is used a significant amount of nucleophilic ring opening is observed [60], presumably as a result of both the lower basicity of phenyllithium (making it relatively more nucleophilic than basic) together with the $BF_3 \cdot Et_2O$ activation of the epoxide. Use of $BF_3 \cdot Et_2O$ as an additive with cyclooctadiene oxide **74** yielded the corresponding unsaturated bicyclic alcohol **75** [Eq. (25)].

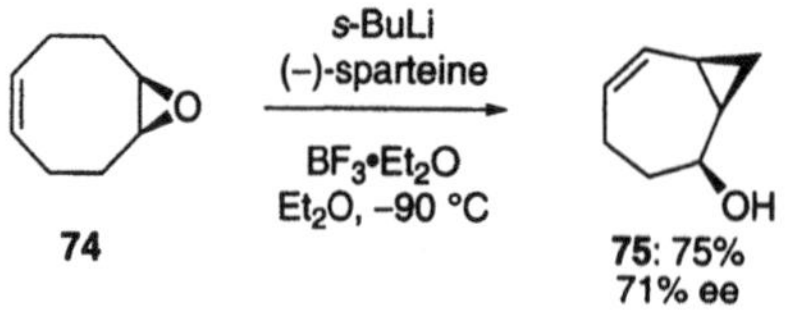

74 → *s*-BuLi, (–)-sparteine, $BF_3 \cdot Et_2O$, Et_2O, –90 °C → **75**: 75%, 71% ee (25)

In order to broaden the scope of asymmetric transannular rearrangement, functionalised *meso*-epoxides derived from medium-sized cycloalkenes have been studied. Protected diol derivatives of cyclooctadiene oxide were used as the substrates; variations in the substitution pattern in the cyclooctene oxide ring resulted in differing product outcomes. Substitution by *t*-butyldimethylsilyloxy groups at the 5- and 6-positions, as in **76**, proved effective in this rearrangement [Eq. (26)] [61,62]. In this case a mixture of bicyclic alcohol **77** from transannular rearrangement, alkene **78** from reductive alkylation and ketone **79** from 1,2-hydride shift was obtained, the ratio being dependent on the reaction conditions, but usually favouring the bicyclic alcohol **77**.

RLi, ligand, Et_2O: **76** → **77**: 12-72%, 32-89% *ee* + **78**: 0-40% + **79**: 0-20% (26)

The use of *s*-BuLi and (–)-sparteine at –78 °C in the rearrangement of **76** yielded 51% of the bicyclic alcohol **77** in 73% ee. This ee is close to the 70% obtained in the unsubstituted system **62** and suggests that the slow interconversion between the two main (enantiomeric) conformers **76a** and **76b** [Eq. (27)] of the disubstituted epoxide **76** (NMR analysis clearly indicates that this compound exhibits a lack of conformational mobility) does not affect the enantioselectivity of the deprotonation-rearrangement. This observation could be explained by the relatively unchanged environment in the approach of the *s*-BuLi/(–)-sparteine complex to the rigid epoxide fragment in either cyclooctene oxide **62** or the substituted cyclooctene oxide **76**. Similar to cyclooctene oxide, it was found with the substituted cyclooctene oxide that *i*-PrLi/(–)-α-isosparteine gave the best yield of the desired bicyclic alcohol **77** (72%), and also the best ee (89%) in Et_2O at –98 °C. The use of a bisoxazoline (*S*,*S*)-**5** [Eq. (3)] allowed access to *ent*-**77** in moderate ee (52%).

76a ⇌ **76b** (27)

The effect of different substitution on this transannular rearrangement was further studied on a range of disubstituted cyclooctene oxides. *cis*-4,7-Di(*t*-butyldimethylsilyloxy)-cyclooctene oxide **80** yielded a mixture of bicyclic alcohol **81**, and alkene **82** arising from reductive alkylation [Eq. (28)] [61,62].

RLi, ligand, Et_2O: **80** → **81**: 31-70% + **82** (28)

The yield of the bicyclic alcohol **81** was dependent on the ligand used. Indeed, in the absence of ligand, the reaction carried out with *s*-BuLi at –78 °C only gave a low isolated yield of **81** (30%). The addition of (–)-sparteine raised it to around 50% (72% ee); the use of *i*-PrLi allowing a slight improvement in both yield

(54%) and ee (80%). As observed in the earlier systems, the best ee (up to 84%) was obtained when the reaction was carried out at –98 °C. It is also noteworthy that the lower temperature allowed a better isolated yield (up to 70%). In contrast with the earlier systems, the use of (–)-α-isosparteine did not give a better ee (82%), and the yield of **81** (48%) was only moderate with this ligand.

In contrast to above, the *trans*-isomer **83** mainly yielded the corresponding bicyclic alcohol **84**, and the allylic alcohol **85** from β-elimination (or α-deprotonation followed by 1,2-hydrogen shift) [Eq. (29)].

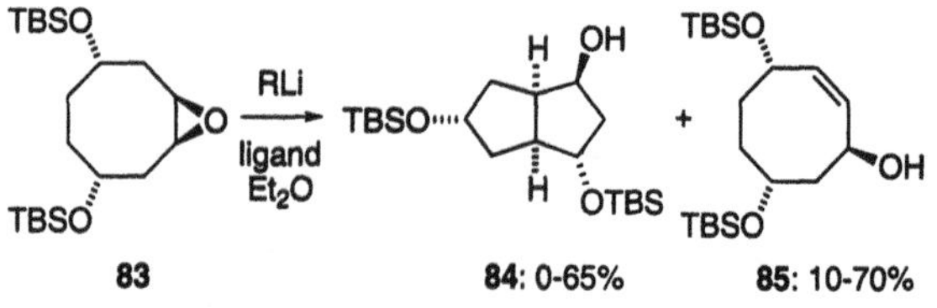

TBSO, TBSO, **83** —RLi, ligand, Et_2O→ **84**: 0-65% + **85**: 10-70% (29)

Here again the ligand exhibited a considerable effect on the deprotonation-rearrangement chemistry. In the absence of ligand, the major product was the allylic alcohol **85** (36%), the bicyclic alcohol **84** being only a minor product (7%) of the reaction. In presence of TMEDA 70% of allylic alcohol **85** was obtained, with no bicyclic alcohol **84** detected. The same observation was made in the presence of bisoxazoline (*S*,*S*)-**5** [Eq. (3)], albeit in reduced yield (34%). Using (–)-sparteine with *i*-PrLi or *s*-BuLi at –78 °C yielded a 45:55 mixture of the two alcohols in favour of allylic alcohol **85** with similar ee (62 to 71%). Moving to (–)-α-isosparteine with *s*-BuLi reversed the product profile. Interestingly, in the latter case the ee of the bicyclic alcohol **84** and the allylic alcohol **85** were significantly different (84 and 60%, respectively), likely implying different mechanisms; yet the ee did not significantly change upon moving from *s*-BuLi to *i*-PrLi. In the presence of (–)-α-isosparteine even the alkyllithium has a strong effect on the reaction conversion and ratio **84:85** (*s*-BuLi: 75% and 87:13; *i*-PrLi: 44% and 72:28).

Proton removal at the *R*-epoxide stereocentre is consistently seen in the enantioselective α-deprotonation-rearrangement of epoxides using the sparteines; this enantioselectivity may be explained by considering a sparteine-RLi-epoxide complex **86**, where the C-H bond on the epoxide *R* stereocentre is held closer to the organolithium than the *S* stereocentre, minimising non-bonded interactions between sparteine and the epoxide (Scheme 14).

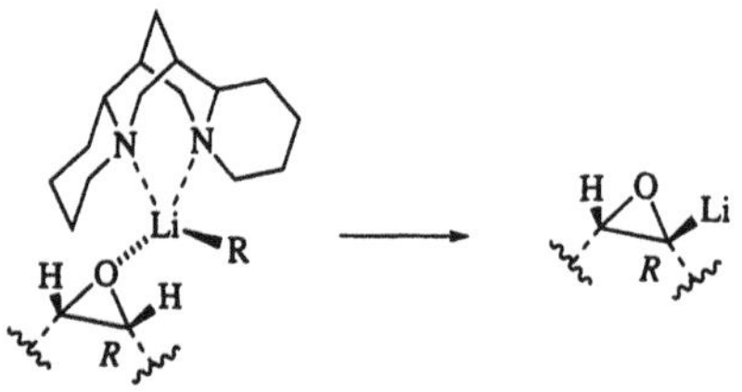

86

Scheme 14.

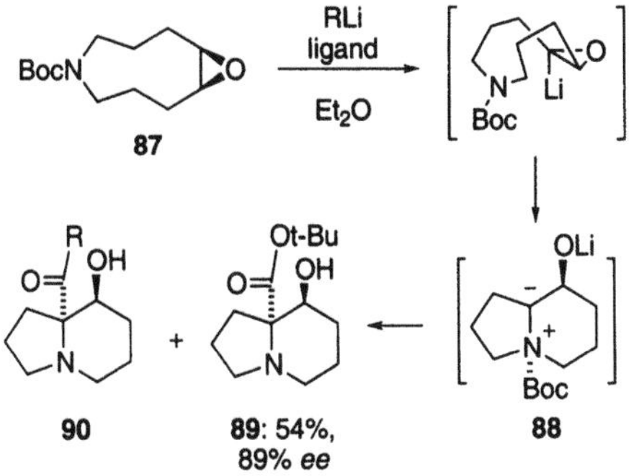

Scheme 15.

The enantioselective transannular desymmetrisation chemistry has also been applied to an azacyclononene oxide **87** (Scheme 15) [62, 63].

The above results are consistent with formation of an intermediate lithiated epoxide which rearranges via an ammonium ylide **88** (Scheme 15). The ketone **90** then arises from addition of the organolithium (present in excess) to the ester **89** (OH=OLi). Using *i*-PrLi in combination with (–)-sparteine at –98 °C yielded 54% of the ester-subsituted indolizidine **89** in 89% ee. A sub-stoichiometric amount of (–)-sparteine (24%) also allowed conversion to proceed to **89** with high ee (82%), albeit at lower rate. Similarly, a sub-stoichiometric amount of (–)-α-isosparteine (24 mol %) proved to be as good as the stoichiometric combination of (–)-sparteine and *i*-PrLi.

Transannular rearrangements of lithiated epoxides are not restricted to medium ring systems. *exo*-Norbornene oxide **91** undergoes a similar rearrangement using an organolithium or a lithium amide, yielding nortricyclanol **92** (Scheme 16) [49, 64].

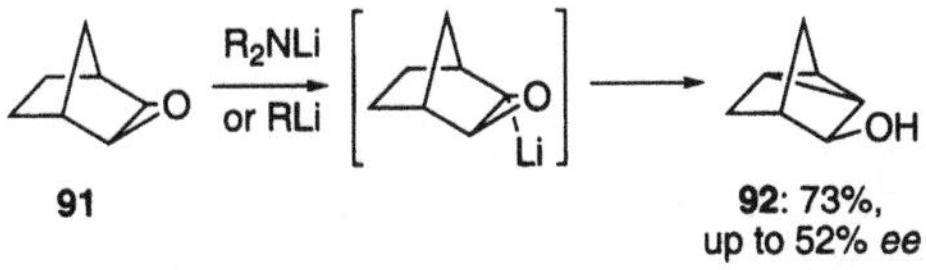

Scheme 16.

In the above case no β-elimination can occur. Reversibility observed during the α-deprotonation of such an epoxide with a lithium amide (vide supra) might result in lowering the ee when using a chiral lithium amide, since reversible deprotonation could compromise the kinetic control in enantioselective deprotonation. Nevertheless deprotonation of *exo*-norbornene oxide **91** with lithium (*S*,*S*)-bis(1-phenyl)ethylamide **11** [Eq. (7)] gave tricyclanol **92** in good yield (73%) and moderate ee (49%) (Scheme 16). When the rearrangement of *exo*-norbornene oxide **91** is carried out with *s*-BuLi in pentane from –78 °C to room temperature in the presence of (–)-sparteine, the ee is only slightly improved (52%), the yield being similar.

The scope of this kind of rearrangement of bicyclic alkene oxides has also been extended to heterobicyclic compounds. The asymmetric rearrangement of *N*-Boc 7-azanorbornene oxide **93** yields azanortricyclanol **94** in good yield and ee [Eq. (30)] [65].

Boc N ... O **93** —MeO–C₆H₃(Me)–Li, (–)-sparteine, Et_2O, –78 °C→ Boc N ... OH **94**: 60%, 77% *ee* (30)

In this case, the use of chiral lithium amides proved to be unsuccessful (lithium amide **11**, Fig. 4, gave alcohol **94** in only 20% yield and 9% ee) but using an aryllithium in combination with (–)-sparteine gave both an improvement in yield (up to 61%) and ee (up to 77%). The use of the diisobutyl-substituted valine-derived bisoxazoline (*S*,*S*)-**5** [Eq. (3)] in combination with PhLi gave a significant improvement in the ee (up to 87%) of *ent*-**94**, albeit with slightly lower yields (up to 51%). This methodology has recently been combined with a subsequent radical rearrangement to achieve the syntheses of epibatidine analogues [66, 67].

4.2.2
Transannular Rearrangement Versus 1,2-Hydride Shift

Both the rearrangements and 1,2-shifts of hydrogens on bicyclic systems shown in Scheme 17 have also been carried out enantioselectively [49, 68].

However, using either lithium (*S*,*S*)-bis(1-phenyl)ethylamide **11** [Eq. (7)] and/or the chiral, non-racemic bislithium bisamide *ent*-**21** [Eq. (8)] only led to moderate yields and ees. Interestingly, the reaction of bicyclo[2.2.2]octadiene oxide **97** with these enantioenriched bases now preferentially proceeds via a 1,2-shift of hydrogen yielding the corresponding ketone **99**, which contrasts with the results obtained in the LDA-induced reaction.

The above examples illustrate that a delicate balance of factors control the fate of the lithiated epoxides.

4.2.3
1,2-Hydride Shifts

Formation of an enolate from a lithiated epoxide could occur by two mechanisms: α-ring opening (vide supra) followed by insertion of the carbene in the LiOC-H bond or electrocyclic β-ring opening (Scheme 18). There is some experimental support for both pathways [69, 70].

Deprotonation of *meso*-cycloheptene oxide diastereoisomers **100** and **102**, possessing a pseudoasymmetric C-5 carbon, led to the enantiomeric ketones

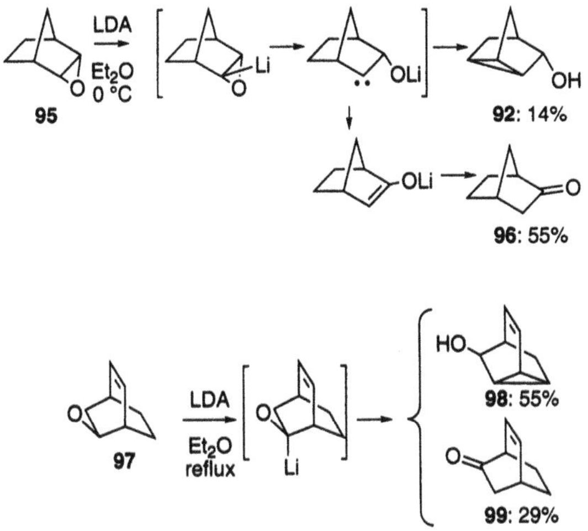

Scheme 17.

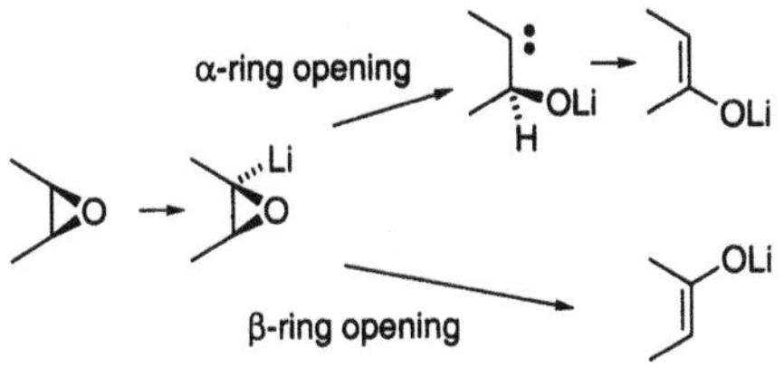

Scheme 18.

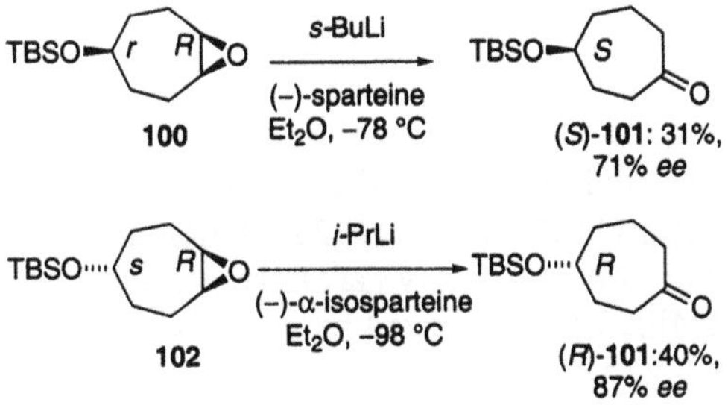

Scheme 19.

(*S*)- and (*R*)-**101**, respectively (Scheme 19) [71]. Ketone (*R*)-**101** is a known intermediate in the synthesis of (*S*)-physoperuvine.

Deprotonation of **100** and **102** in presence of the sparteines could be assumed to occur at the *R*-configured epoxide terminus (as observed in all medium ring systems, see Scheme 14); if this is the case, the above results indicate that the base-induced rearrangement of epoxides **100** and **102** to ketones (*S*)- and (*R*)-**101** proceeds solely by α-ring opening.

4.3
Electrophile Trapping of Lithiated Epoxides

Although electrophile trapping of lithiated epoxides is clearly not a rearrangement process, it is included here to enable comparison with the other asymmetric processes of lithiated epoxides previously discussed in this chapter. The development of α-lithiated epoxides as nucleophiles has mainly relied on the presence of anion-stabilising groups on the lithiated carbon. Starting with an enantioenriched epoxide usually leads to an enantioenriched subsituted epoxide (see examples below). Enantioselective generation and electrophile trapping of a lithiated epoxide is a recent development, covered toward the end of this section.

4.3.1
Lithiated Epoxides Bearing Anion-Stabilising Groups

It has recently been shown that an α-trifluoromethyl epoxide, (*S*)-**103**, can be deprotonated, the electron-withdrawing effect of the trifluoromethyl group promoting formation (and stabilisation) of the intermediate lithiated epoxide **104**, which can subsequently be trapped with a range of electrophiles: carbonyl, triphenylsilyl, triphenyltin and methyl compounds (Scheme 20) [72]. Lithiation-substitution proceeds without loss of configuration.

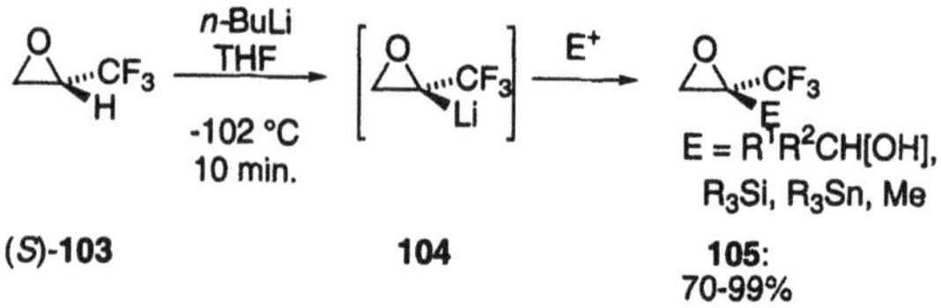

CF_3, H; *n*-BuLi, THF, -102 °C, 10 min.; CF_3, Li; E^+; CF_3, E

E = R^1R^2CH[OH], R_3Si, R_3Sn, Me

(*S*)-**103** **104** **105**: 70-99%

Scheme 20.

An optically active silyl-stabilised lithiated epoxide has been used in the synthesis of (+)-cerulenin **111** (Scheme 21) [73]. Starting from the epoxysilane **107**, obtained in high enantiomeric purity via Sharpless epoxidation of **106**, deprotonation by *s*-BuLi·TMEDA was followed by electrophile trapping using (4*E*,7*E*)-nonadienal **109** to allow a concise preparation of **110**, which is a key intermediate in the (+)-cerulenin synthesis.

The first example of enantioselective deprotonation-electrophile trapping of a *meso*-epoxide is shown in Eq. (31) [37, 74]. This reaction was favoured by both the stabilisation of the lithiated epoxide by the adjacent phenyl group, and the use of TMSCl as an internal electrophile.

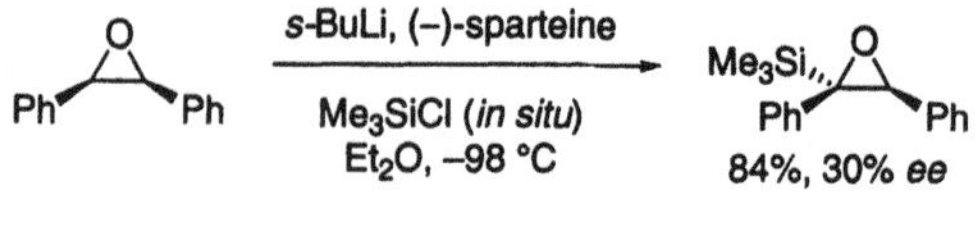

Ph, Ph; *s*-BuLi, (–)-sparteine; Me_3SiCl (*in situ*), Et_2O, –98 °C; Me_3Si, Ph, Ph; 84%, 30% *ee* (31)

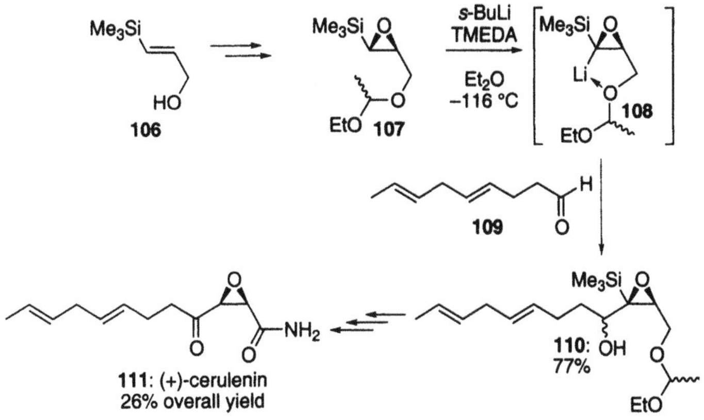

Scheme 21.

Florio et al. have recently reported that benzylic deprotonation of optically active styrene oxides and subsequent trapping of the corresponding α-lithiated epoxides provides a stereospecific route to substituted styrene oxides [Eq. (32)] [75].

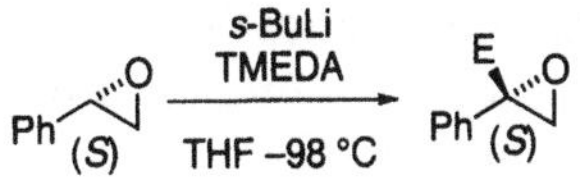

(32)

4.3.2 *Unstabilised and Destabilised Lithiated Epoxides*

Recently, hydrogen-metal exchange has proven to be useful for the preparation and electrophile trapping of non-stabilised (H-substituted) and destabilised lithiated epoxides. Treatment of terminal epoxides **112** with a diamine·*s*-BuLi complex leads to the desired lithiated epoxides **113** (Scheme 22). Deprotonation was observed to occur at the methylene position, the proton abstracted being *trans* to the bulkiest substituent of the oxirane. The lithiated epoxide so generated was stable enough to react with an external deuterium source or TMSCl in situ to give **114**; other external electrophiles did not allow the trapping of the preformed anions [76].

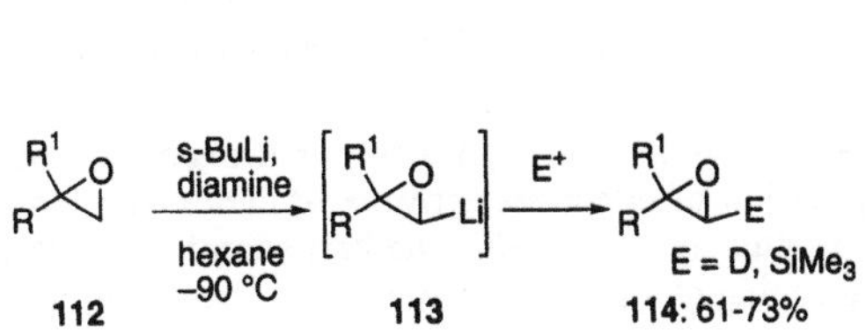

Scheme 22.

The above methodology nevertheless required the use of very specific diamines containing the diazabicyclononane structural motif (shown in bold in Fig. 4).

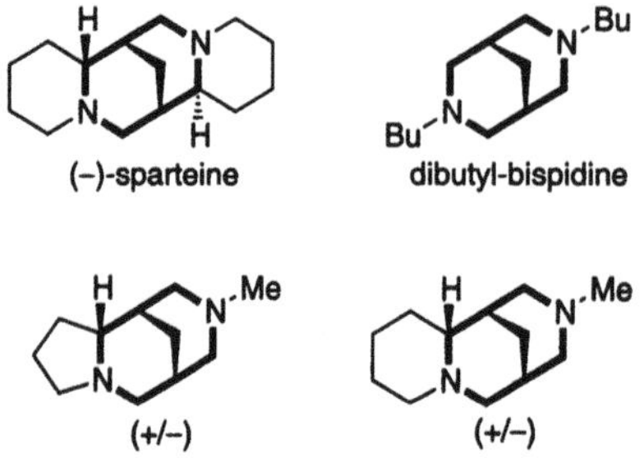

(–)-sparteine

dibutyl-bispidine

(+/–)

(+/–)

Fig. 4.

Very recently, deprotonation-electrophile trapping of "simple" *meso*-epoxides was applied to medium-sized *meso*-cycloalkene oxides [77]. When cyclooctene oxide **62** is treated with *s*-BuLi in the presence of a diamine at –90 °C, the lithiated epoxide is stable enough to be trapped with a wide range of electrophiles, allowing the creation of carbon-heteroatom or carbon-carbon bonds [Eq. (33)]. The deprotonation is a symmetry-breaking step, and enantioselective deprotonation was successfully achieved in the presence of (–)-sparteine, leading to a range of enantioenriched functionalised epoxides **115** (in up to 86% ee).

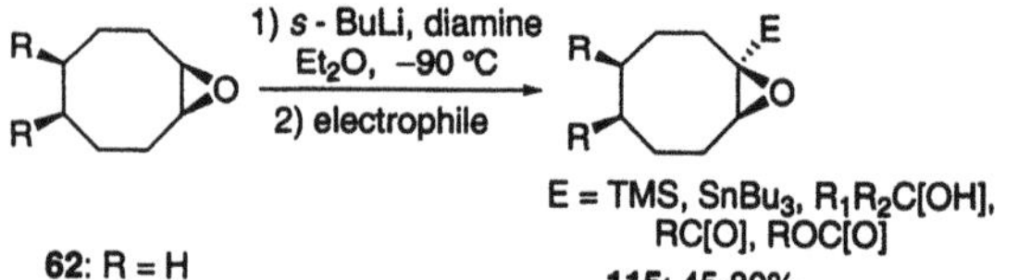

1) *s*-BuLi, diamine Et_2O, –90 °C; 2) electrophile (33)

62: R = H
76: R = OTBS

E = TMS, $SnBu_3$, $R_1R_2C[OH]$, RC[O], ROC[O]
115: 45-80%

5 Concluding Remarks

The enantioselective Wittig and epoxide rearrangements have been widely explored in the last few years. They are of interest since such processes can, in a single step, induce asymmetry and molecular complexity that would be difficult to generate otherwise. Advances in substrate scope, reaction efficiency and catalytic process constitute significant challenges for the future.

Note added in proof. A total synthesis of (–)-xialenon A using bicyclic alcohol **77** [Eq. (26)] represents the first application of enantioselective α-lithiation transannular C-H insertion of epoxides in natural product synthesis [78]. The first enantioselective generation – intermolecular nucleophile trapping of a lithium carbenoid has been described, via enantioselective α-deprotonation of the epoxides of oxa- and aza-bicyclic alkenes [e.g. **93** Eq. (30)] using alkyllithiums in the presence of (–)-sparteine or bisoxazoline **5** [Eq. (3)] [79]. Insertion of a second equivalent of the organolithium into the lithiated epoxide, followed by elimination of the oxa- or aza-bridge leads to cyclic unsaturated diols and amino alcohols in a regio-, stereo- and enantiocontrolled fashion (up to 87% ee).

References

1. Wittig G, Löhmann L (1942) Liebigs Ann Chem 550:260
2. (a) Schöllkopf U (1970) Angew Chem Int Ed Engl 9:763; (b) Marshall JA (1991) In: Trost BM, Fleming I (eds), Comprehensive organic synthesis. Pergamon Press, London, Vol. 3, p 975; (c) Tomooka K, Nakai T (1996) J Synth Org Chem Jpn 54:1000; (d) Tomooka K, Yamamoto H, Nakai T (1997) Liebigs Ann Recl 1275
3. Mazaleyrat J-P, Welvart Z (1983) Nouv J Chim 7:491
4. (a) Tomooka K, Yamamoto K, Nakai T (1999) Angew Chem Int Ed Engl 38:3741; (b) Tomooka K (2001) J Synth Org Chem Jpn 59:322
5. Rautenstrauch V, Büchi G, Wuest H (1974) J Am Chem Soc 96:2576
6. Miyata O, Koizumi T, Ninomiya I, Naito T (1996) J Org Chem 61:9078
7. (a) Hoppe D (1984) Angew Chem Int Ed Engl 23:932; (b) Zhang P, Gawley RE (1993) J Org Chem 58:3223; (c) Superchi S, Sotomayor N, Miao G, Joseph B, Campbell MG, Snieckus V (1996) Tetrahedron Lett 37:6061
8. Tomooka K, Shimizu H, Inoue T, Shibata H, Nakai T (1999) Chem Lett 759
9. Tomooka K, Igarashi T, Nakai T (1994) Tetrahedron 50:5927
10. (a) Nakai T, Mikami K (1994) Org React 46:105; (b) Nakai T, Tomooka K (1997) Pure Appl Chem 69:595
11. (a) Marshall JA, Lebreton J (1987) Tetrahedron Lett 28:3323; (b) Marshall JA, Lebreton J (1988) J Am Chem Soc 110:2925
12. Marshall JA, Lebreton J (1988) J Org Chem 53:4108
13. Gibson SE, Ham P, Jefferson GR (1998) Chem Commun 123
14. Kang J, Cho WO, Cho HG, Oh HJ (1994) Bull Korean Chem Soc 15:732
15. (a) Kawasaki T, Kimachi T (1998) Synlett 1429; (b) Kawasaki T, Kimachi T (1999) Tetrahedron 55:6847
16. Tsubuki M, Kamata T, Nakatani M, Yamazaki K, Matsui T, Honda T (2000) Tetrahedron Asymmetry 11:4725
17. Capriati V, Florio S, Ingrosso G, Granito C, Troisi L (2002) Eur J Org Chem 478
18. Manabe S (1998) Chem Pharm Bull 46:335
19. Manabe S (1997) Chem Commun 737
20. (a) Tomooka K, Komine N, Nakai T (1998) Tetrahedron Lett 39:5513; (b) Tomooka K, Komine N, Nakai T (2000) Chirality 12:505
21. Felkin H, Tambuté A (1969) Tetrahedron Lett 821
22. (a) Felkin H, Frajerman C (1977) Tetrahedron Lett 3485; (b) Briére R, Chérest M, Felkin H, Frajerman C (1977) Tetrahedron Lett 3489
23. (a) Tomooka K, Yamamoto H, Nakai T (2000) Angew Chem Int Ed Engl 39:4500; (b) Tanabe Y, Tomooka K (2001) Annual Meeting of the Chemical Society of Japan, Tokyo, Japan
24. Kikuchi T, Seki N, Endou K, Tanabe Y, Tomooka K (2002) Annual Meeting of the Chemical Society of Japan, Tokyo, Japan
25. (a) Colvin EW (1981) Silicon in organic synthesis, Butterworths, London; (b) Brook MA (2000) Silicon in organic, organometallic, and polymer chemistry, Wiley, New York
26. Kleinfeld SH, Wegelius E, Hoppe, H (1999) Helv Chim Acta 82:2413
27. Hammerschmidt F, Völlenkle H (1986) Liebigs Ann Chem 2053
28. (a) Hammerschmidt F, Hanninger A (1995) Chem Ber 128:823; (b) Hammerschmidt F, Schmidt S (2001) Phosphorus Sulfur Silicon Relat Elem 174:101
29. Satoh T (1996) Chem Rev 96:3303
30. Hodgson DM, Gras E (2002) Synthesis 1625
31. Erden I (1996) In: Katritzky AR, Rees CW, Scriven EFV (eds), Comprehensive heterocyclic chemistry II, vol 1A. Pergamon, Oxford p 97

32. Jacobsen EN (2000) Acc Chem Res 33:421
33. Schaus SE, Brandes BD, Larrow JF, Tokunaga M, Hansen KB, Gould AE, Furrow ME, Jacobsen EN (2002) J Am Chem Soc 124:1307
34. Crandall JK, Apparu M (1983) Org React 29:345
35. Seebach D (1988) Angew Chem Int Ed Engl 27:1624
36. Collum DB (1993) Acc Chem Res 26:227
37. Hodgson DM, Gibbs AR, Lee GP (1996) Tetrahedron 52:14361
38. Hodgson DM, Gibbs AR, Drew MGB (1999) J Chem Soc Perkin Trans 1 3579
39. Denmark SE, Barsanti PA, Wong KT, Stavenger RA (1998) J Org Chem 63:2428
40. Magnus A, Bertilsson SK, Andersson PG (2002) Chem Soc Rev 31:223
41. Cope AC, Tiffany BD (1951) J Am Chem Soc 73:4158
42. Morgan KM, Gajewski JJ (1996) J Org Chem 61:820
43. Eisch JJ, Galle JE (1976) J Am Chem Soc 98:4646
44. Boche G, Lohrenz JCW (2001) Chem Rev 101:697
45. Morgan KM, Gronert S (2000) J Org Chem 65:1461
46. Ramirez A, Collum DB (1999) J Am Chem Soc 121:11114
47. Cope AC, Trumbull PA, Trumbull ER (1958) J Am Chem Soc 80:2844
48. Cope AC, Lee H-H, Petree HE (1958) J Am Chem Soc 80:2849
49. Hodgson DM, Lee GP, Marriott RE, Thompson AJ, Wisedale R, Witherington J (1998) J Chem Soc Perkin Trans 1 2151
50. Boeckman RKJ (1977) Tetrahedron Lett 49:4281
51. Hoppe D, Hense T (1997) Angew Chem Int Ed Engl 36:2282
52. Hodgson DM, Lee GP (1996) Chem Commun 1015
53. Beak PJ, Kerrick ST, Wu S, Chu J (1994) J Am Chem Soc 116:3231
54. Hodgson DM, Lee GP (1997) Tetrahedron Asymmetry 8:2303
55. Gallagher DJ, Wu S, Nikolic NA, Beak PJ (1995) J Org Chem 60:8148
56. Zschage O, Hoppe D (1992) Tetrahedron 48:5657
57. Kang J, Cho WO, Cho HG (1994) Tetrahedron Asymmetry 5:1347
58. Kang J, Kim JI, Lee JH (1994) Bull Kor Chem Soc 15:865
59. Alexakis A, Vrancken E, Mangeney P (2000) J Chem Soc Perkin Trans 1 3354
60. Alexakis A, Vrancken E, Mangeney P (1998) Synlett 1165
61. Hodgson DM, Cameron ID (2001) Org Lett 3:441
62. Hodgson DM, Cameron ID, Christlieb M, Green R, Lee GP, Robinson LA (2001) J Chem Soc Perkin Trans 1 2161
63. Hodgson DM, Robinson LA (1999) Chem Commun 309
64. Hodgson DM, Wisedale R (1996) Tetrahedron Asymmetry 7:1275
65. Hodgson DM, Maxwell CR, Matthews IR (1999) Tetrahedron Asymmetry 10:1847
66. Hodgson DM, Maxwell CR, Matthews IR (1998) Synlett 1349
67. Hodgson DM, Maxwell CR, Wisedale R, Matthews IR, Carpenter KJ, Dickenson AH, Wonnacott S (2001) J Chem Soc Perkin Trans 1 3150
68. Hodgson DM, Marriott RE (1997) Tetrahedron Asymmetry 8:519
69. Thies RW, Chiarello RH (1979) J Org Chem 44:1342
70. Yanagisawa A, Yasue K, Yamamoto H (1994) J Chem Soc Chem Commun 2103
71. Hodgson DM, Robinson LA, Jones ML (1999) Tetrahedron Lett 40:8637
72. Yamauchi Y, Katagiri T, Uneyama K (2002) Org Lett 4:173
73. Mani NS, Townsend CA (1997) J Org Chem 62:636
74. Witherington J (1994) PhD thesis, University of Reading
75. Capriati V, Florio S, Luisi R, Salomone A (2002) Org Lett 4:2445
76. Hodgson DM, Norsikian SLM (2001) Org Lett 3:461
77. Hodgson DM, Gras E (2002) Angew Chem Int Ed 41:2376
78. Hodgson DM, Galano JM, Christlieb M (2002) Chem Commun 2436
79. Hodgson DM, Maxwell CR, Miles TJ, Paruch E, Stent MAH, Matthews IR, Wilson FX, Witherington J (2002) Angew Chem Int Ed 41:4313

Topics Organomet Chem (2003) 5: 251–286
DOI 10.1007/b10338

Enantioselective Synthesis by Lithiation to Generate Planar or Axial Chirality

Jonathan Clayden

Department of Chemistry, University of Manchester, Oxford Road, Manchester M13 9PL, UK.
E-mail: j.p.clayden@man.ac.uk

Enantioselective lithiation can be used to introduce and control new elements of planar or of axial chirality. The principal classes of compounds displaying these stereochemical features are ferrocenes, arenechromium tricarbonyl complexes, biaryls, atropisomeric amides and allenes. Methods for the enantioselective lithiation of these compound classes, relying on either substrate (chiral auxiliary) or reagent (chiral base) control will be reviewed.

Keywords. Ferrocenes, Arenechromium tricarbonyls, Atropisomers, Chiral lithium amides, (–)-Sparteine

1 Generating Planar Chirality by Lithiation

Planar chiral compounds usually (and for the purpose of this review, always) contain unsymmetrically substituted aromatic systems. Chirality arises because the otherwise enantiotopic faces of the aromatic ring are differentiated by the coordination to a metal atom – commonly iron (in the ferrocenes) or chromium (in the arenechromium tricarbonyl complexes). Withdrawal of electrons by the metal centre means that arene-metal complexes and metallocenes are more readily lithiated than their parent aromatic systems, and the stereochemical features associated with the planar chirality allow lithiation to be diastereoselective (if the starting material is chiral) or enantioselective (if only the product is chiral).

1.1 Planar Chirality in Ferrocenes

Chiral ferrocenes form a class of powerful and useful chiral ligands, and a number of methods are available for their enantioselective synthesis. Ferrocenes are readily deprotonated by alkyllithiums (perhaps too readily – it is difficult, though possible, to control the monolithiation of ferrocene itself) and the introduction of a second substituent via directed metallation [1] is a simple and powerful way to make disubstituted ferrocenes with planar chirality from simple, achiral monosubstituted ferrocenes [2]. Chiral 1,2-disubstituted ferrocenes **1** for this reason are much more readily available than chiral 1,3-disubstituted ferrocenes **2**. Control of absolute planar stereochemistry during the lithiation of a monosubstituted ferrocene **3** may be achieved either by reagent control (using a chiral base to remove selectively only one of two enantiotopic protons H^A and H^B of **3** with a simple achiral substituent R) or by substrate control (starting with a ferrocene **3** lacking planar chirality but bearing a chiral substituent R, which directs a base towards removal of only one of two diastereotopic protons H^A and H^B).

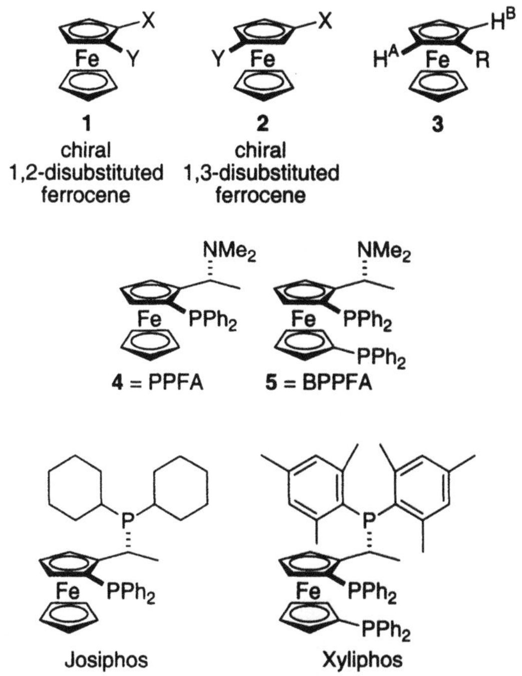

X Y Fe 1 chiral 1,2-disubstituted ferrocene
X Y Fe 2 chiral 1,3-disubstituted ferrocene
H^B H^A Fe R 3

NMe_2 Fe PPh_2 4 = PPFA
NMe_2 Fe PPh_2 PPh_2 5 = BPPFA

P Fe PPh_2 Josiphos
P Fe PPh_2 PPh_2 Xyliphos

Scheme 0.

1.1.1
Diastereoselective Lithiation of Chiral Ferrocenes

Nozaki pioneered the investigation of both routes to enantiomerically pure ferrocenes in the late 1960s [3], and after one early example of a diastereoselective acylation [4], most early attempts to control planar chirality made use of ferrocenes bearing chiral and enantiomerically pure aminoalkyl substituents. The amino groups could readily be made by reductive amination or substitution; they also facilitated both resolution and lithiation, acting as metallation directing groups [1]. Many ferrocenes used as chiral ligands (such as PPFA **4** and BPPFA **5**) retain these aminoalkyl substituents, and others (such as Josiphos or Xyliphos) carry groups derived from them by simple substitution chemistry [5]. More recently, in the interests of versatility, fully removable auxiliaries have been used to direct diastereoselective lithiation, including those based on oxazolines, hydrazones, acetals and sulfoxides, each of which is discussed below.

1.1.1.1
Lithiation Directed by Aminoalkyl Groups

In 1969 Nozaki and coworkers published the first diastereoselective lithiation of a ferrocene **6** (Scheme 1) [3]. Ugi improved [6] upon these reactions by shifting the stereocontrolling influence closer to the ring [7, 8]. His lithiation of **9**, which

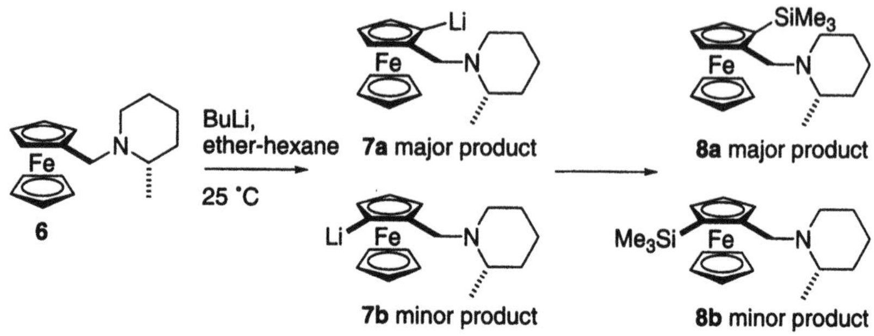

Scheme 1.

achieved 96:4 diastereoselectivity over the new chiral plane, is under kinetic control, and is governed by the relative favourability of the transition state leading to **10a**, which keeps the α-methyl group clear of the lower portion of the ferrocene sandwich (Scheme 2).

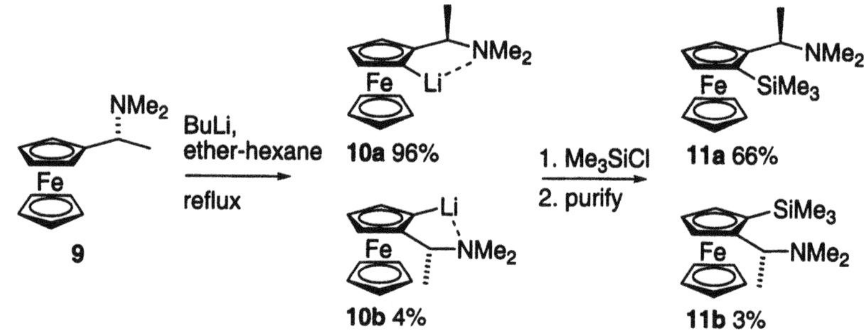

Scheme 2.

The stereochemistry of the almost diastereoisomerically pure organolithium **10** was proved by an X-ray crystal structure of its anisaldehyde adduct. Compound **10** reacts with a range of electrophiles, including, importantly, $ClPPh_2$, which generates the ligand PPFA **4** after removal of a small amount of the minor diastereoisomer by recrystallisation [9]. Addition of a second equivalent of alkyllithium can be used to form a "dianion", allowing a one-pot double functionalisation - reaction of **14** (R=Me), for example, gives the ligand BPPFA **5** (Scheme 3) [9–14].

Despite the resolution which was required to produce the enantiomerically pure starting materials (which fortunately is highly efficient - recrystallisation of the mother liquors allows isolation of both enantiomers) [15], Ugi's lithiation provided the basis for a rapid growth in the use of enantiomerically pure, planar chiral ferrocenes which has continued since. Several reviews [5] have covered applications of enantiomerically pure ferrocenes as chiral ligands, which until the 1990s were all made using Ugi's method [16].

The diastereoselectivity of Ugi's lithiation can be improved upon by replacing the methyl group α to nitrogen by isopropyl [17]. Complete (>99:1 selectivity)

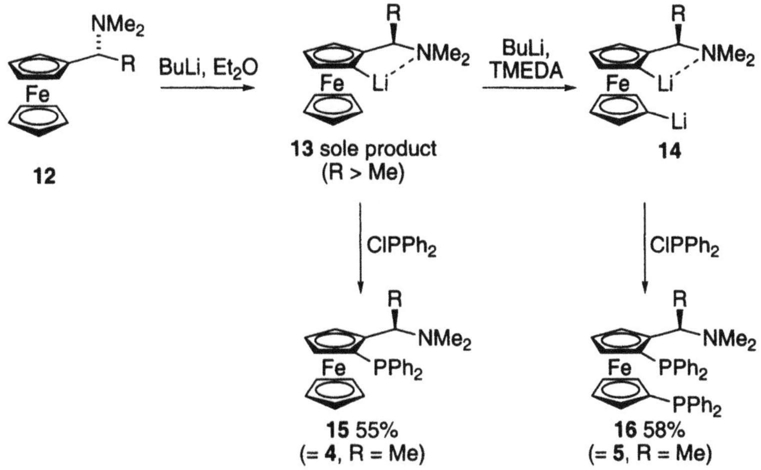

Scheme 3.

is observed in the lithiation of **12** (R=*i*-Pr). Attempts to lithiate compounds **12** with even bigger α-substituents (and avoid the need for a resolution in the synthesis of the enantiomerically pure starting material) failed.

The resolution required for the synthesis of **9** or **12** can be avoided by making them by asymmetric reduction [18] or by asymmetric alkylation of an aldehyde [19]. The amines **12** (R=Et or *n*-Bu) formed in this way are lithiated with diastereoselectivity similar to, or greater than, that achieved with **9**. α-Ethyl- and α-butylphosphines **15** incidentally may show even higher selectivity than the more widely used α-methylphosphine ligand PPFA **4**.

1.1.1.2
Lithiation Directed by Acetals

Although much mileage has been made out of ligands which retain the stereogenic centre of the aminoalkyl side-chain in compounds derived from **10** and **13**, more versatile methods based on removable auxiliaries are clearly desirable. The first of these to be reported was Kagan's use of the acetal **18** [20], made by condensing aldehyde **17** with a malic acid-derived triol followed by *O*-methylation (Scheme 4). Lithiation of **18** with *t*-BuLi in hexane-ether at –78 °C (higher temperatures yield lower selectivities, though completion of the deprotonation at 25 °C is necessary for good yields) gives organolithium **19** with complete diastereoselectivity, probably via a transition state involving chelation of Li by the OMe group [20, 21]. The reaction of this organolithium with electrophiles yields products such as **20** with >99:1 diastereoselectivity and in 80–90% yield. Acid hydrolysis allows clean removal of the diol auxiliary, allowing the synthesis of versatile formylferrocenes such as **21** of ee=95% with planar chirality only.

Attempts to make C_2-symmetric ferrocenes by double lithiation of a bis-acetal met with only limited success [22]. A second lithiation of the ferrocenylacetal **19** leads to functionalisation of the lower ring of the ferrocene, in contrast

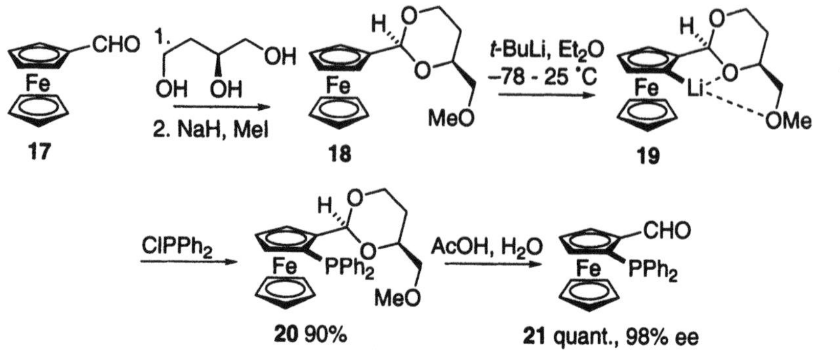

Scheme 4.

with the second adjacent lithiation of the oxazolines described below. This can be used to advantage if, for example, the first-formed aldehyde **22** is protected in situ by addition of the lithiopiperazine **23** [23], directing *t*-BuLi to the lower ring (Scheme 5) [24]. The same strategy can be used to introduce further functionalisation to products related to **24**. For example, silane **25**, produced in enantiomerically pure form by the method of Scheme 4, may be converted to the ferrocenophane **26** by lithiopiperazine protection, lithiation and functionalisation (Scheme 6) [24].

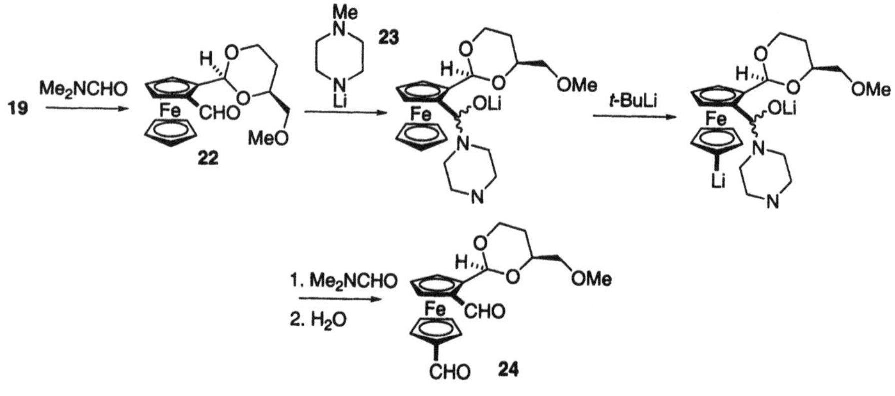

Scheme 5.

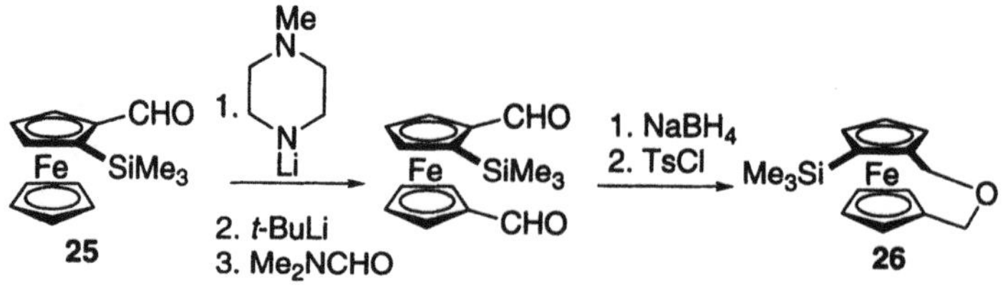

Scheme 6.

1.1.1.3
Lithiation Directed by Oxazolines

Early in 1995, the groups of Richards [25], Uemura [26] and Sammakia [27] simultaneously published results showing that chiral oxazolines attached to aromatic rings are capable of directing diastereoselective lithiation. The best conditions for the diastereoselective lithiation of **27**, derived from either valine or *tert*-leucine, are *s*-BuLi in TMEDA-hexane at −78 °C. Organolithium **28** is formed with a diastereoselectivity of >500:1 [28], and electrophilic quench gives a single diastereoisomer of the silane **29** (Scheme 7). Similar results have been obtained with related oxazolines bearing a further chiral substituent [29].

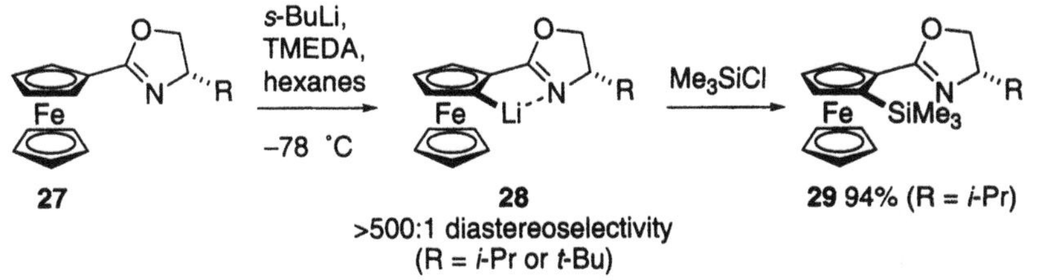

Scheme 7.

Lithiation of the conformationally constrained ferrocenyloxazoline **30** (Scheme 8) provided useful information about the mechanism by which lithiation of **27** achieves diastereoselectivity [30]. The product **32** must have arisen from an organolithium **31** in which the oxazoline nitrogen coordinates to the lithium (Scheme 8). It is likely that stereoselectivity in the lithiation of **27** results from a similar intermediate **28**, in which the oxazoline R group surprisingly points towards the ferrocene nucleus. Presumably, while this is not the ground state of the starting material, lithiation in this conformation is faster since the organolithium's approach is unhindered.

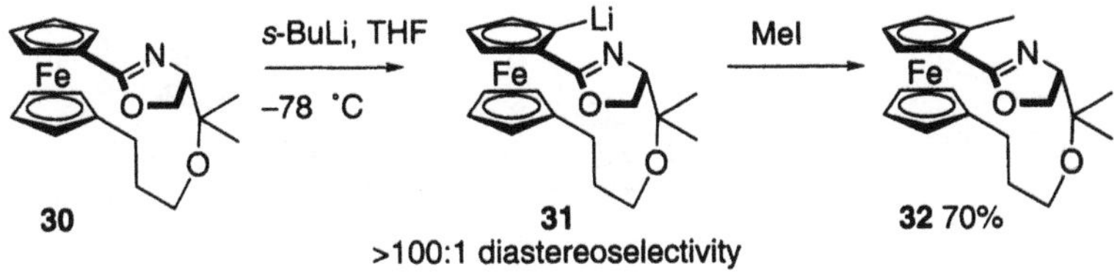

Scheme 8.

Although oxazolines can be used as auxiliaries and later removed [27, 31], they have also been retained in target molecules which have then been used as ligands for a variety of asymmetric transformations. Ferrocenes carrying oxazoline and phosphine coordination sites [25, 26, 32–34], oxazoline and amine coordination sites [35], and ferrocene bis-oxazolines [36–38] have been synthesised by the method of Scheme 7.

A ferrocenyloxazoline with only one adjacent position available for deprotonation will lithiate at that position irrespective of stereochemistry. This means that the same oxazoline can be used to form ferrocenes with either sense of planar chirality. The synthesis of the diastereoisomeric ligands **33** and **35** illustrates the strategy (Scheme 9), which is now commonly used with other substrates to control planar chirality by lithiation (see below). Ferrocene **33** is available by lithiation of **27** directly, but diastereoselective silylation followed by a second lithiation (best carried out in situ in a single pot) gives the diastereoisomeric phosphine **35** after deprotection by protodesilylation [25, 32, 34].

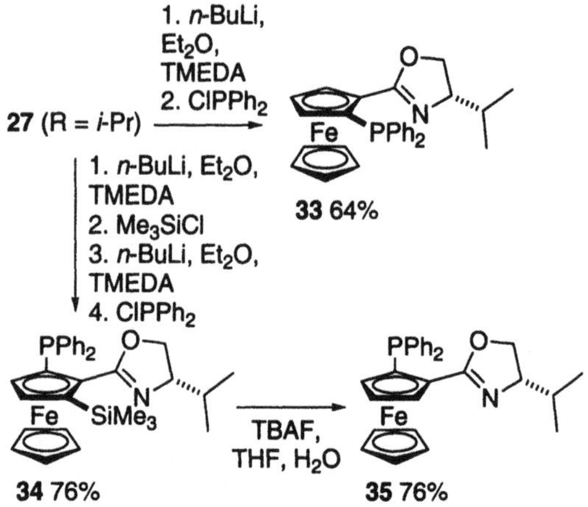

Scheme 9.

Double lithiation of bisoxazoline **36** is possible with either *s*-BuLi [38] or *t*-BuLi [36], and interestingly each base gives different planar diastereoselectivity. Double lithiation with *s*-BuLi, quenching the dianion with $ClPPh_2$, leads to the C_2-symmetric bisphosphine **37** as the major product (Scheme 10). With *t*-BuLi, the diastereoisomeric bisphosphine **38** is isolated as the sole product.

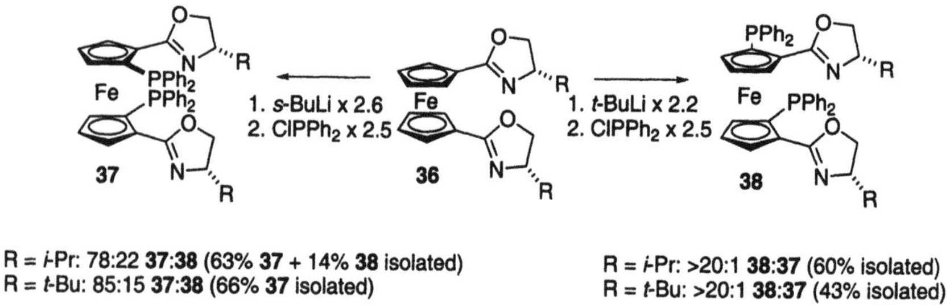

R = *i*-Pr: 78:22 **37**:**38** (63% **37** + 14% **38** isolated)
R = *t*-Bu: 85:15 **37**:**38** (66% **37** isolated)

R = *i*-Pr: >20:1 **38**:**37** (60% isolated)
R = *t*-Bu: >20:1 **38**:**37** (43% isolated)

Scheme 10.

Bis-oxazoline ligands can also be produced by oxidative coupling of the copper derivative of diastereoisomerically pure **28** (Scheme 11) [37]. Further lithiations of the product **39**, which was produced as single diastereoisomer, occur (as

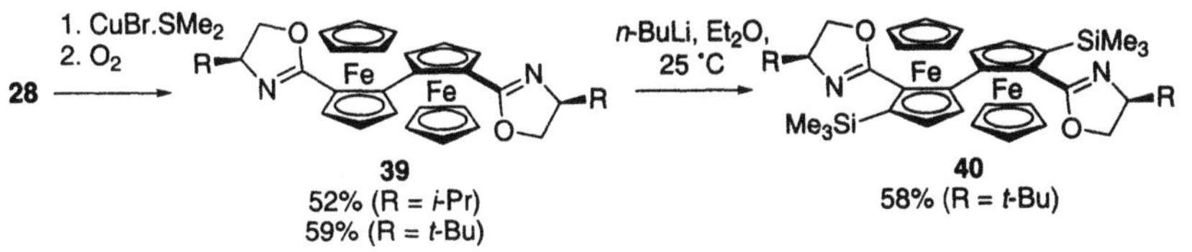

28 → (1. $CuBr.SMe_2$ 2. O_2) → **39** 52% (R = *i*-Pr) 59% (R = *t*-Bu) → (*n*-BuLi, Et_2O, 25 °C) → **40** 58% (R = *t*-Bu)

Scheme 11.

in Scheme 9) at the second site adjacent to the oxazoline, giving for example **40**, despite the (presumably) less favourable stereochemistry of the lithiation step. Bisoxazolines **40** direct the asymmetric copper-catalysed cyclopropanation of styrene using diazoacetate.

In removing the oxazoline auxiliary from the products, Richards has demonstrated the use of the diastereoselective ferrocenyloxazoline lithiation in the synthesis of conformationally constrained amino acid derivatives (Scheme 12) [31]. Amination of **27** was achieved by nitration, reducing to the amino group after removal of the oxazoline under standard conditions. Using the trick of silylating the more reactive diastereotopic site, it was possible to make either enantiomer of **43** from the same oxazoline starting material.

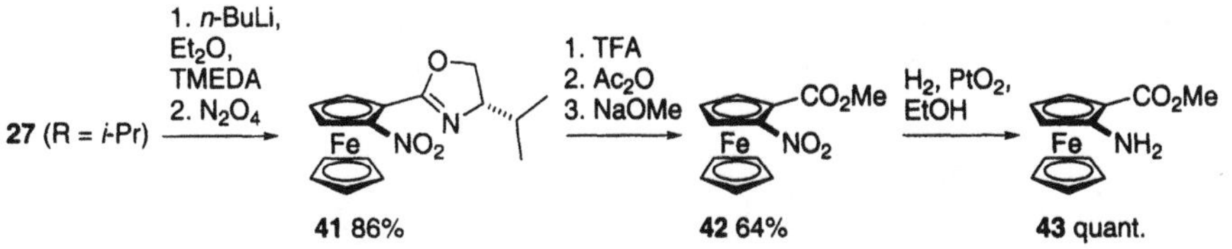

27 (R = *i*-Pr) → (1. *n*-BuLi, Et_2O, TMEDA 2. N_2O_4) → **41** 86% → (1. TFA 2. Ac_2O 3. NaOMe) → **42** 64% → (H_2, PtO_2, EtOH) → **43** quant.

Scheme 12.

1.1.1.4
Lithiation Directed by Other Diamine and Amino Alcohol Derivatives

A number of alternatives to oxazolines based on other diamine or amino alcohol derivatives have been proposed, and in several cases good control over planar stereoselectivity can be achieved. The earliest, published before any work on ferrocenyloxazolines, drew on Nozaki's early studies [3,39] on stereoselective lithiation of aminomethylferrocenes, and made use of the proline-derived amino ether **45** (Scheme 13) [40]. Substitution of **44** gave **46**, whose lithiation proceeded with 93:7 stereoselectivity with *n*-BuLi in ether at −78 °C and with 99:1 stereoselectivity with *s*-BuLi in ether at −78 °C. Reaction of **47** with $ClPPh_2$ gave **48**; the enantiomer is available by silylation, re-lithiation, phosphination and deprotection in the manner of Scheme 9. Removal of the prolinol auxiliary is achieved by acetylation and hydrolysis to **49** [41].

The same proline-derived ring system features in Enders' RAMP and SAMP chiral hydrazone auxiliaries, and Enders [42–45] has shown than RAMP and

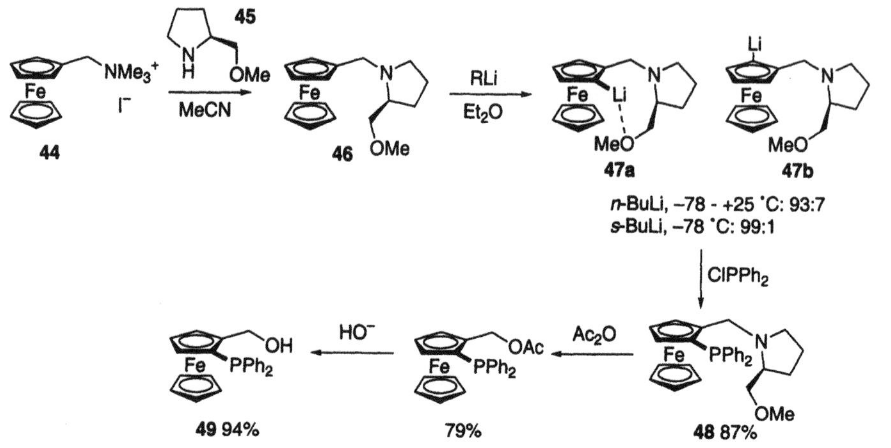

Scheme 13.

SAMP hydrazones derived from acylferrocenes will direct the diastereoselective lithiation of the ferrocene ring (Scheme 14). For example, treating benzoylferrocene **50** with (*S*)-*N*-amino-*O*-methylprolinol **51** generates **52** after silica-catalysed equilibration to the more stable *E*-hydrazone [42]. Lithiation of **52** gives organolithium **53** selectively: after electrophilic quench, products **54** were obtained in 80–95% yield and with <2% of a minor diastereoisomer. Removal of the auxiliary is achieved by oxidative or reductive means.

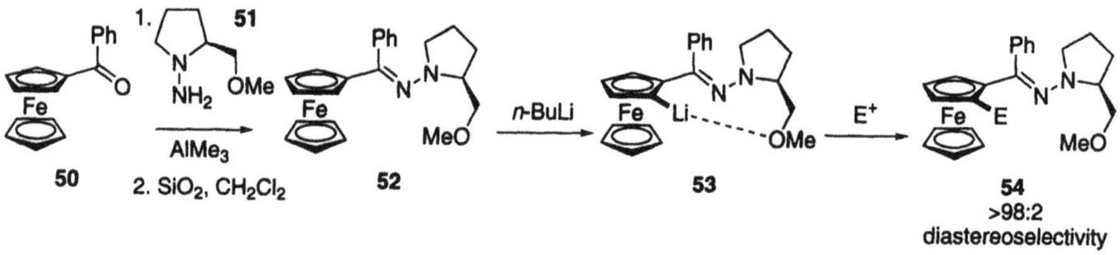

Scheme 14.

Similar chemistry is possible starting from hydrazones bearing acidic α protons: an initial diastereoselective enolisation and electrophilic functionalisation of the hydrazone can be followed by derivatisation which is stereoselective in the planar sense [43, 44].

An acyclic methoxyamine, *O*-methylephedrine **55**, fulfils a similar role in the diastereoselective lithiation of **56**. Substitution with **55** gives **56** from **44**, and lithiation of **56** typically gives products **57** with >90:10 diastereoselectivity (Scheme 15) [46]. Auxiliary removal is possible by quaternisation and substitution with dimethylamine (giving **58**) or other nucleophiles, producing potential ligands with planar chirality only. Usefully, the products formed by this method are enantiomeric with those formed by most other auxiliary methods when the more readily available enantiomers of the starting materials are used.

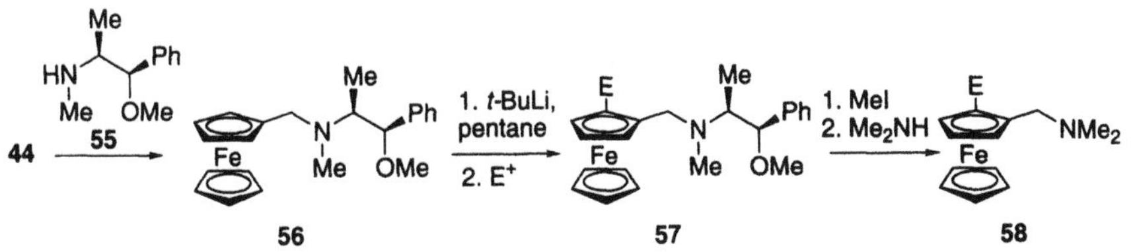

Scheme 15.

Aminomethylferrocene **60**, this time without further methoxy substituents, also lithiates diastereoselectively (Scheme 16) [47, 48], and similar results may be obtained with simple chiral pyrrolidines [49]. Treatment of **44** with the binaphthylamine **59** yields **60**. Lithiation generates a 9:1 mixture of diastereoisomeric organolithiums, which give the phosphine **61** (Scheme 16) [47, 48]. Attempts to obtain reversed planar diastereoselectivity by using silylation to block the more reactive lithiation site failed.

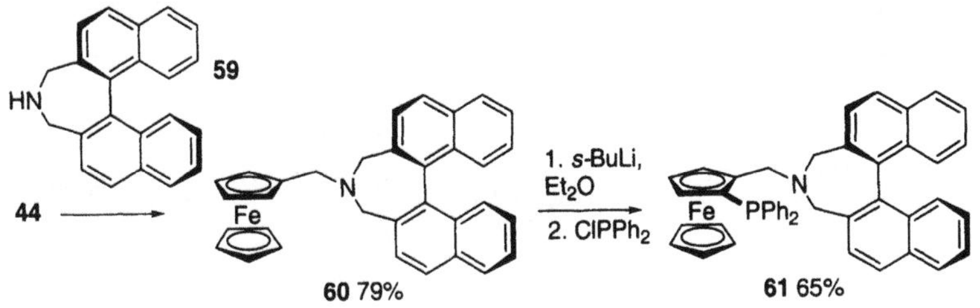

Scheme 16.

Some success has been achieved by a method which bridges the gap between the use of a chiral base and the use of an auxiliary to functionalise ferrocenes stereoselectively. Double protection of the dialdehyde **62** by addition of the chiral amine **63**, and lithiation with *t*-BuLi, gives a mixture of lithiated species which leads to **64** and **65**, in proportions dependent on the amount of *t*-BuLi employed (Scheme 17). Though the yields of each are not high, both products **64** and **65** are obtained in very high enantiomeric excess [24].

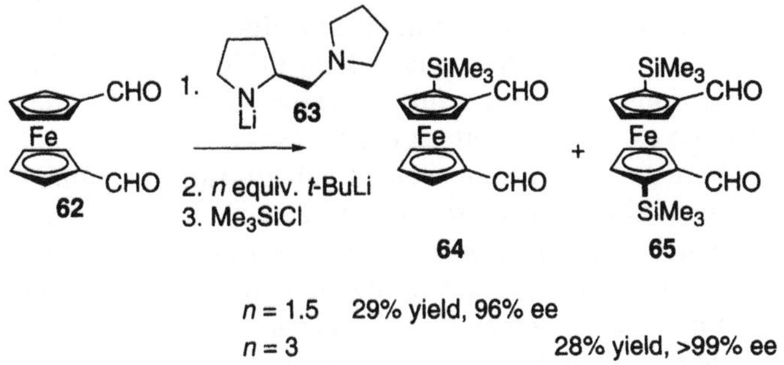

Scheme 17.

1.1.1.5
Lithiation Directed by Sulfoxides and Sulfoximines

Probably the most versatile of all the auxiliary methods for asymmetric functionalisation of ferrocenes is based upon sulfoxides. First reported by Kagan in 1993 [50], the strategy (Scheme 18) entails the formation of a ferrocenyl sulfoxide in enantiomerically pure form and its diastereoselective lithiation. For example, the *t*-butylferrocenyl sulfoxide **69** may be formed either with full enantiomeric purity by stereospecific substitution of the sulfite **68** or in 90% ee by Sharpless-Kagan oxidation of the sulfide **67**. Treatment of the sulfoxide **69** with *n*-BuLi at 0 °C leads to diastereoselective lithiation with >98:2 selectivity, and subsequent additions of electrophiles, such as MeI, produces diastereoisomer **71** as the major product [50–52]. Evidence from the crystal structure of the product suggests that the lithiation is directed by the orientation of the sulfoxide oxygen with the *t*-butyl group pointing away from the ferrocene nucleus. The stereochemistry of the reaction is under kinetic control – higher temperatures lead to some erosion of stereoselectivity [51]. A second lithiation of **69** functionalises the second Cp ring and can give, via **72**, the diphosphine **73** [50]. The planar chiral ferrocenyl *t*-butyl sulfoxides have been used as chiral auxiliaries [51] and as chiral ligands [52].

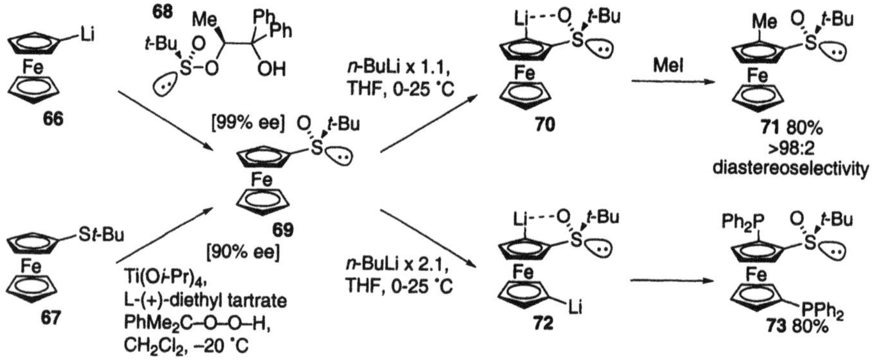

Scheme 18.

Similar deprotonations are possible using the *p*-tolyl sulfoxide **74** made by substitution of menthyl toluenesulfinate (Scheme 19) [51, 53]. This substitution is prone to loss of stereospecificity, since the product is able to undergo inversion by further reaction with another molecule of lithioferrocene, although careful control of conditions [53] ensures fully invertive substitution. The ability of the sulfur to undergo electrophilic attack enhances the value of the tolyl sulfoxides by allowing them to undergo subsequent substitution reactions by sulfoxide-lithium exchange. Lithiation (with LDA, to avoid nucleophilic substitution at sulfur) and quench of **74** gives products **75**. Treatment with *t*-BuLi then generates an organolithium which reacts quite generally with electrophiles. The sequence shown in Scheme 19 therefore allows a ferrocene to be elaborated enantioselectively with any pair of groups which can be introduced by successive electrophilic attack. Among the molecules produced in this way are the interest-

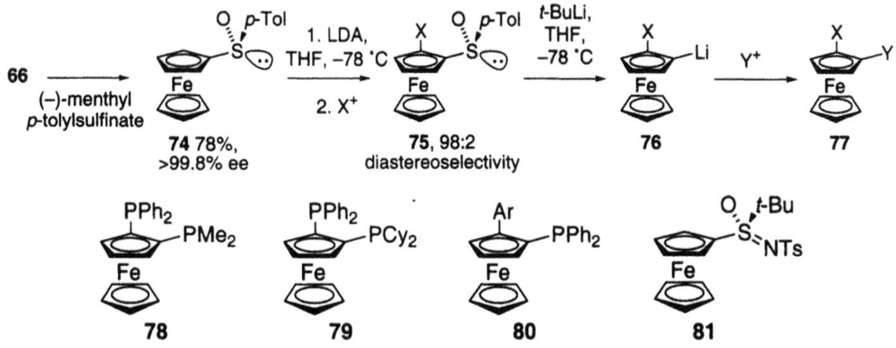

Scheme 19.

ing planar chiral bis-phosphines **78**, **79** [54] and **80** [55], which direct a number of catalytic transformations with moderate to good enantioselectivity.

Sulfoximinoferrocenes [56] and ferrocenyl sulfonates [57] can also be lithiated diastereoselectively, with the sulfoximine oxygen atom of **81** directing the deprotonation.

1.1.2
Enantioselective Lithiation of Achiral Ferrocenes

1.1.2.1
Chiral Lithium Amide Bases

The diastereoselective lithiation of **74** shows that ferrocenes bearing electron-withdrawing directors of lithiation are sufficiently acidic to allow deprotonation with lithium amide bases. By replacing LDA with a chiral lithium amide, enantioselectivity can be achieved in some cases. The phosphine oxide **82**, for example, is silylated in 54% ee by treatment with *N*-lithiobis(α-methylbenzyl)amine **83** in the presence of Me_3SiCl (Scheme 20) [58].

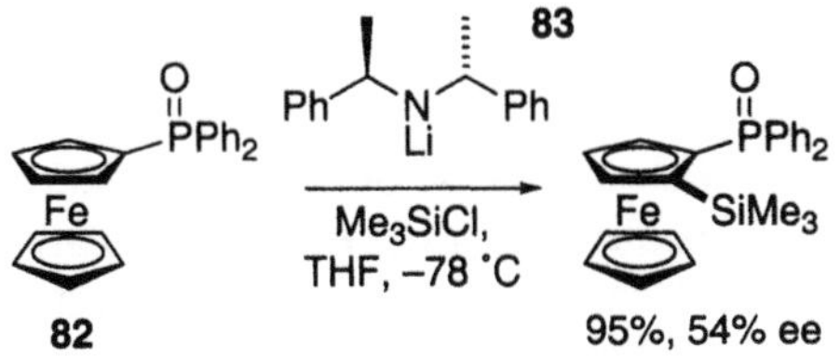

Scheme 20.

1.1.2.2
Alkyllithium-(–)-Sparteine Complexes

More generally successful however has been the use of alkyllithium/chiral tertiary amine [and in particular *s*-BuLi-(–)-sparteine] combinations [59]. One of the very first examples of asymmetric lithiation involved ferrocene: Nozaki's [3, 39] (–)-sparteine-promoted functionalisation of isopropylferrocene with moderate asymmetric induction. For useful results in this area, however, a lithiation-directing group is required. A breakthrough came when Snieckus lithiated the amide **84** with *n*-BuLi in the presence of (–)-sparteine **85** at –78 °C in Et_2O: reaction of the organolithium **86** with electrophiles generated products **87** in generally excellent yield and up to 99% ee (Scheme 21) [60, 61]. Further lithiation of **87** led to substitution on the lower, unfunctionalised ring.

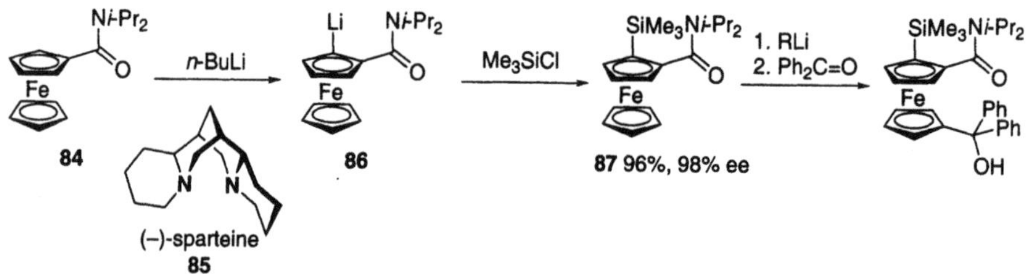

Scheme 21.

Lithiation of the dicarboxamide **88** is similarly diastereoselective, and after two successive *n*-BuLi-(–)-sparteine lithiations C_2-symmetric products such as **90** can be made with high ee (Scheme 22) [62, 63]. Attempted double lithiation of **88** in one pot fails with *n*-BuLi; with *s*-BuLi-(–)-sparteine the major product is the *meso*-isomer of **90** [63]. The amidophosphine **89** has been used as a chiral ligand for Pd chemistry; the amides can be reduced to amino groups with $BH_3 \cdot THF$ [60].

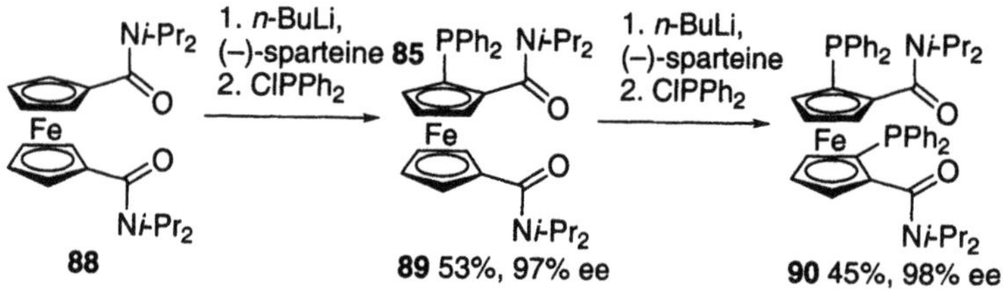

Scheme 22.

Contemporaneously with Snieckus, Uemura [64] showed that ferrocene **92** bearing an aminomethyl group may also be lithiated enantioselectively by alkyllithiums: in these cases, better results are obtained with the C_2-symmetric amine **91** than with (–)-sparteine (Scheme 23).

1.2
Planar Chirality in Arenechromium Tricarbonyl Complexes

Unlike the ferrocenes, the arene complexes of chromium, in particular the arenechromium tricarbonyls, have seen much less use in asymmetric catalysis. This is beginning to change, however [65], and a number of synthetic transfor-

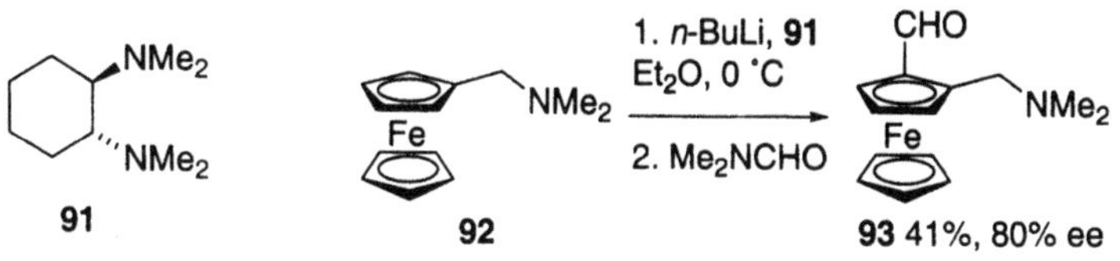

91 92 1. n-BuLi, **91** Et_2O, 0 °C 2. Me_2NCHO **93** 41%, 80% ee

Scheme 23.

mations of arenechromium tricarbonyls owe their existence to the formation of planar chiral chromium complexes by asymmetric lithiation processes [66]. The relative importance of substrate-controlled chiral auxiliary methods and reagent-controlled chiral base methods in the synthesis of planar chiral arenechromium tricarbonyls is essentially the inverse of the situation with ferrocenes: the chiral lithium amide base chemistry is much more well developed with the chromium compounds because of the greater acidity of the arene protons relative to the Cp protons of ferrocene, making them removable even by LDA [67]. We shall deal with the use of auxiliaries to direct planar stereoselective lithiation before moving on to chiral base methods.

1.2.1 *Diastereoselective Lithiation of Chiral Arenechromium Tricarbonyl Complexes*

1.2.1.1 Chiral Alkoxy, Aminoethyl and Sulfinyl Substituents

Early examples of diastereoselective lithiation of arenechromium tricarbonyls involved arenes bearing chiral amino substituents, following the precedent of Ugi's ferrocene chemistry (see above). Lithiation of **94** with t-BuLi (n-BuLi is less selective) gives organolithium **95** which may be quenched with electrophiles to give single diastereoisomers of products such as the methylated **96** in high yield (Scheme 24) [68–72]. Lithiation proceeds when the methyl group can lie *anti* to the chromium in the chelated intermediate. Additions of **95** to aldehydes proceed with diastereoselectivity at the new hydroxy-bearing centre; either diastereoisomer of the products **98** can be formed by choosing between this reaction and an alternative diastereoselective nucleophilic addition to aldehyde **97**. Interestingly, decomplexation and recomplexation provides a third diastereoisomer **98c**. Similar chemistry is possible using ephedrine-derived arenechromium tricarbonyls [73].

As a chiral auxiliary, the aminoethyl group is a poor candidate as it is hard to remove. Davies demonstrated that removable chiral alkoxy groups will direct diastereoselective lithiation provided LDA is used as the base [74] (alkyllithiums lead to substitution reactions with chromium-complexed ethers) [71]. Chromium-promoted substitution of **99** with **100** gives **101**, whose treatment with LDA gives **102** completely regio- and stereoselectively (Scheme 25). Quenching with DMF yields the aldehyde **103**. This aldehyde reacts diastereoselectively with PhMgBr, and decomplexation and Birch reduction of the auxiliary yield the otherwise difficult to obtain diol **105**. The enantiomer **105b** is available by acylation of **102** to give **106** followed by reduction to **104b** and deprotection.

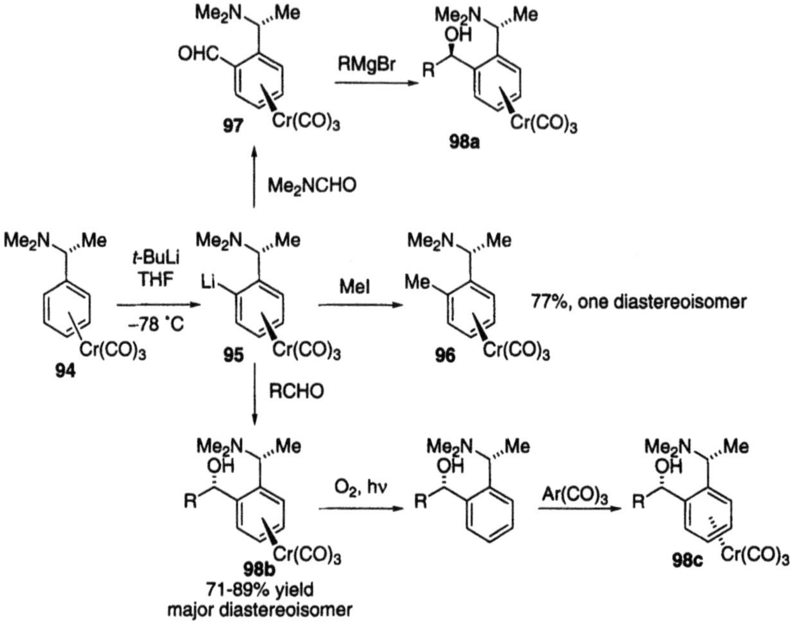

Scheme 24.

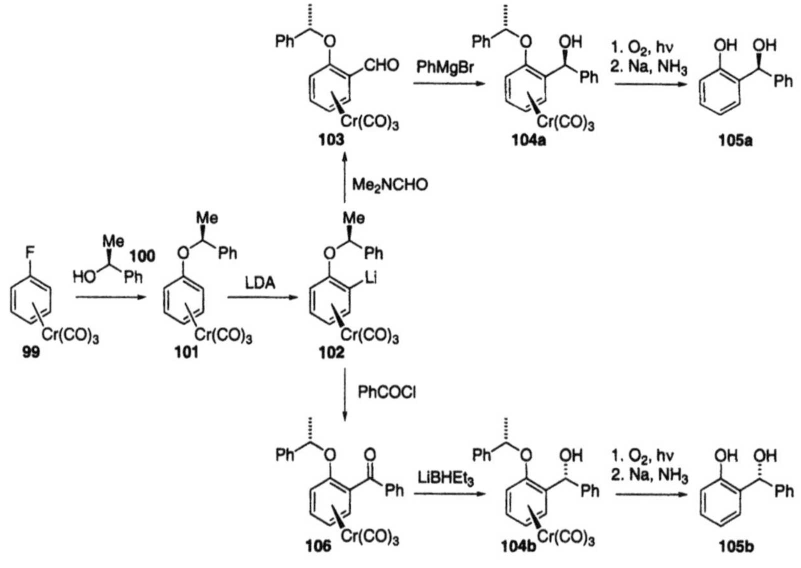

Scheme 25.

Some headway has been made using sulfoxides to direct the lithiation of arenechromium tricarbonyls in the manner of Kagan's work with ferrocenes [75, 76]. Diastereoselectivities in the lithiation-quench of **106** are excellent, though yields are poor with most electrophiles (Scheme 26). Diastereoselectivity reverses on double lithiation, because the last-formed anionic site in **108** is the most reactive.

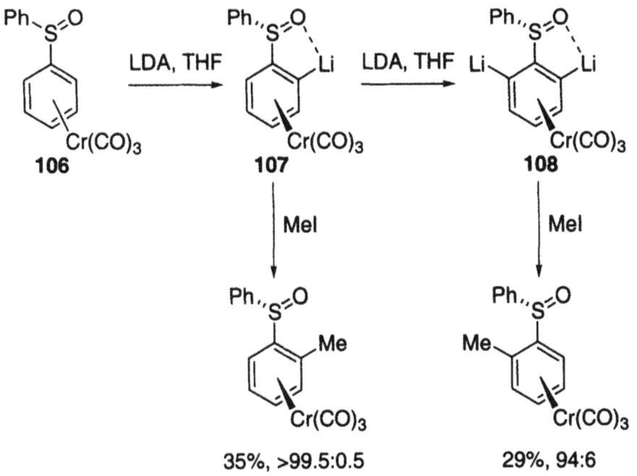

Scheme 26.

1.2.1.2
Chiral Acetals and Aminals

In 1991, Green showed that slow addition of *n*-BuLi to a solution of the tartrate-derived acetal **109** leads to diastereoselective lithiation and hence allows formation of the complexes **111** (Scheme 27) [77, 78]. Similar acetals **112** and **113** (R=H) performed much less successfully. Formation of the organolithium **110** required an excess of alkyllithium base, and its stereoselectivity appeared to be under thermodynamic control arising from equilibration of various lithiated species [78]. Unfortunately, removal of the auxiliary from **111** and its analogues proved problematic.

The malic acid-derived auxiliary which gives good results in the ferrocene series (see above) also looks promising among the chromium complexes, and the six-membered acetal of **114** is much more easily hydrolysed than the tartrate-derived acetals of Scheme 27 [79, 80]. Lithiation and bromination of **114** gives, after hydrolysis of the acetal, the complex **116** in 90% ee, increasing to >99% after recrystallisation (Scheme 28). Compound **116** is an intermediate in a formal synthesis of (–)-steganone (see Scheme 45).

Among derivatives of acetophenone, the acetal **113** (R=Me) performs the best [81]. Lithiation with *t*-BuLi in THF and electrophilic quench gives **117a** in 60–85% yield and with about 95:5 diastereoselectivity (Scheme 29). Switching to Et_2O as the solvent leads to a precipitate which reacts with completely reversed diastereoselectivity, giving **117b**.

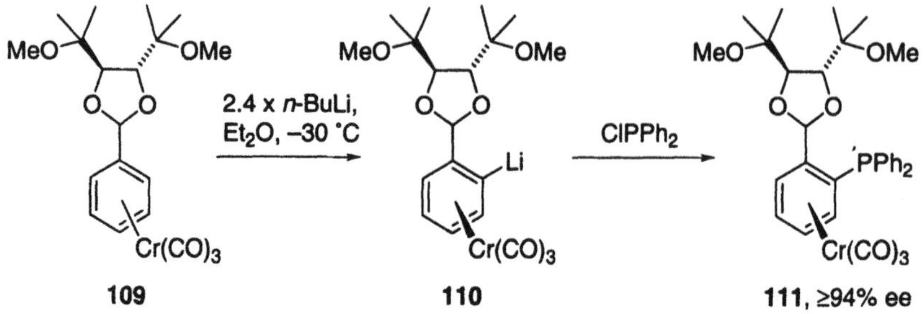

Scheme 27.

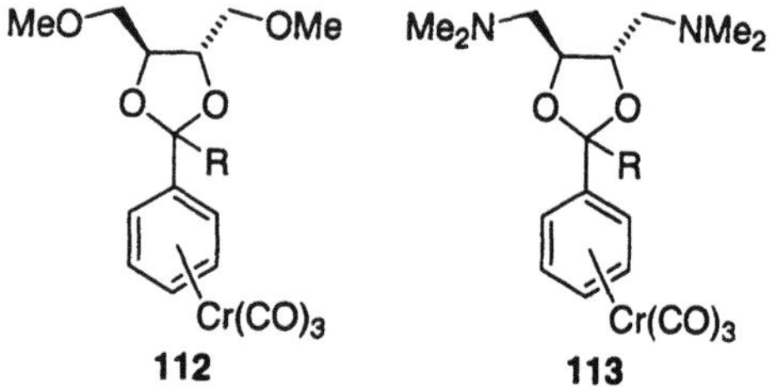

Scheme 27a.

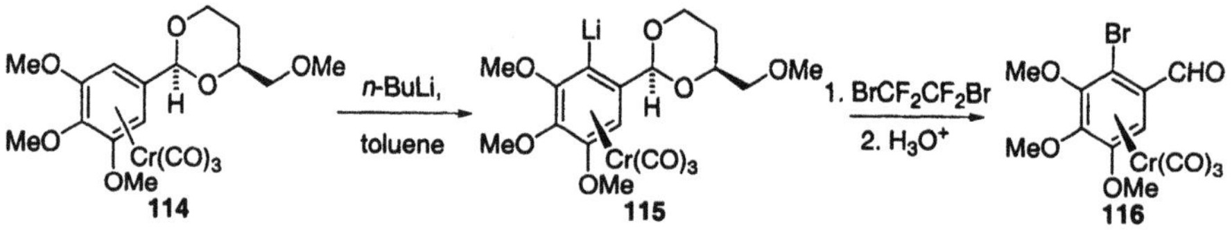

Scheme 28.

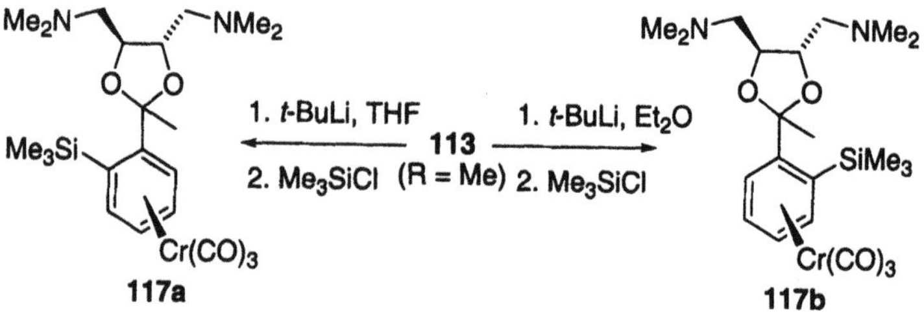

Scheme 29.

The problem of auxiliary removal is overcome when acetals are replaced with the much more labile aminals, formed by reaction of benzaldehydechromium tricarbonyl **118** with diamines, and readily cleaved with mild acid. The best choice of diamine is **119**, because the aminal **120** lithiates with good to excellent regioselectivity in the *ortho*-position (Scheme 30): the same could not be said for diamines lacking further lithium-coordinating side-chains [82]. Treatment of **120** with three equivalents of *n*-BuLi in THF at –30 °C followed by electrophilic quench with 1,2-dibromoethane, for example, yields the product **121** with no trace of its diastereoisomer (>99.5:0.5 selectivity) in 64% isolated yield.

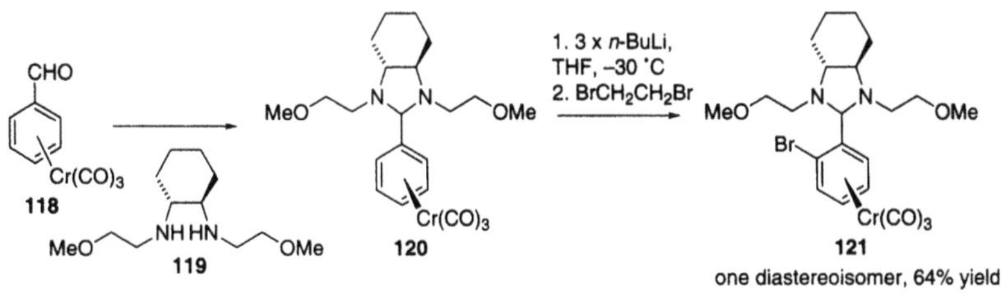

Scheme 30.

Even greater ease of deprotection is achieved with the in situ protection strategy for the lithiation of aldehydes developed by Comins [23]. Alexakis showed that a lithiodiamine **122** related to **119** achieved in situ protection of benzaldehydechromium tricarbonyl at the same time as directing regio- and stereoselective lithiation (Scheme 31) [82]. Treatment of **118** with a slight excess of the lithium amide followed by three equivalents of *n*-BuLi and then the electrophile (Me_3SiCl) gave, after work-up, the product **124** in 72% ee and 68% yield. NMR examination of the intermediate **123** suggests that stereoselectivity is controlled by diastereoselective formation of the new α-aminoalkoxide centre.

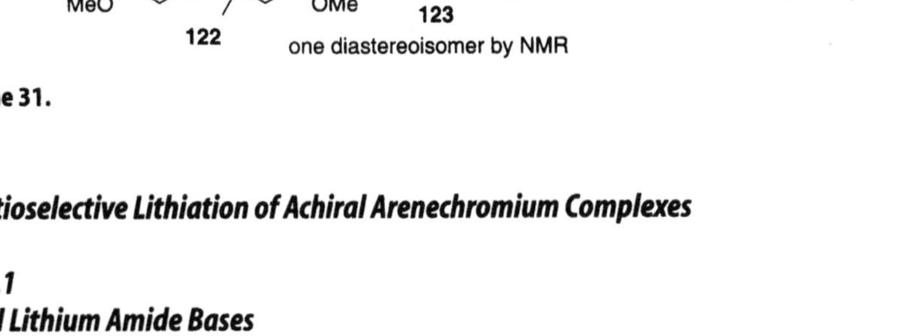

Scheme 31.

1.2.2
Enantioselective Lithiation of Achiral Arenechromium Complexes

1.2.2.1
Chiral Lithium Amide Bases

Arenechromium tricarbonyls are considerably more acidic than ferrocenes. Complexation to $Cr(CO)_3$ allows even electron-rich ring systems such as anisole to be deprotonated by lithium amides [66]. In 1994, Simpkins showed that chiral lithium amides [83] could be used to achieve this transformation enantioselectively [84]. Anisole complex **125** was treated with the chiral base **83** in the presence of Me_3SiCl, giving the silane **126** in 83% yield and 84% ee accompanied by small amounts of a doubly silylated by-product (Scheme 32). Other related deprotonations were less successful, either in terms of yield (**127** gives silylated product of 84% ee but only in 36% yield, with benzylic deprotonation being a problem) or ee (**128–130** give varying amounts of product in <50% ee).

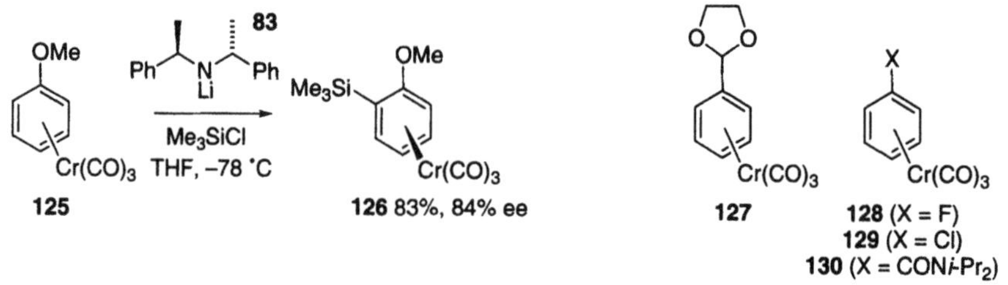

Scheme 32.

Early attempts to vary the electrophile or to use an "external quench" – i.e. to add the electrophile after lithiation was complete – were unsuccessful. However, further studies [85, 86] on this reaction produced some intriguing and valuable insights into its mechanism and allowed some of these problems to be solved. It became clear that racemisation of the lithiated intermediate **132** by proton exchange with the amine **133** was the process which had previously destroyed enantioselectivity when electrophiles were added after completion of the lithiation. For example, an authentic sample of **132** could be formed by a second lithiation-quench of **126**, desilylating with TBAF, and then submitting the product to tin-lithium exchange (Scheme 33). Compound **132** turned out to be fully configurationally stable in the absence of **133**, but configurationally labile in its presence. The lithiation of **125** turns out to be much faster in the presence of LiCl, and this acceleration is sufficient to allow complete metallation before significant racemisation has occurred, and therefore to extend the range of electrophiles which can be used with **132** (for example, benzaldehyde successfully gives **134**). Preferred conditions for the metallation are shown in Scheme 34. Other bases failed to improve the enantioselectivity [87].

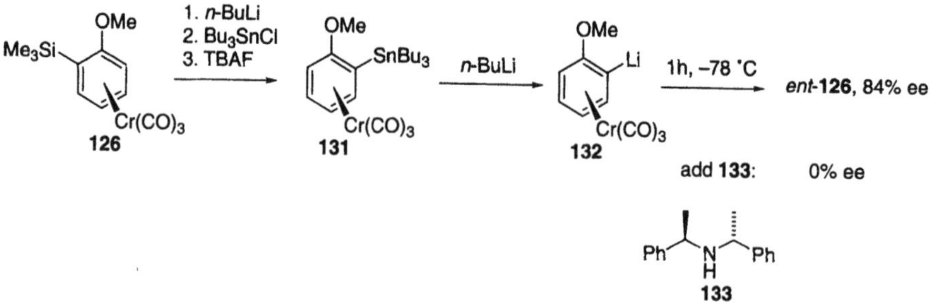

Scheme 33.

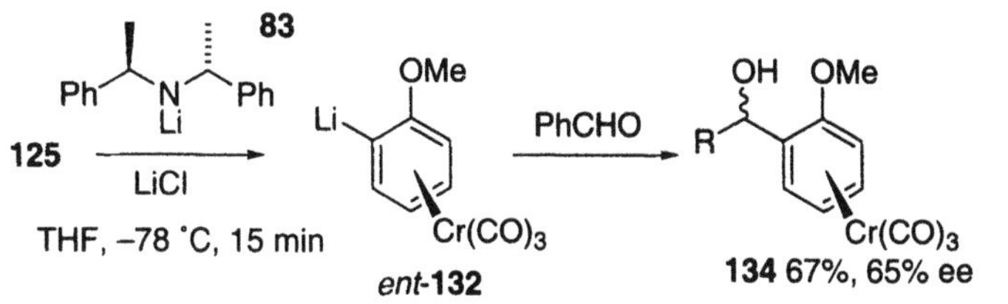

Scheme 34.

The enantioselective lithiation of anisolechromium tricarbonyl was used by Schmalz in a route towards the natural product (+)-ptilocaulin [88, 89]. In situ lithiation of **125** with *ent*-**83** gave *ent*-**126** in an optimised 91% ee (reaction carried out at −100°C over 10 min). A second, substrate-directed lithiation with BuLi alone, formation of the copper derivative, and a quench with crotyl bromide, gave **135**. The planar chirality and reactivity of the chromium complex was then exploited in a nucleophilic addition of dithiane, which generated the ptilocaulin precursor **136** (Scheme 35). The stereochemistry of compound **126** has also been used to direct dearomatising additions, yielding other classes of enones [90].

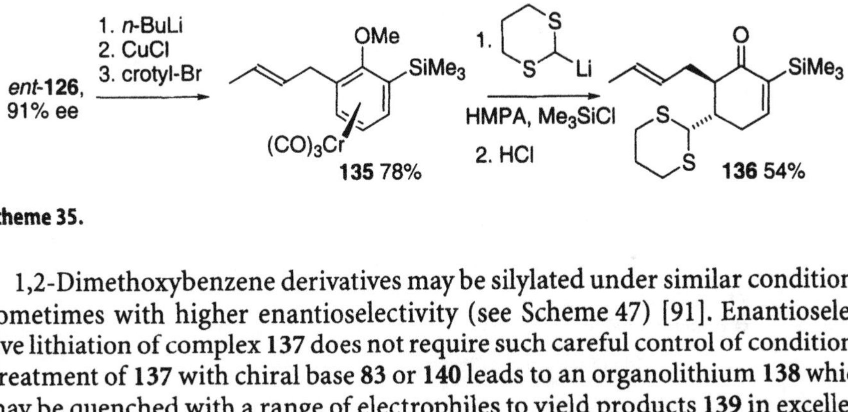

Scheme 35.

1,2-Dimethoxybenzene derivatives may be silylated under similar conditions, sometimes with higher enantioselectivity (see Scheme 47) [91]. Enantioselective lithiation of complex **137** does not require such careful control of conditions. Treatment of **137** with chiral base **83** or **140** leads to an organolithium **138** which may be quenched with a range of electrophiles to yield products **139** in excellent yield and with ees in the range 65–75%, recrystallisable to >90% (Scheme 36) [92, 93]. The intermediate **138**, on warming, undergoes rearrangement to a phenoxide trapped as **141**, but in only 54% ee due to concurrent racemisation. The application of other bases to this deprotonation has also been studied [87].

This reaction has been used in the synthesis of some phosphine ligands: **139** (E=Br) of 62% ee was recrystallised to provide material of 95–97% ee. Suzuki coupling gave biarylchromium complexes **142**, and the chromium's electron-withdrawing ability was exploited to turn the carbamate into a leaving group: addition of Ph_2PLi gave **143** (Scheme 37) [94].

The chromium tricarbonyl complex **144** of triphenylphosphine oxide is lithiated and silylated (in situ quench) by **83** and Me_3SiCl to give **145** (X=Si) in 90% yield and 73% ee (Scheme 38) [95, 96]. Interestingly, the sense of asymmetric induction is reversed from the anisole result. With two equivalents of chiral base, bis-silyl compounds **147** could be formed.

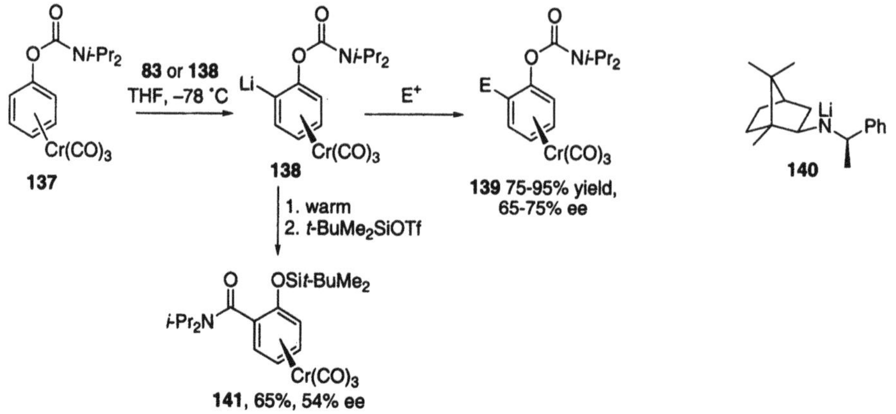

Scheme 36.

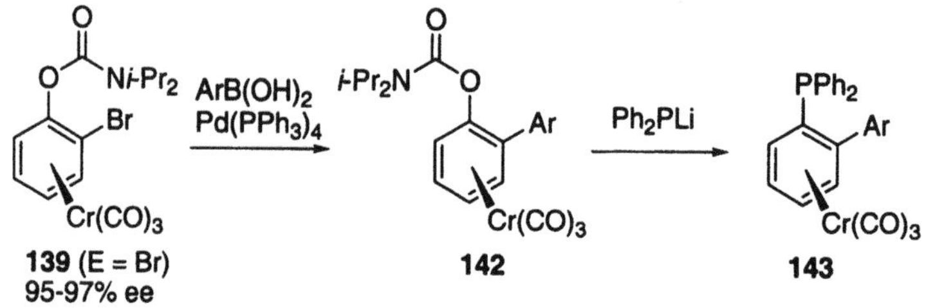

Scheme 37.

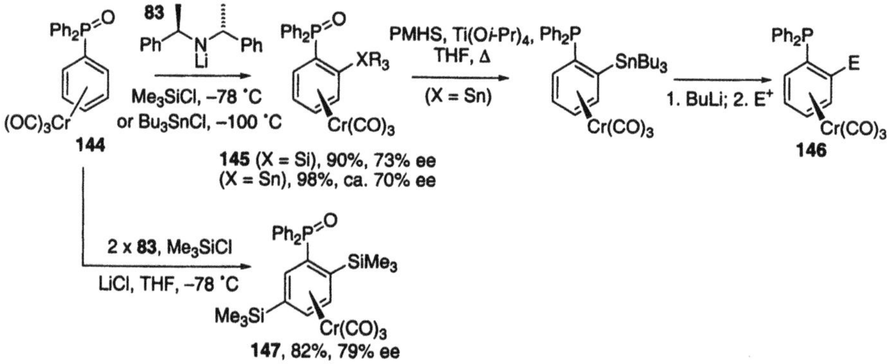

Scheme 38.

Adding electrophiles "externally" (after complete lithiation) failed, and in order to make potential ligands based upon the arenechromium skeleton, it was necessary to use Bu_3SnCl in an internal quench procedure. Reduction of **145** (X= Sn) yielded a phosphine without decomplexation, and tin-lithium exchange can lead to a variety of products **146** in about 70% ee.

Alkoxyalkyl-substituted arenechromium complexes have been lithiated enantioselectively in the benzylic position without control over planar chirality [97,

98], and the tendency for lithium amides to lithiate at a benzylic position rather than on the ring itself poses a problem with some substrates. In the lithiation of **148**, an element of planar stereochemistry is introduced during a benzylic lithiation (Scheme 39) [86, 99]. Exploiting the accelerating effect of added LiCl on such lithiations, treatment of **148** with **83** and LiCl followed by MeI gave the ether **150** in 79% ee. Oxidative decomplexation yields the known lactone **151**.

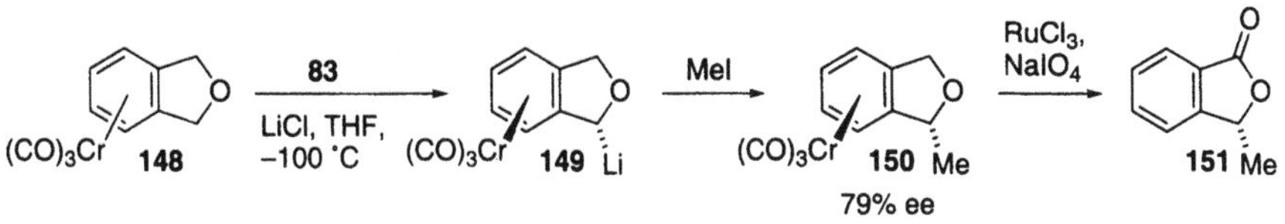

Scheme 39.

The chiral base **83** turns out not to be the best choice for enantioselective lithiation of the sulfur analogue **152** (Scheme 40): the bis-lithium amide **153** in the presence of LiCl at -100°C gives better yields and enantioselectivity [96, 100]. The base **153** often turns out to be a good choice as an alternative to **83** for reactions that fail to give good enantioselectivity [98].

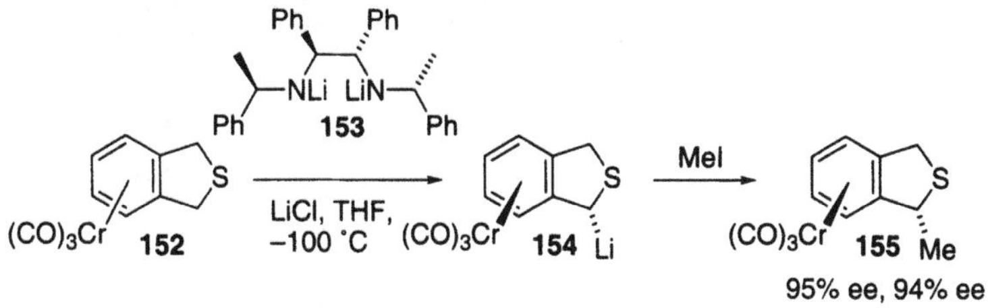

Scheme 40.

Attempts to deprotonate the benzaldehyde derivative **127** enantioselectively met with some success (see above), but the reaction is complicated by benzylic lithiation [87, 92]. Better results are obtained with the benzaldimine **156**, which is lithiated by **83** with good enantioselectivity and whose product is easily hydrolysed back to the aldehyde **157** (Scheme 41).

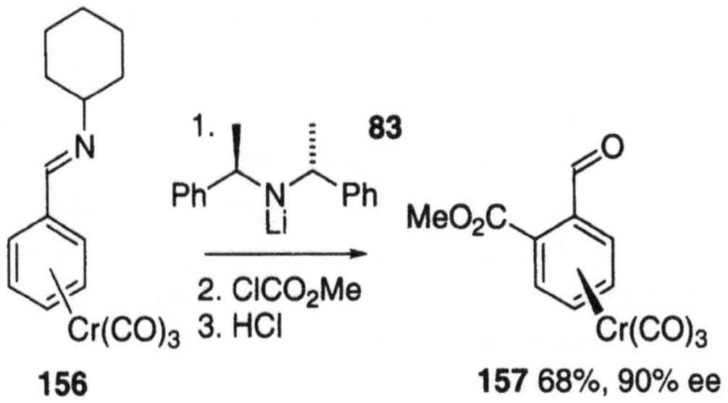

Scheme 41.

1.2.2.2
Chiral Alkyllithium-Diamine Complexes

An alternative approach to the asymmetric synthesis of arenechromium tricarbonyls is to use achiral alkyllithiums in the presence of a chiral ligand – the diamine (–)-sparteine **85**, for example. In a study of the relative efficiencies of a range of diamines, Uemura showed that the best ligand for introducing enantioselectivity into the lithiation of **158** and **161** was the diamine **159** (Scheme 42) [101]. (–)-Sparteine **85** performed relatively poorly with **158**.

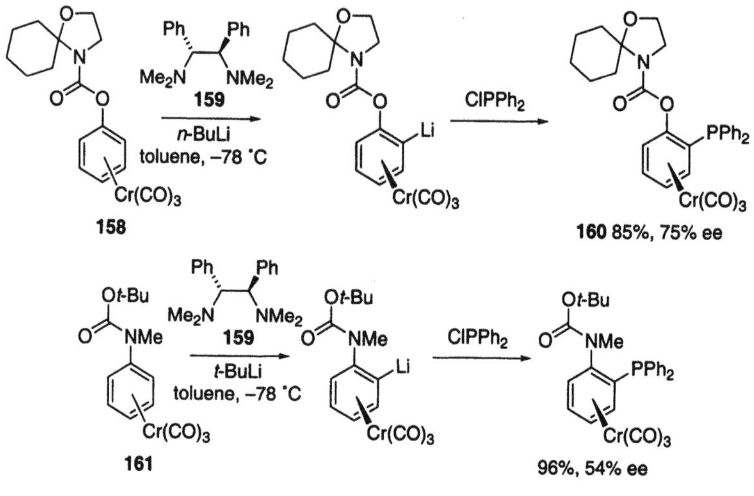

Scheme 42.

By careful optimisation, Widdowson was able to show that methoxymethyl ethers of phenols are better substrates for alkyllithium-diamine controlled enantioselective deprotonation, and (–)-sparteine **85** is then also the best ligand among those surveyed: the BuLi-(–)-sparteine complex deprotonates **162** to give, after electrophilic quench, compounds such as **164** in 58% yield and 92% ee (Scheme 43) [102]. Deprotonation of the anisole complex **125** under these conditions gave products of opposite absolute stereochemistry with poor ee.

A problem with (–)-sparteine **85** is its lack of availability in both enantiomeric forms. Reversed selectivity in the generation of planar chirality has been achieved by second lithiations (see Schemes 26 and 34) and a remarkable modification of this strategy works with arenechromium tricarbonyls. By using excess BuLi (sometimes *t*-BuLi is required) in the presence of sparteine **85**, a doubly lithiated species **165** may be formed from **162** [103]. The formation of the doubly lithiated species may be confirmed by double deuteration with excess D_2O. However, other electrophiles react selectively only once and give products of opposite absolute stereochemistry from those formed after monolithiation, albeit in rather low yield. Presumably, the first lithiation, which is directed by (–)-sparteine, produces an organolithium whose complexation with (–)-sparteine remains favourable. The second lithiation must produce a less stable organolith-

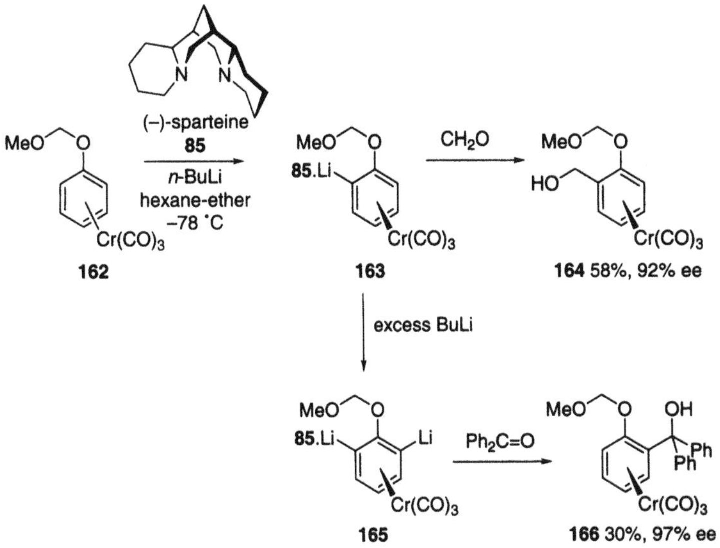

(–)-sparteine
85
n-BuLi
hexane-ether
–78 °C
MeO
O
Cr(CO)3
162
85.Li
163
CH2O
HO
164 58%, 92% ee
excess BuLi
Li
165
Ph2C=O
OH
Ph
166 30%, 97% ee

Scheme 43.

ium – one which cannot form a stabilising complex with (–)-sparteine. This second formed "anion" is the first to react, and monofunctionalisation achieves the required reversal of stereochemistry (Scheme 43).

Green has deprotonated arenechromium tricarbonyls bearing acidic benzylic protons with chiral bases in which the organolithium itself is chiral. Menthyllithium **167** (R=H) performed variably in terms of ee, but 8-phenylmenthyllithium **167** (R=Ph) gave good yields and ees [in the opposite enantiomeric series from **167** (R=H)] in the deprotonation of **127** (Scheme 44) [104]. Unlike most chiral bases, **167** (R=Ph) gives better results in THF than in ether.

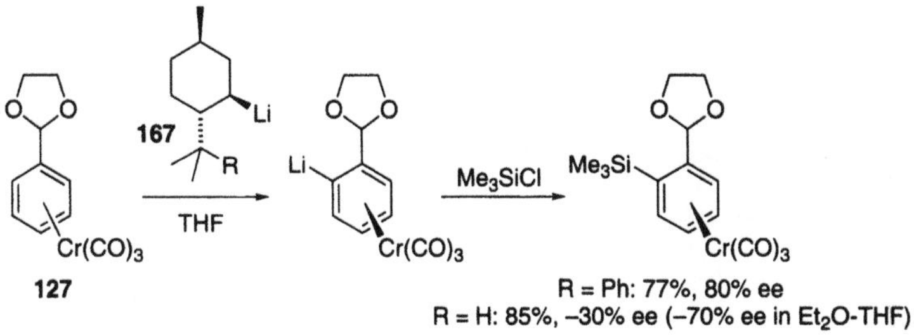

167
Li
R
THF
Cr(CO)3
127
Me3SiCl
Me3Si
R = Ph: 77%, 80% ee
R = H: 85%, –30% ee (–70% ee in Et2O-THF)

Scheme 44.

2
Generating Axial Chirality by Lithiation

The use of enantioselective lithiation to generate axial chirality directly is in its infancy. However, several strategies are available for the conversion of planar or central chirality into axial chirality, and several research groups have reported success in using enantioselective lithiation first to generate planar or central chirality which can itself be used to control a new stereogenic axis.

2.1 Axial Chirality in Biaryls

2.1.1 *Via Planar Chirality*

Uemura pioneered the atroposelective Suzuki coupling of planar chiral haloarenechromium tricarbonyls with areneboronic acids to yield biarylchromium tricarbonyls with defined planar and axial chirality [105]. A valuable feature of these reactions is that the major atropisomer about the new aryl-aryl bond generated in the coupling is typically the one bearing the larger group *syn* to chromium. This also turns out to be thermodynamically the less stable of the two atropisomers, so thermal equilibration allows the axial stereochemistry of the product to be inverted, and choice of conditions allows either atropisomer to be formed [106]. Schemes 45 and 46 show this method in action in the synthesis of two biaryl natural products, (–)-steganone [79] and *O,O'*-dimethylkorupensamine A [107, 108]. In the first, an acetal auxiliary is used to direct the diastereoselective lithiation and bromination, yielding **116** (see Scheme 28). Coupling with boronic acid **168** in refluxing methanol yields the more thermodynamically stable of two relatively easily interconverted atropisomeric biaryls. Further steps convert the axial chirality of **169** into the axially chiral natural product.

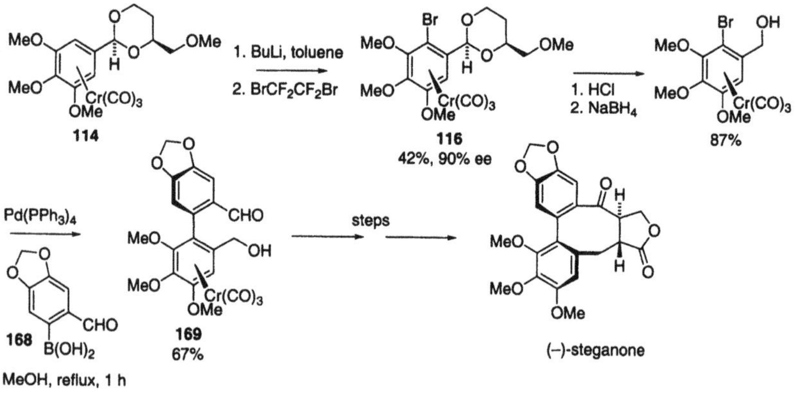

Scheme 45.

The protected diol side-chain of **170** is introduced by asymmetric dihydroxylation and directs diastereoselectivity in the formation of **172** by lithiation. The most acidic position of **170**, between the two methoxy groups, is first protected by silylation. Suzuki coupling of **173** with the boronic acid **174** gives the kinetic product **175** – the more severe hindrance to bond rotation in this compound

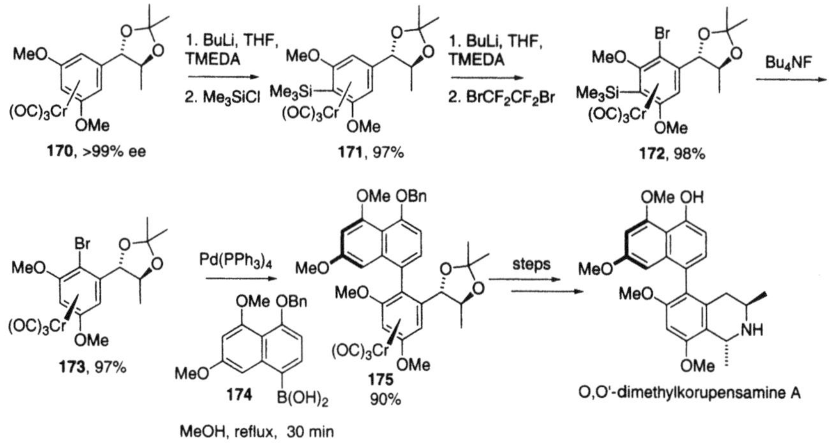

Scheme 46.

does not allow equilibration to the more stable atropisomer of the biaryl under the conditions of the reaction.

Couplings can also be carried out by simple nucleophilic substitution reactions of arenechromium tricarbonyls [109]. For example, in the synthesis of biaryl **182**, asymmetric lithiation of **176** using in situ silylation provides the complex **179**. Nucleophilic substitution by the tolyl Grignard reagent **180** yields **181** as a single atropisomer in 68% yield, and decomplexation gives the biaryl **182** in 92% yield (Scheme 47).

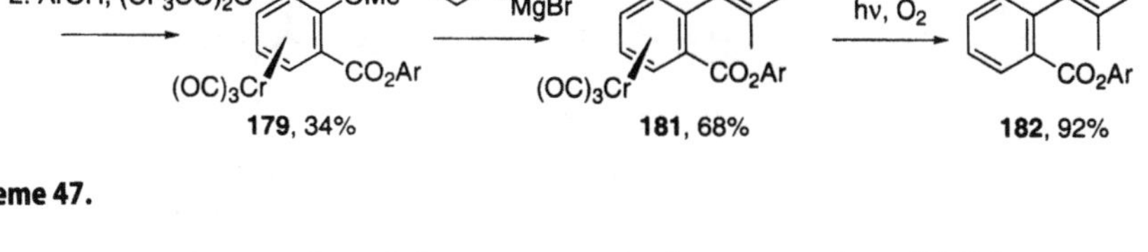

Scheme 47.

Widdowson has exploited the asymmetric deprotonation of **183** in a synthesis of a protected version **191** of the biaryl component of vancomycin, actinoidinic acid (Scheme 48) [110, 111]. One of the rings derives from an arenechromium tricarbonyl with stereochemistry controlled by asymmetric lithiation. The most readily lithiated position of **183**, between the two methoxy groups, first needed blocking. Enantioselective lithiation and chlorination of **184** gave **186** (TMEDA was needed to displace sparteine from **185** and restore reactivity towards a poor electrophile). Suzuki coupling of **187** with the boronic acid **188** transfers planar

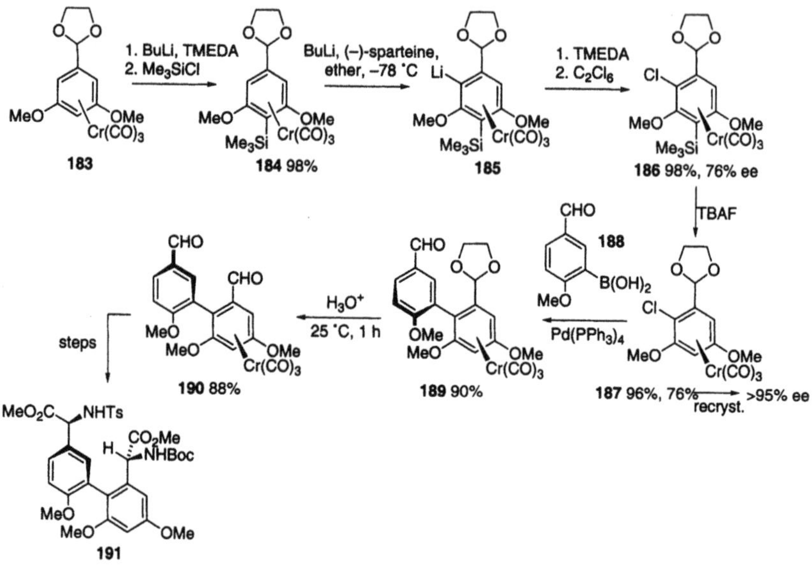

Scheme 48.

to axial stereochemistry and gives a single axial stereoisomer **189.** Deprotection of the aldehyde lowers the barrier to Ar-Ar rotation and allows relaxation to the thermodynamically preferred stereoisomer **190.**

2.1.2
By Direct Lithiation

The only example of direct control of axial chirality in biaryls is an early report by Raston of the lithiation of the achiral biaryl **192** using *n*-BuLi-(–)-sparteine. The product **194** of unknown configuration was obtained in moderate ee (Scheme 49) [112].

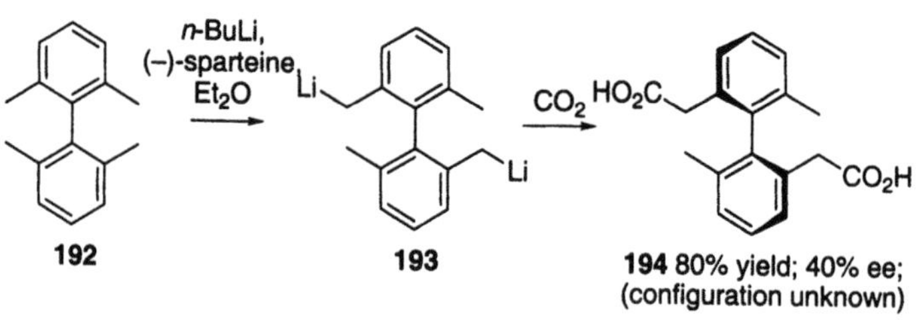

Scheme 49.

2.2
Axial Chirality in Other Atropisomers

During the late 1990s, several research groups investigated the stereoselective reactions of compounds with slowly rotating bonds other than Ar-Ar [113].

Principal among these are amides: either benzamides, exemplified by **197**, or anilides, exemplified by **200** (Scheme 50). Given the known lithiation directing ability of amide groups [1, 114], these compounds are clearly prime candidates for asymmetric synthesis using enantioselective lithiation.

2.2.1
Via Planar Chirality

As with atropisomeric biaryls, axial chirality in atropisomeric amides may be introduced by stereochemical control in the atroposelective reactions of planar chiral complexes [115]. Enantioselective lithiation was reported in this context by Uemura, who showed that the achiral complexes **195**, **198**, **201** and **204** are deprotonated enantioselectively by treatment with chiral lithium amide bases (Scheme 50) [116–118]. The stereogenic C-C and C-N axes in these compounds are orientated such that the larger NR_2 and acyl groups, respectively, are directed away from the chromium. A range of chiral lithium amides was investigated, and by careful selection it was possible to obtain products **196**, **199**, **202** and **205**

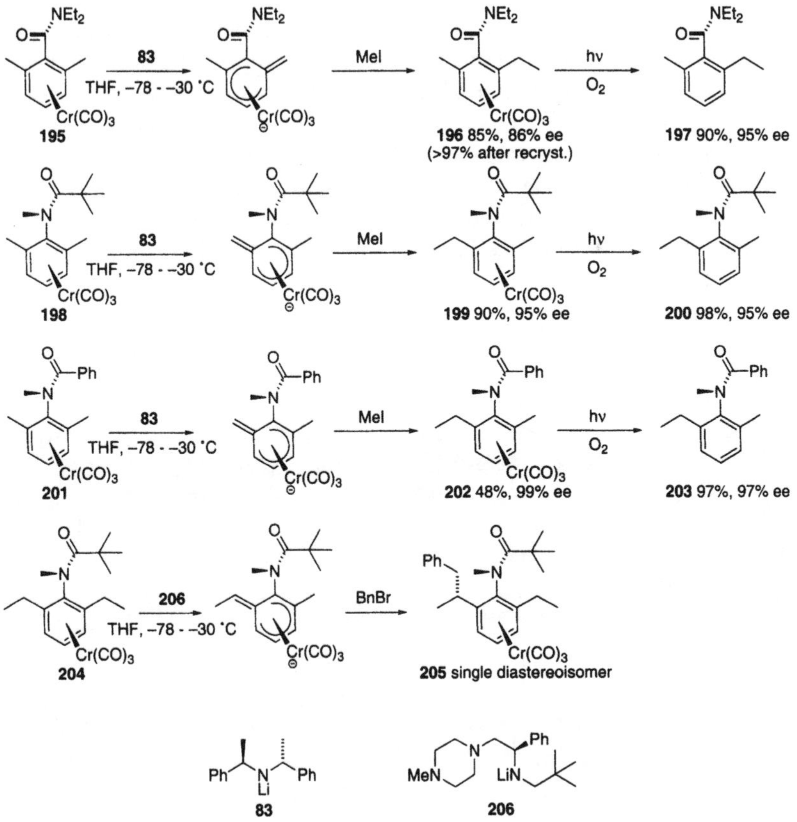

Scheme 50.

with excellent yield and enantioselectivity. Decomplexation returns atropisomers with no loss of absolute stereochemistry, though products such as **197** racemise slowly on standing. Even with the same base, the anilides **198** and amides **195** give products in opposite enantiomeric series. The diethyl-substituted anilide **204** could be functionalised enantio- and diastereoselectively to yield such products as **205** with control over chiral plane, axis and centre.

2.2.2
Via Central Chirality

A stereogenic centre constructed by the (–)-sparteine-directed lithiation and electrophilic quench of benzamides reported by Beak [119] can function as a temporary cache of absolute stereochemistry during the enantioselective formation of atropisomeric amides [120]. Lithiation of **207** and reaction with Me_3SiCl in the presence of (–)-sparteine **85** gives silane **208** in >97% ee after two recrystallisations from petroleum ether (Scheme 51). 1-Silylethyl-bearing amides with freely rotating amide axes (such as **208**) have been shown to adopt a conformation approximating to that shown (Scheme 51), in which the amide substituents and the silyl group are able to avoid close non-bonded interactions [121]. The absolute stereochemistry of the silyl-bearing centre therefore translates itself, under simple thermodynamic control, into absolute stereochemistry in the form of the preferred conformation of an amide axis. Trapping the preferred *conformer* of **208** as a single *atropisomer* allowed the absolute stereochemistry of the silicon-bearing centre to be translated completely into axial stereochemistry. Rotation of the amide group was blocked with a second *ortho*-substituent on the benzamide ring. The amide **208** was treated with *s*-BuLi to *ortho*-lithiate it, and the organolithium was quenched with a range of electrophiles including $ClPPh_2$, which gave the phosphine **209**. Finally, removal of any thermodynamic confor-

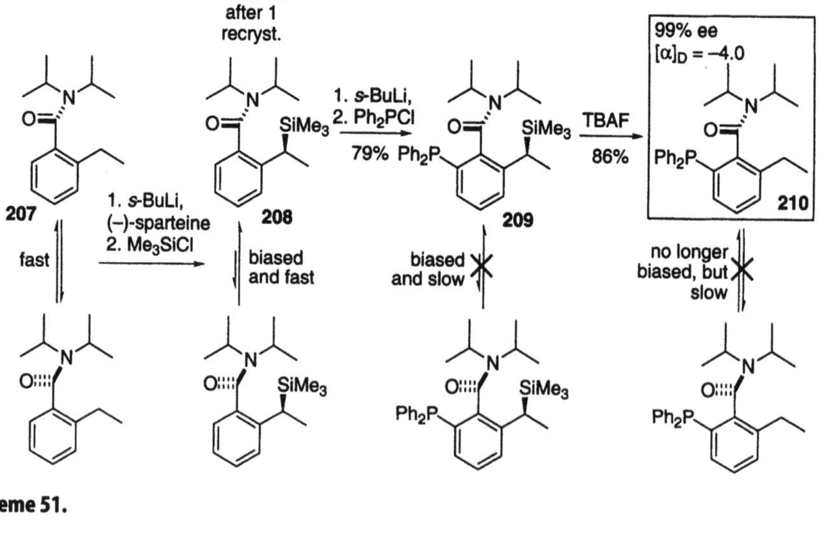

Scheme 51.

mational bias by desilylation gave **210**, whose Ar-CO rotation is slow and is therefore obtained as a single enantiomerically enriched atropisomer [120].

The "thermodynamic" strategy employed in the asymmetric synthesis of **210** is outlined in Scheme 52. The starting amide **207** is a pair of rapidly interconverting (small $\Delta G^{\ddagger}_{rac}$) enantiomeric (*a*R and *a*S) conformers about the Ar-CO bond. Asymmetric construction of the stereogenic centre in **208** perturbs the populations of the conformers, favouring *a*R over *a*S. This thermodynamic perturbation of populations is then fixed kinetically, by introduction of the substituent R into **209**. Now the thermodynamic perturbation – the stereogenic centre – can be removed, and the kinetic barrier to re-establishment of equilibrium (large $\Delta G^{\ddagger}_{rac}$) allows the enantiomeric excess to be retained in the products **210**.

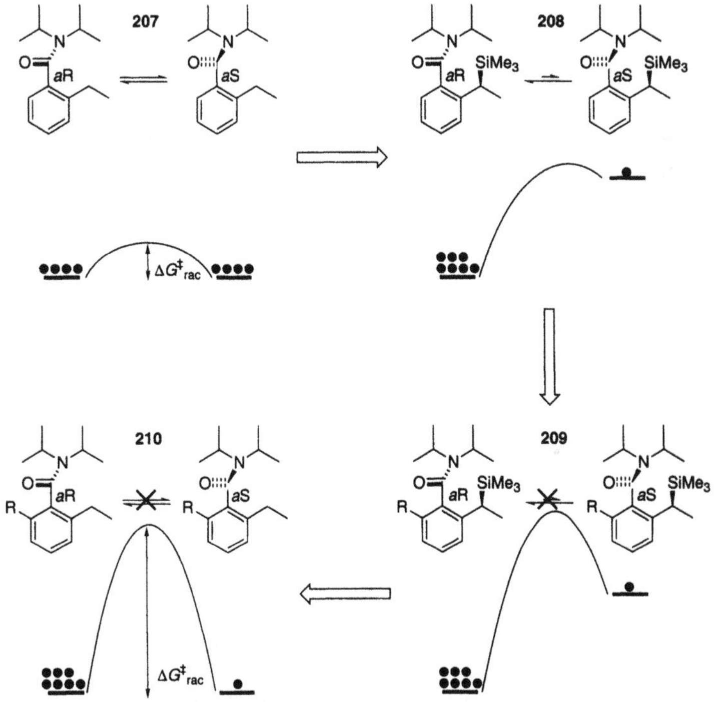

Scheme 52.

2.2.3
By Direct Lithiation

Deprotonation of uncomplexed compounds **211** similar in structure to **195** is also possible, but the yields and enantioselectivities obtained are poorer than for the chromium-complexed analogues (Scheme 53). The range of bases available for use in this reaction is limited by the loss of the acidifying effect of chromium complexation. However, with an internal electrophilic quench it was pos-

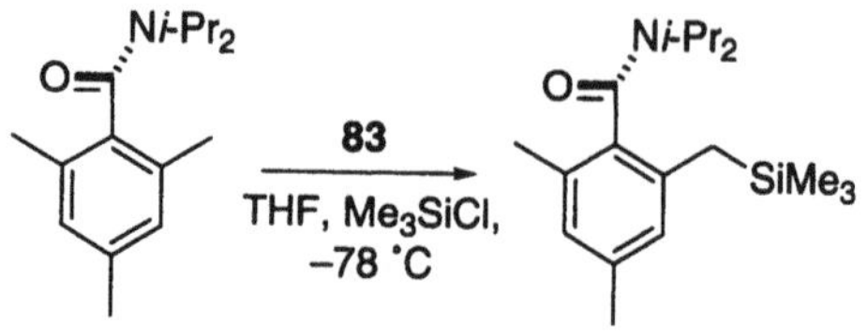

Scheme 53.

sible to form the axially chiral benzamide **212** with the same sense of asymmetric induction as **197** in 89% ee using lithium amide **83** [122].

s-BuLi-(–)-sparteine failed to yield any useful asymmetric induction in the deprotonation of **211**. However, in 1996, Beak and Curran reported an asymmetric *ortho*-lithiation using *s*-BuLi-(–)-sparteine, which deprotonates **213** with moderate enantioselectivity [119]. Low yields of products **214** are obtained (Scheme 54 is representative) and the reaction is almost certainly a kinetic resolution of the starting amide, which at –78 °C exists as a mixture of enantiomeric atropisomers [123].

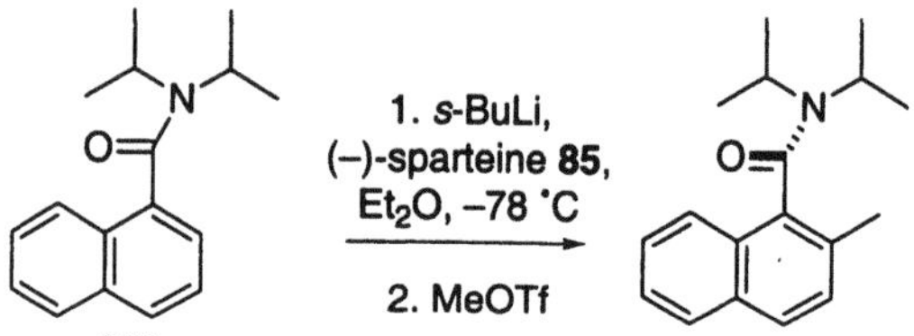

Scheme 54.

2.3 Axial Chirality in Allenes

Axial chiral allenes may be formed from "anions" in a number of methods, and it is perhaps surprising that the methods of asymmetric organolithium chemistry have been applied to relatively few cases. An early example was the demonstration by Nozaki and Noyori (Scheme 55) that decomposition of *gem*-dibromocyclopropanes **215** to allenes by halogen-lithium exchange to the lithium carbenoid **216** in the presence of (–)-sparteine **85** yielded optically active allenes **217** of indeterminate enantiomeric excess [124, 125].

Recently, Hoppe has shown that asymmetric deprotonation of propargyl carbamates **218** can lead to enantiomerically enriched alkynes **220** [126]. Treatment of **218** with *n*-BuLi in the presence of (–)-sparteine **85** gives a crystalline organolithium complex **219** which reacts with electrophiles to yield enantiomerically enriched alkynes **220**. Invertive transmetallation with $ClTi(O\text{-}i\text{-}Pr)_3$ generates

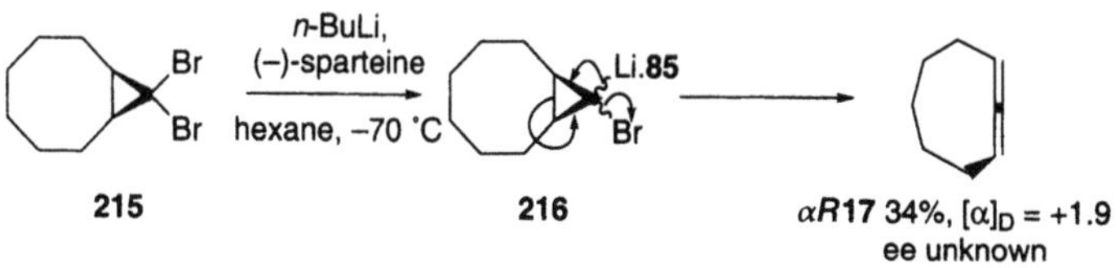

Scheme 55.

the propargyltitanium species **221** whose reactions with electrophiles give enantiomerically enriched allenes with control over axial chirality. For example, simple protonation gives **222** in 88% ee; reactions with aldehydes give single diastereoisomers of **223** in >95% ee (Scheme 56).

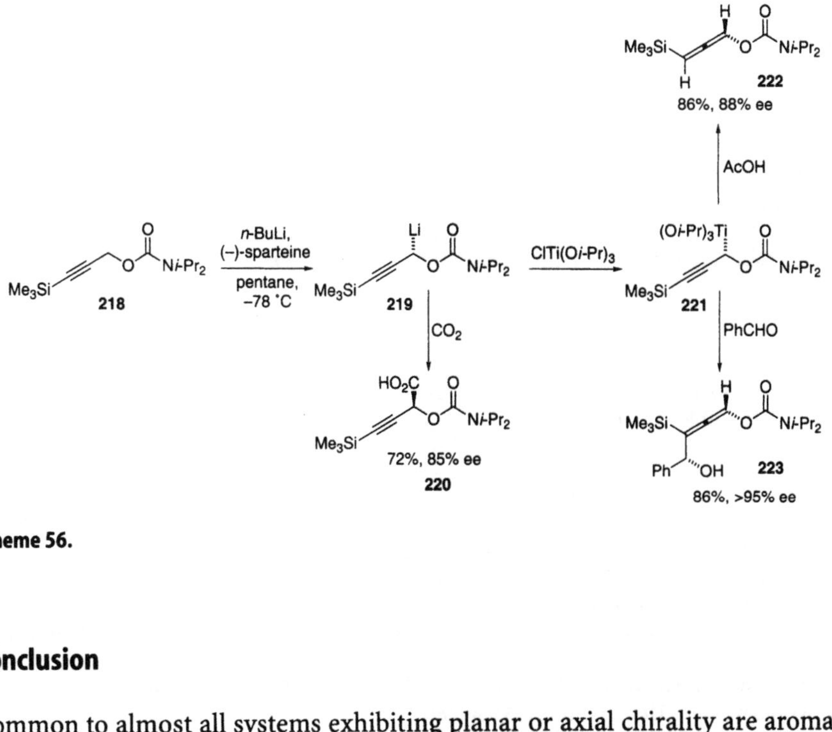

Scheme 56.

3 Conclusion

Common to almost all systems exhibiting planar or axial chirality are aromatic rings, allenes at least having π-systems and spiro compounds excepted. The range of methods available for making aryl-, benzyl- and vinyllithiums [127] therefore suits their enantioselective synthesis by organolithium chemistry. Some areas nonetheless have received much more attention (metal complexes) than others (biaryls and allenes), and there are still many opportunities for further advances.

Note added in proof. The diastereoselective lithiation of *N,N*-dimethylferrocenylethylamine) **9** (Scheme 2) and sparteinemediated enantioselective lithiation of (diisopropylamido)ferrocene **84** (Scheme 21) using MeLi were modeled through an assumed reversible adduct formation at the amine N or amide O, followed by an irreversible ring lithiation; selectivity results from ring lithiation via the adduct conformer with the shortest C-H_{ring}---H_3C-Li interaction [128]. Clayden and coworkers have reported that *ortho*-lithiation and reaction with (-)-menthyl *p*-toluensulfinate introduces a sulfoxide substituent *ortho* to the stereogenic Ar-CO axis of an aromatic amide [129]. The sulfoxide exerts a powerful conformational bias on the axis, such that after rapid equilibration at ambient temperature essentially only one of two diastereoisomeric Ar-CO atropisomers is populated. Sulfoxide-lithium exchange by treatment with *t*-BuLi regenerates the *ortho*-lithiated amide in an enantiomerically pure and conformationally stable form, from which rapid electrophilic trapping generates highly enantiomerically enriched atropisomeric tertiary aromatic amides.

References

1. Gschwend HW, Rodriguez HR (1979) Org Reac 26:1
2. Slocum DW, Rockett BW, Hauser CR (1965) J Am Chem Soc 87:1241
3. Aratani T, Gonda T, Nozaki H (1969) Tetrahedron Lett:2265
4. Falk H, Schlögl K (1965) Monatsh Chem 96:1065
5. Richards CJ, Locke AJ (1998) Tetrahedron: Asymmetry 9:2377
6. Gokel G, Hoffmann P, Kleimann H, Klusacek H, Marquarding D, Ugi I (1970) Tetrahedron Lett:1771
7. Marquarding D, Klusacek H, Gokel G, Hoffmann P, Ugi I (1970) J Am Chem Soc 92:5389
8. Battelle LF, Bau R, Gokel G, Okayawa RT, Ugi IK (1973) J Am Chem Soc 95:482
9. Hayashi T, Mise T, Fukushima M, Kagotani M, Nagashima N, Hamada Y, Matsumoto A, Kawakami S, Konishi M, Yamamoto K (1980) Bull Chem Soc Jpn 53:1138
10. Appleton TD, Cullen WR, Evans JV, Kim T-J, Trotter J (1985) J Organomet Chem 279:5
11. Hayashi T, Yamazaki A (1991) J Organomet Chem 413:295
12. Pastor SD (1988) Tetrahedron 44:2993
13. Naiini AA, Lai C-K, Ward DL, Brubaker CH (1990) J Organomet Chem 390:73
14. Butler IR, Cullen WR, Rettig SJ (1987) Can J Chem 65:1452
15. Gokel G, Ugi I (1972) J Chem Ed 1972:49
16. Beyer L, Richter R, Seidelmann O (1998) J Organomet Chem 561:199
17. Deus N, Robles D, Herrmann R (1990) J Organomet Chem 386:253
18. David DM, Kane-Maguire LAP, Pyne SG (1990) Chem Commun 888
19. Ohno A, Yamane M, Hayashi T, Oguni N, Hayashi M (1995) Tetrahedron: Asymmetry 6:2495
20. Riant O, Samuel O, Kagan HB (1993) J Am Chem Soc 115:5835
21. Riant O, Samuel O, Flessner T, Taudien S, Kagan HB (1997) J Org Chem 62:6733
22. Iftime G, Daran J-C, Manoury E, Balavoine GGA (1998) J Organomet Chem 565:115
23. Comins DL (1992) Synlett:615
24. Balavoine GGA, Daran J-C, Iftime G, Manoury E, Moreau-Bossuet C (1998) J Organomet Chem 567:191
25. Richards CJ, Damalidis T, Hibbs DE, Hursthouse MB (1995) Synlett:74
26. Nishibayashi Y, Uemura M (1995) Synlett:79
27. Sammakia T, Latham H, Schaad DR (1995) J Org Chem 60:10
28. Sammakia T, Latham HA (1995) J Org Chem 60:6002
29. Manoury E, Fossey JS, Ait-Haddou H, Daran J-C, Balavoine GGA (2000) Organometallics 19:3836
30. Sammakia T, Latham HA (1996) J Org Chem 61:1629
31. Salter R, Pickett TE, Richards CJ (1998) Tetrahedron: Asymmetry 9:4239
32. Richards CJ, Mulvaney AW (1996) Tetrahedron: Asymmetry 7:1419
33. Nishibayashi Y, Segawa K, Arikawa Y, Ohe K, Hidai M, Uemura S (1997) J Organomet Chem 545:381
34. Ahn KH, Cho C-W, Baek H-H, Park J, Lee S (1996) J Org Chem 61:4937
35. Sebesta R, Toma S, Salisova M (2002) Eur J Org Chem:692

36. Park J, Lee S, Ahn KH, Cho C-W (1995) Tetrahedron Lett 36:7263
37. Kim S-G, Cho C-W, Ahn KH (1997) Tetrahedron: Asymmetry 8:1023
38. Zhang W, Adachi Y, Hirao T, Ikeda I (1996) Tetrahedron: Asymmetry 7:451
39. Aratani T, Gonda T, Nozaki H (1970) Tetrahedron 26:5453
40. Ganter C, Wagner T (1995) Chem Ber 128:1157
41. Xiao L, Mereiter K, Weissensteiner W, Widhalm M (1999) Synthesis 1354
42. Enders D, Peters R, Lochtman R, Runsink J (1997) Synlett 1462
43. Enders D, Peters R, Lochtman R, Raabe G (1999) Angew Chem Int Ed 38:2421
44. Enders D, Peters R, Lochtman R, Raabe G, Runsink J, Bats JW (2000) Eur J Org Chem 3399
45. Enders D, Klumpen T (2001) J Organomet Chem 637:698
46. Xiao L, Kitzler R, Weissensteiner W (2001) J Org Chem 66:8912
47. Widhalm M, Mereiter K, Bourghida M (1998) Tetrahedron: Asymmetry 9:2983
48. Widhalm M, Nettekoven U, Mereiter K (1999) Tetrahedron: Asymmetry 10:4369
49. Farrell A, Goddard R, Guiry PJ (2002) J Org Chem 67:4209
50. Rebière F, Riant O, Ricard L, Kagan HB (1993) Angew Chem Int Ed Engl 32:568
51. Lagneau NM, Chen Y, Robben PM, Sin H-S, Takasu K, Chan J-S, Robinson PD, Hua DH (1998) Tetrahedron 54:7301
52. Priego J, Mancheño OG, Cabrera S, Carretero JC (2002) J Org Chem 67:1346
53. Riant O, Argouarch G, Guillaneux D, Samuel O, Kagan HB (1998) J Org Chem 63:3511
54. Argouarch G, Samuel O, Riant O, Daran J-C, Kagan HB (2000) Eur J Org Chem 2893
55. Pedersen HL, Johanssen M (1999) Chem Commun 2517
56. Bolm C, Kesselgruber M, Muñiz K, Raabe G (2000) Organometallics 19:1648
57. Metallinos C, Snieckus V (2002) Org Lett 4:1935
58. Price DA, Simpkins NS (1995) Tetrahedron Lett 36:6135
59. Hoppe D, Hense T (1997) Angew Chem Int Ed Engl 36:2282
60. Tsukazaki M, Tinkl M, Roglans A, Chapell BJ, Taylor NJ, Snieckus V (1996) J Am Chem Soc 118:685
61. Bonini BF, Comes-Franchini M, Fochi M, Mazzanti G, Ricci A, Alberti A, Macciantelli D, Marcaccio M, Roffia S (2002) Eur J Org Chem 543
62. Laufer RS, Veith U, Taylor NJ, Snieckus V (2000) Org Lett 2:629
63. Jendralla H, Paulus E (1997) Synlett 471
64. Nishibayashi Y, Arikawa Y, Ohe K, Uemura S (1996) J Org Chem 61:1172
65. Bolm C, Muniz K (1999) Chem Soc Rev 28:51
66. Berger A, Djukic JP, Michon C (2002) Coord Chem Rev 225:215
67. Gilday JP, Negri JT, Widdowson DA (1989) Tetrahedron 45:605
68. Blagg J, Davies SG, Goodfellow CL, Sutton KH (1987) J Chem Soc Perkin Trans 1 1805
69. Davies SG, Goodfellow CL (1987) J Organomet Chem 370:C5
70. Heppert JA, Aubé J, Thomas-Miller ME, Milligan ML, Takusagawa F (1988) Organometallics 7:2581
71. Heppert JA, Aubé J, Thomas-Miller ME, Milligan ML, Takusagawa F (1990) Organometallics 9:727
72. Uemura M, Miyake R, Nakayama K, Shiro M, Hayashi Y (1993) J Org Chem 58:1238
73. Coote SJ, Davies SG, Goodfellow CL, Sutton KH, Middlemiss D, Naylor A (1990) Tetrahedron: Asymmetry 1:817
74. Davies SG, Hume WE (1995) Chem Commun 251
75. Davies SG, Loveridge T, Clough JM (1995) Chem Commun 817
76. Davies SG, Loveridge T, Teixeira MFCC, Clough JM (1999) J Chem Soc Perkin Trans 1 3405
77. Kondo Y, Green JR, Ho J (1991) J Org Chem 56:7199
78. Kondo Y, Green JR, Ho J (1993) J Org Chem 58:6182
79. Uemura M, Daimon A, Hayashi Y (1995) Chem Commun 1943
80. Watanabe T, Shakadou M, Uemura M (1999) Inorg Chim Acta 296:80
81. Aubé J, Heppert JA, Milligan ML, Smith MJ, Zenk P (1992) J Org Chem 57:3563
82. Alexakis A, Kanger T, Mangeney P, Rose-Munch F, Perrotey A, Rose E (1995) Tetrahedron: Asymmetry 6:47
83. Gibson SE, Reddington EG (2000) Chem Commun 989

84. Price DA, Simpkins NS, MacLeod AM, Watt AP (1994) J Org Chem 59:1961
85. Price DA, Simpkins NS, MacLeod AM, Watt AP (1994) Tetrahedron Lett 35:6159
86. Ewin RA, MacLeod AM, Price DA, Simpkins NS, Watt AP (1997) J Chem Soc Perkin Trans 1 401
87. Pache S, Botuha C, Franz R, Kündig EP, Einhorn J (2000) Helv Chim Acta 83:2436
88. Schmalz H-G, Schellhaas K (1996) Angew Chem Int Ed Engl 35:2146
89. Schellhaas K, Schmalz H-G, Bats JW (1998) Chem Eur J 4:57
90. Quattropani A, Anderson G, Bernardinelli G, Kündig EP (1997) J Am Chem Soc 119:4773
91. Schmalz H-G, Schellhaas K (1995) Tetrahedron Lett 36:5515
92. Kündig EP, Quattropani A (1994) Tetrahedron Lett 35:3497
93. Quattropani A, Bernardinelli G, Kündig EP (1999) Helv Chim Acta 82:90
94. Nelson S, Hilfiker MA (1999) Org Lett 1:1379
95. Ariffin A, Blake AJ, Li W-S, Simpkins NS (1997) Synlett 1453
96. Ariffin A, Blake AJ, Ewin RA, Li W-S, Simpkins N (1999) J Chem Soc Perkin Trans 1 3177
97. Courton ELM, Gibson (née Thomas) SE, Schneider MJ, Smith MH (1996) Chem Commun 839
98. Gibson S, O'Brien P, Rahimian E, Smith MH (1999) J Chem Soc Perkin Trans 1 909
99. Ewin RA, Simpkins NS (1996) Synlett 317
100. Ariffin A, Blake AJ, Ewin RA, Simpkins NS (1998) Tetrahedron: Asymmetry 9:2563
101. Uemura M, Hayashi Y, Hayashi Y (1994) Tetrahedron: Asymmetry 5:1427
102. Wilhelm R, Sebhat IK, White AJP, Williams DJ, Widdowson DA (2000) Tetrahedron Asymmetry 11:5003
103. Tang Y-L, Widdowson DA, Wilhelm R (2001) Synlett 1632
104. Siwek MJ, Green JR (1999) Chem Commun 2359
105. Kamikawa K, Uemura M (2000) Synlett 938
106. Watanabe T, Kamikawa K, Uemura M (1995) Tetrahedron Lett 36:6695
107. Watanabe T, Uemura M (1998) Chem Commun 871
108. Kamikawa K, Watanabe A, Daimon A, Uemura M (2000) Tetrahedron 56:2325
109. Kamikawa K, Uemura M (1996) Tetrahedron Lett 37:6359
110. Wilhelm R, Widdowson DA (2001) Org Lett 3:3079
111. Tan YL, White AJP, Widdowson DA, Wilhelm R, Williams DJ (2001) J Chem Soc Perkin Trans 1 3269
112. Engelhardt LM, Leung W-P, Raston CL, Salem G, Twiss P, White AH (1988) J Chem Soc Dalton Trans 2403
113. Clayden J (1997) Angew Chem Int Ed Engl 36:949
114. Snieckus V (1990) Chem Rev 90:879
115. Koide H, Uemura M (1998) Chem Commun 2483
116. Hata T, Koide H, Taniguchi N, Uemura M (2000) Org Lett 2:1907
117. Hata T, Koide H, Uemura M (2000) Synlett 1145
118. Koide H, Hata T, Uemura M (2002) J Org Chem 67:1929
119. Thayumanavan S, Beak P, Curran DP (1996) Tetrahedron Lett 37:2899
120. Clayden J, Johnson P, Pink JH, Helliwell M (2000) J Org Chem 65:7033
121. Clayden J, Pink JH, Yasin SA (1998) Tetrahedron Lett 39:105
122. Clayden J, Johnson P, Pink JH (2001) J Chem Soc Perkin Trans 1 371
123. Ahmed A, Bragg RA, Clayden J, Lai LW, McCarthy C, Pink JH, Westlund N, Yasin SA (1998) Tetrahedron 54:13277
124. Nozaki H, Aratani T, Noyori R (1968) Tetrahedron Lett 2087
125. Nozaki H, Aratani T, Toraya T, Noyori R (1971) Tetrahedron 27:905
126. Schutz-Fademrecht C, Wibbeling B, Fröhlich R, Hoppe D (2001) Org Lett 3:1221
127. Clayden J (2002) Organolithiums: selectivity for synthesis. Pergamon, Oxford
128. Howell JAS, Yates PC, Fey N, McArdle P, Cunningham D, Parsons S, Rankin DWH (2002) Organometallics 21:5272
129. Clayden J, Mitjans D, Youssef LH (2002) J Am Chem Soc 124:5266

Topics Organomet Chem (2003) 5: 287–310
DOI 10.1007/b10334

Enantioselective Carbolithiations

Jean F. Normant

Laboratoire de Chimie des Organoéléments, Université P & M Curie,4 Place Jussieu, 75252 Paris, Cedex 05, France. *E-mail: Normant@ccr.jussieu.fr*

The addition of RLi to unactivated C=C bonds is much more restricted than the addition to activated, electron-poor C=C bonds. The known examples are briefly surveyed, and the enantioselective version is detailed: in an intermolecular way (cinnamyl and non-phenylated substrates), and in an intramolecular way (cyclizations from an enantiomerically enriched organolithium derivative, or from a racemic one, under the influence of a chiral ligand).

Keywords. Carbometallation, Sparteine, Carbocyclizations

Abbreviations

Bn	benzyl
Cby	2,2,4,4-tetramethyl-1,3-oxazolidin-3-ylcarbonyl
de	diastereomeric excess
dr	diastereoisomeric ratio
ee	enantiomeric excess
er	enantiomeric ratio

HMPA	hexamethylphosphoric triamide
LDBB	lithium di-*tert*-butylbiphenylide
THF	tetrahydrofuran
TMEDA	tetramethylethylenediamine
rt	room temperature

1 General Aspects of Carbolithiations

1.1 Introduction

The preceding chapters have described chiral lithium reagents, and the enantioselective addition of organolithiums to polar carbon-heteroatom double bonds and C=C bonds conjugated with an electron-withdrawing group. In this chapter, we consider the enantioselective addition of RLi reagents to unactivated C=C bonds[1, 2]. This subject has been tackled only recently: the major reason is that carbolithiations, in the racemic series, were restricted to a limited amount of substrates. A major limitation comes from the fact that the desired (homologated) reagent B is very similar to the starting one A, (both possess an sp^3-C-Li bond) and polymerization may occur [Eq. (1)]. It must be recalled that the anionic polymerization of alkenes [3, 4] was initially performed with organoalkali metals.

$$\underset{A}{RLi} \xrightarrow{R_2R_3C=CR_1R_4} \underset{B}{R-CR_2R_1-CR_3R_4-Li} \xrightarrow{R_2R_3C=CR_1R_4} R(-CR_2R_1-CR_3R_4-)_2Li \;\ldots \qquad (1)$$

This drawback is circumvented if the addition occurs on a triple bond [Eq. (2)], where a less reactive sp^2-C-Li bond is formed, or on an intramolecular double bond [Eq. (3)], where entropy factors are favorable. In this latter case, a new sp^3 carbon has been created.

$$RLi + R_1-C\equiv C-R_2 \longrightarrow (R)(R_1)C=C(Li)(R_2) \qquad (2)$$

$$\text{(3)}$$

We shall then consider mostly cases concerning Eqs. (1) and (3), and first recall which substrates allow such carbolithiations in the racemic series.

1.2
Intramolecular Carbolithiations

Intramolecular carbolithiation is now a common process to generate cyclic substrates, and is particularly well suited for the creation of five-membered rings via a *5-exo-trig* cyclization [5, 6] (Scheme 1). Cascade reactions can be performed when starting with a diethylenic organolithium reagent [7].

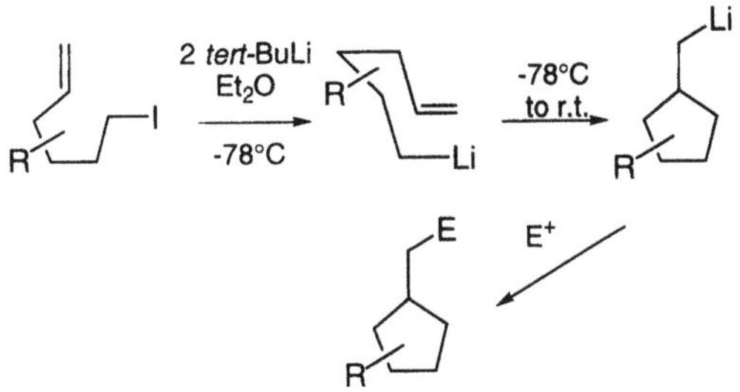

2 *tert*-BuLi
Et_2O
-78°C
-78°C
to r.t.
E^+

Scheme 1.

This cyclization is efficient if the alkene moiety is terminal, or bears a phenyl, trimethylsilyl, cyclopropyl [8], or sulfide [9] group, but not if it is internal [10]. The reagent is usually prepared by iodine/ lithium exchange in a low polarity solvent (pentane-ether), so that coordination of the lithium atom to the π bond [6, 11] is favored. The starting iodoalkene must be primary, or else Wurtz-type or elimination products reduce yields [12]. However, δ-ethylenic benzyllithiums (even tertiary ones) generated from Se/Li exchange, undergo the reaction [13]. δ-Ethylenic vinyllithiums where lithium is located on carbon 2 undergo cyclization, with comparable rates [14] and lead to exomethylene ring systems (see Scheme 2).

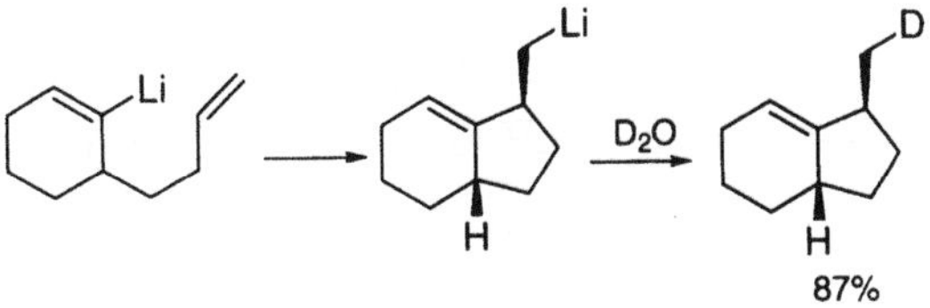

D_2O
87%

Scheme 2.

δ-Ethylenic allyllithiums also cyclize, with possible further proton transfer, or further internal nucleophilic substitution [15]. Finally, β-methylpyrrolidines are accessible via such *5-exo-trig* cyclizations from α-lithiomethyl allylamines [16]. Four-membered rings can be prepared via a *4-exo-trig* cyclization [17, 18], but with lower yields, since the cyclobutylcarbinyllithium thus formed tends to reopen [18, 19] (Scheme 3), unless it is stabilized by a neighboring alkoxy moiety [18] (Scheme 4).

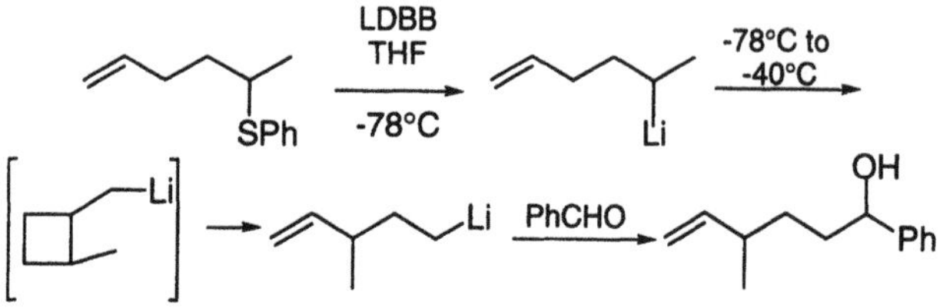

Scheme 3.

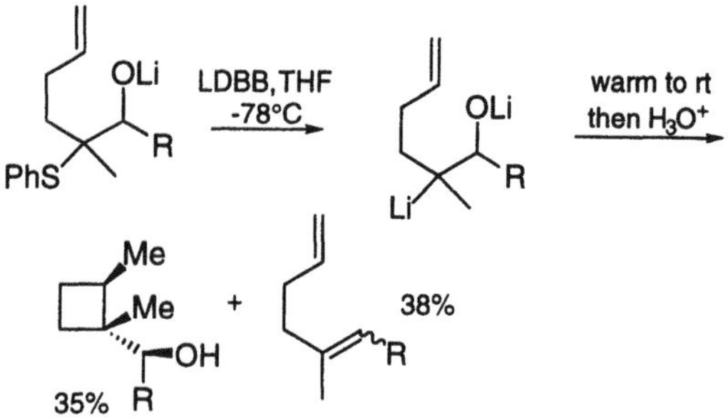

Scheme 4.

In a similar way, cyclopropylcarbinyllithiums, which can be formed by cyclization of homoallyllithiums undergo 1,2-vinyl rearrangements [19, 20], unless a lithioalkoxy group, initially β to lithium accelerates the ring closure, and stabilizes the cyclic product [19a] (Scheme 5). An efficient similar stabilization is observed if the starting lithioalkene is derived from a vinyl thioether [19b].

Cyclizations of a 6-*exo-trig* nature are possible, but are much slower than the 5-*exo-trig* ones [12], and require the presence of TMEDA (Scheme 6), except in cases where the lithiated position is benzylic [21] or aza-allylic [22]

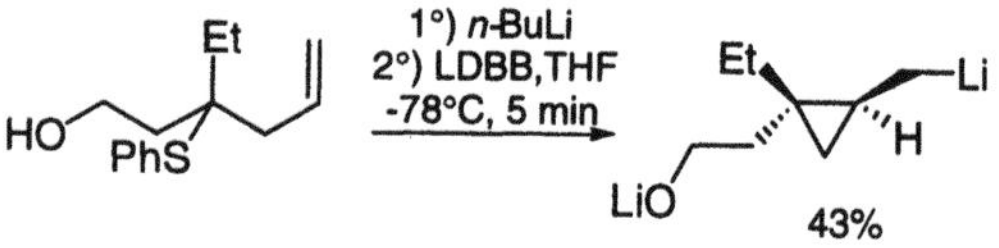

Scheme 5.

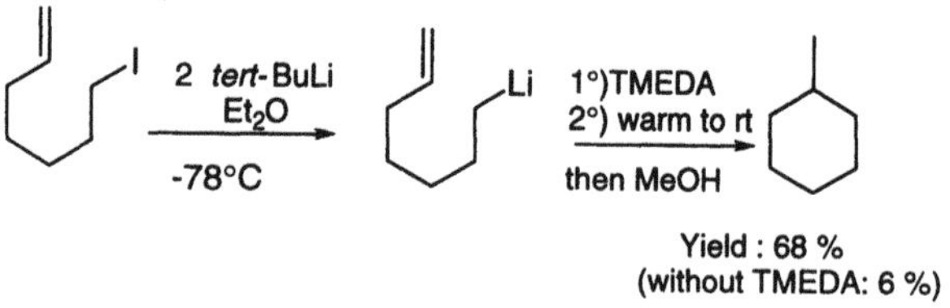

Scheme 6.

1.3
Intermolecular Carbolithiations

The addition of RLi to unactivated alkenes [23] is difficult. Wittig [24], in his pioneering work, showed that it was promoted by a donor group, vicinal to the unsaturation, and allylic alcohols were found to be good substrates [25]. More recently, addition of *n*-butyllithium (2.2 equivalents) to cinnamyl alcohol, either *Z* or *E*, has been performed with good yields [26], in the presence of TMEDA. The benzyllithium thus obtained has been trapped by methyl iodide (Scheme 7) to give a *syn* diastereomer with an excellent diastereoselectivity (dr=98:2). This steric outcome is interpreted by a facile epimerization of the formed benzyllithium, whose lithium atom is chelated in a five-membered ring. In this ring, both substituents prefer a *trans* relationship. The subsequent trapping takes place with a formal inversion of configuration, and a cyclic dilithiated intermediate has been proposed (Scheme 7).

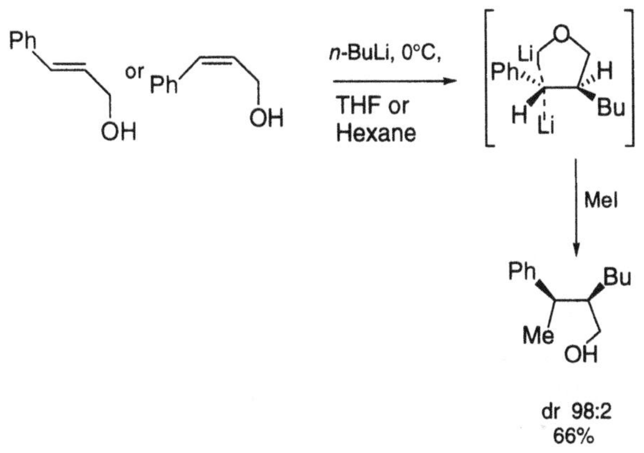

Ph, OH or Ph, OH — *n*-BuLi, 0°C, THF or Hexane → [Li, O, Ph, H, H, Li, Bu] — MeI → Ph, Bu, Me, OH

dr 98:2
66%

Scheme 7.

However, if one starts from a cinnamylamine, either *Z* or *E*, butyllithiation under the same conditions, followed by trapping of the intermediate by electrophiles, now leads to the *anti* product [27](Scheme 8), via retention of configuration.

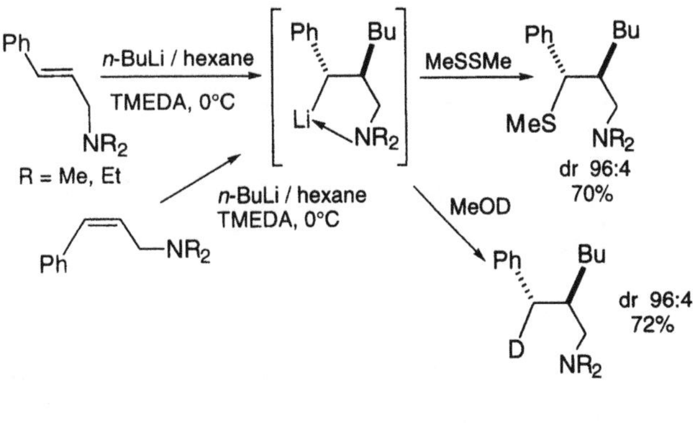

Ph, NR_2 (R = Me, Et) — *n*-BuLi / hexane, TMEDA, 0°C → [Ph, Bu, Li, NR_2] ← *n*-BuLi / hexane, TMEDA, 0°C — Ph, NR_2

→ MeSSMe → Ph, Bu, MeS, NR_2 — dr 96:4, 70%

→ MeOD → Ph, Bu, D, NR_2 — dr 96:4, 72%

Scheme 8.

This steric outcome has been interpreted [28, 29]. The presence of an allylic hydroxy or amino group is not necessary, in fact, and styrene itself adds various tertiary, secondary, and even primary alkyllithiums if ether is used as a solvent (and not THF which promotes the known polymerization) [30]. MeLi, PhLi, or vinyllithiums do not react, even in the presence of TMEDA, but in the presence of this additive, they will react on *ortho*-allyloxystyrenes, and mediate the migration of the allyl moiety to a benzylic position [31, 32] (Scheme 9).

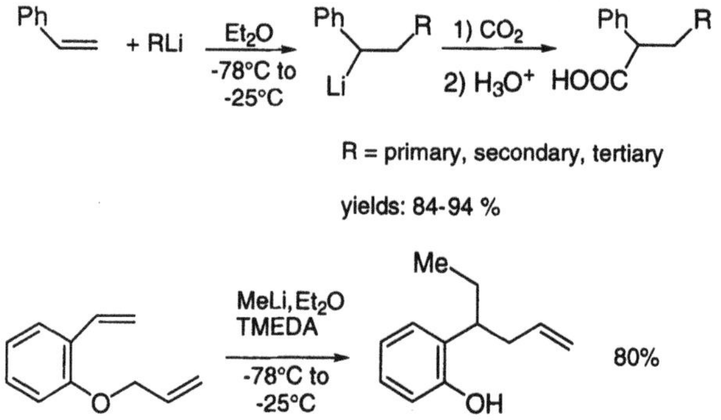

Scheme 9.

Styrenes, *ortho*-substituted by a γ-unsaturated chain, undergo a "cascade reaction" corresponding to a double addition sequence, to give disubstituted tetralins [33] (Scheme 10), whereas the addition of a β-unsaturated alkyllithium to styrene leads to *trans*-1,2-disubstituted cyclopentanes [34] (Scheme 11), via successive inter- and intramolecular additions.

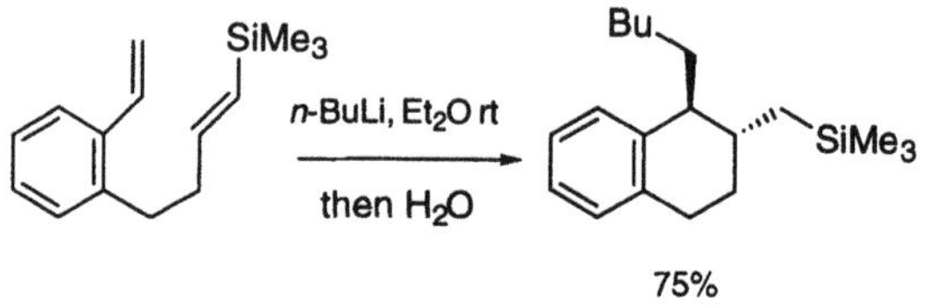

Scheme 10.

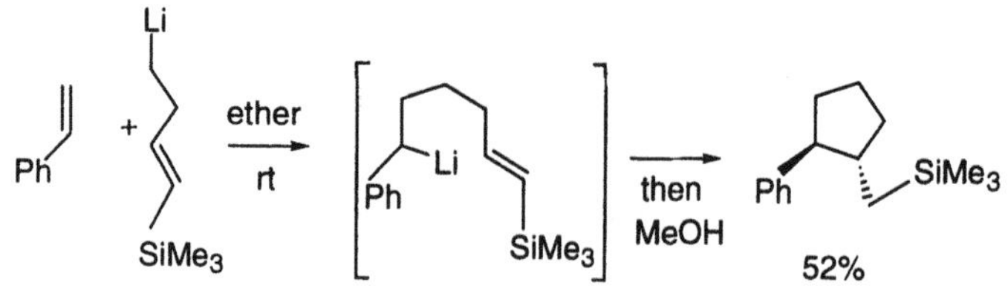

Scheme 11.

α-Aryl-*O*-vinyl carbamates also add regioselectively a variety of organolithium reagents, and the resulting benzyllithium can be trapped at –78 °C, whereas

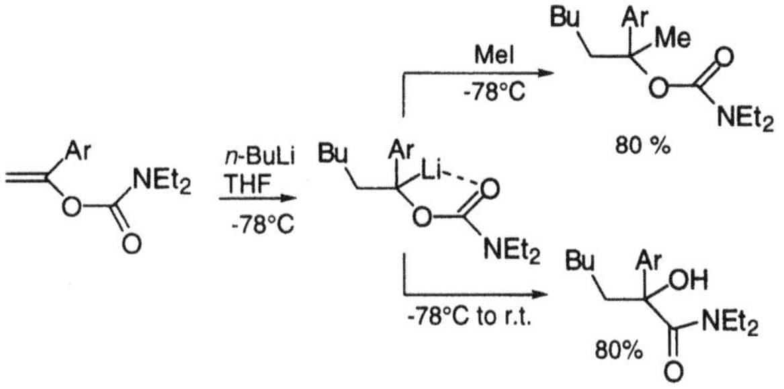

Scheme 12.

on warming to room temperature, it undergoes a [1,2]-Wittig rearrangement to form a 2-aryl-2-hydroxyamide [35] (Scheme 12).

Very few non-arylated alkenes undergo intermolecular carbolithiation. This is the case for diallylamines once they have been converted to a dianionic intermediate which promotes the desired addition of *n*-butyllithium [36] (Scheme 13), and of 2-azaallyllithiums which react with vinylsilanes [37] by an anionic [3+2]-cycloaddition process followed by elimination (Scheme 14).

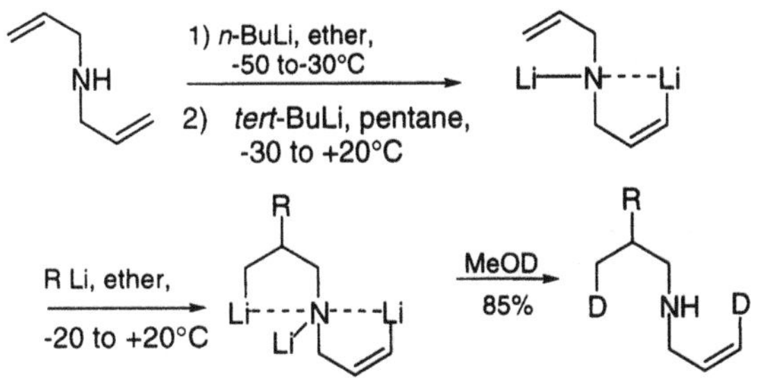

Scheme 13.

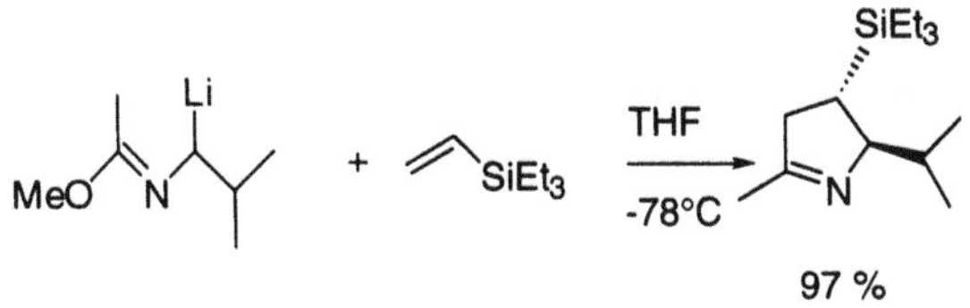

Scheme 14.

2
Enantioselective Intermolecular Carbolithiations

All substrates described above should be, in principle, amenable to an enantioselective version of carbolithiation, under the influence of either a chiral ligand for lithium, which would promote a face selection on the ethylenic unsaturation, or of a chiral group already present on the substrate (diastereoselection). The first efficient results, according to the former process, came from the (–)-sparteine-mediated carbolithiation of (*E*)-cinnamyl alcohol [38], in hexane or cumene (but not in an ethereal solvent), by primary and secondary alkyllithium reagents (Scheme 15). For butylation in cumene, the hydrolyzed product was obtained in 82% yield, with an ee of 80%.

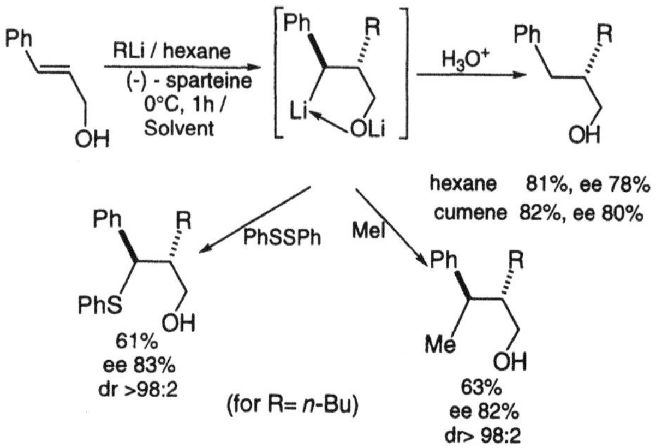

Scheme 15.

As discussed above, for the racemic series [26–29], a further quench of the intermediate benzyllithium by electrophiles delivered highly diastereoselectively (>98:2) the substitution products, derived from a formal inversion of configuration (Scheme 15). Two chiral centers have thus been created. Curiously enough, this high diastereoselection remains valid, if one starts from (*Z*)-cinnamyl alcohol, but the enantioselection is now reversed, although with slightly lower ees (70%). Thus, although sparteine is only available in the (–) form, both enantiomers of 2-benzylhexan-1-ol can be prepared, by switching from the *E* to the *Z* form of cinnamyl alcohol [38] (Scheme 16). If the allylic alcohol is not substituted, as is the case with allyl alcohol, addition occurs but without any induction.

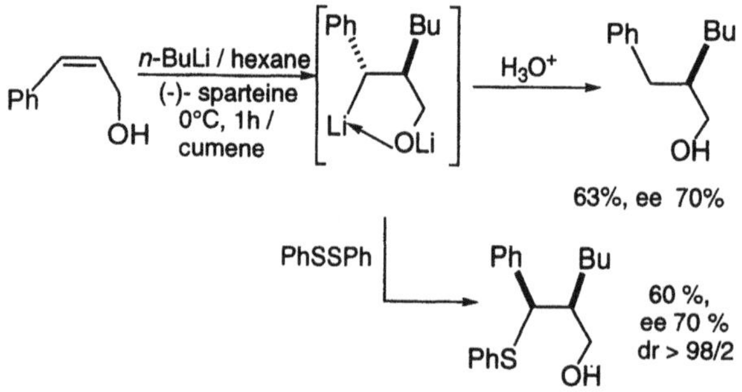

Scheme 16.

The presence of a free alcohol is not a requirement, and cinnamyl ethers, or amines can be used as well [38]. In all cases, an interesting reagent for the trapping of the intermediate benzyllithiums is zinc bromide, which gives access to the organo zinc species: This metal-metal exchange takes place also with inversion of configuration, thus the intermediate lithium reagents shown in Schemes 9–15, 9–16, (phenyl and butyl groups *anti* to each other) give the corresponding zinc reagents where these groups are now *syn*. These kinetically formed species are configurationally stable, up to –30 °C for hours, and can be trapped by electrophiles, or they can be epimerized to the thermodynamically favored diastereomer (phenyl and butyl groups *anti* to each other) by warming to room temperature or higher [27] (Scheme 17).

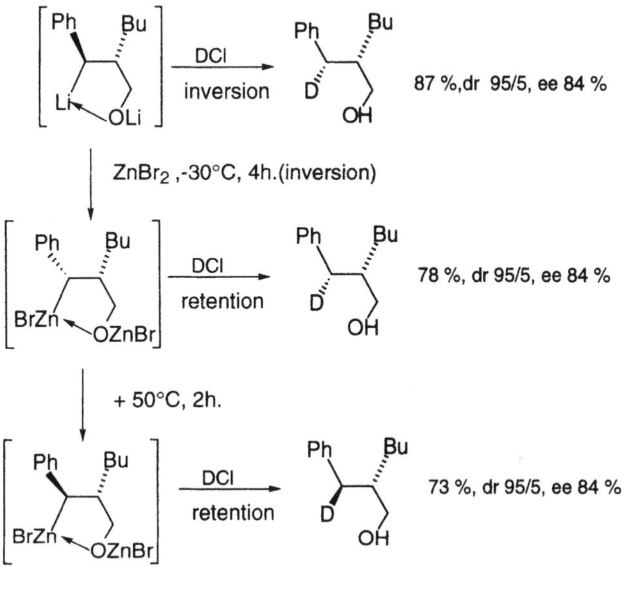

DCl, inversion — 87 %, dr 95/5, ee 84 %

$ZnBr_2$, -30°C, 4h. (inversion)

DCl, retention — 78 %, dr 95/5, ee 84 %

+ 50°C, 2h.

DCl, retention — 73 %, dr 95/5, ee 84 %

Scheme 17.

The necessity of using a hydrocarbon as the solvent in the reactions discussed thus far raised the problem of using a "home-made" lithium reagent, since these derivatives are best made from a halogen/lithium exchange in an ethereal solvent. This difficulty may be circumvented if pentane, hexane, or cumene is added to the RLi,LiX reagent, prepared in ether, but a larger amount of sparteine has to be added, since the diamine coordinates efficiently with the lithium halide. The best way is to start from an alkyl chloride, so that LiCl already precipitates in ether, and to pump off this solvent after the addition of cumene: under these conditions, a 1:1 ratio of sparteine: RLi can be used [27].

Sparteine, as a tertiary diamine (like TMEDA), is necessary for the addition to proceed, and one may surmise that a catalytic amount of it should be efficient if this ligand moves from the formed benzyllithium to the starting alkyllithium. This is indeed the case, and in the presence of 5% of (–)-sparteine, butyllithium

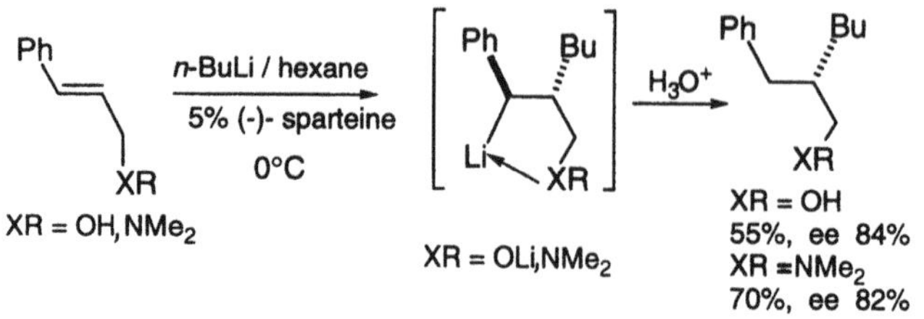

Scheme 18.

adds to (*E*)-cinnamyl alcohol or (*E*)-cinnamyldimethylamine with ees of 84 and 82% [38] (Scheme 18)

In order to improve the fair enantiosectivity described above, the coordination between the chiral organometallic complex, the electron-donating oxygen, and the π-bond of the styryl moiety were reinforced by switching from cinnamyl alcohol to its mixed acetal derived from 2-methoxypropene. For this latter case the carbolithiation takes place at –50 °C (instead of 0 °C with cinnamyl alcohol) and primary or secondary alkyllithiums (in hexane or cumene) give the addition product in 70–80% yield, with ees of 94 and 90%, respectively [39, 40] (Scheme 19). With *n*-butyllithium, the use of 10% sparteine led to 67% yield (with an ee of 92%), whereas use of 1% sparteine gave 50% yield (with an ee of 85%) [39].

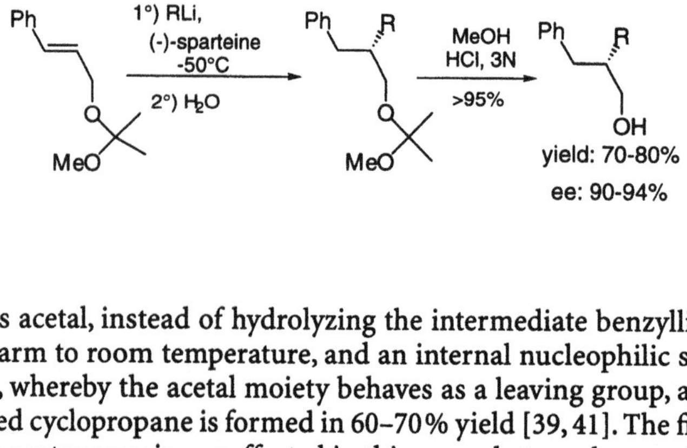

Scheme 19.

With this acetal, instead of hydrolyzing the intermediate benzyllithium, one can let it warm to room temperature, and an internal nucleophilic substitution takes place, whereby the acetal moiety behaves as a leaving group, and a *trans*-disubstituted cyclopropane is formed in 60–70% yield [39, 41]. The first-formed stereogenic center remains unaffected in this second step, whereas the benzylic lithiated carbon is able to epimerize [38, 39], leading to the more stable *trans*-cyclopropane [42–44] (Scheme 20).

According to these preliminary results, it was believed that the observed enantioselectivities were due to the presence of a heteroatom (O,N) in the allylic position. Later on, it was shown that homologated substrates with a hydroxy function in the homo, or bishomo position [45], also undergo carbolithiation, with only slight erosion of ees (Scheme 21).

The question was then raised whether such functionalities were indispensable. Since styrenes add organolithiums (Sect, 1.3), β-alkylstyrenes were tested in

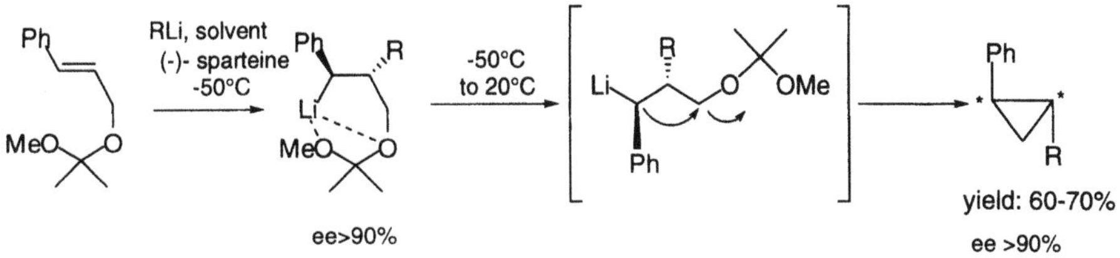

Scheme 20.

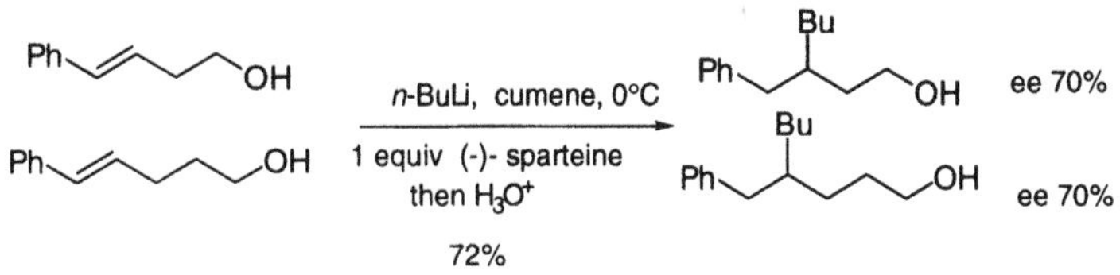

Scheme 21.

a sparteine-mediated carbolithiation. With one equivalent of (–)-sparteine, in hexane, at –15 °C, for 4 hours, various primary alkyllithiums add to (*E*)-β-methylstyrene in good yields, and good enantioselectivities (ee: 76–85%) (Scheme 22) [45].

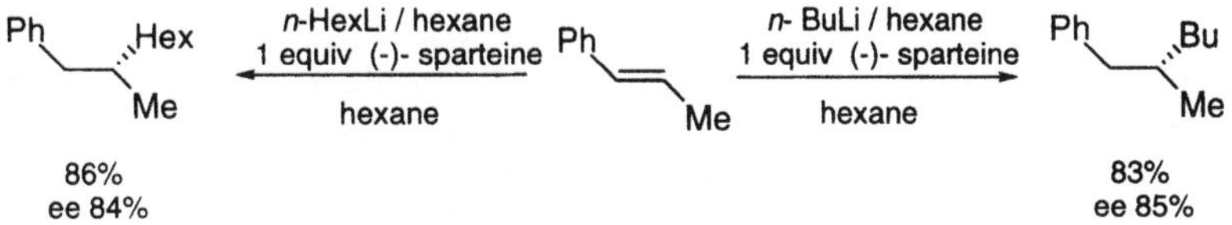

Scheme 22.

In this case also, a catalytic amount (10%) of sparteine can be used, but a lower ee of 70% is obtained. As in the cinnamyl series, the geometry of the alkene is crucial. If (*Z*)-β-methylstyrene is reacted with *n*-BuLi in the presence of 1 equivalent of (–)-sparteine, the addition is much slower, and requires 10 h at 0 °C to give 50% of the adduct [which is the enantiomer of the one obtained from (*E*)-β-methylstyrene], but with a low ee of 28%. According to this result, a kinetic resolution has been performed [45], in the addition of *n*-BuLi to β-ethylstyrene as a mixture of the two isomers (*E*:*Z*= 90:10): The butylated adduct is obtained in 87% yield with 78% ee, whereas the (*Z*)-styrene is recovered to the extent of 7%. (Scheme 23).

Stilbenes also add *n*-BuLi in the presence of (–)-sparteine. Yields are generally good, but the stereoselectivity is very low (<30%). Interestingly, *ortho*-methoxystilbene reacts with a good regioselectivity, which is opposite to that observed with *ortho*-hydroxystilbene, but both give poor enantioselectivities [46] (Scheme 24).

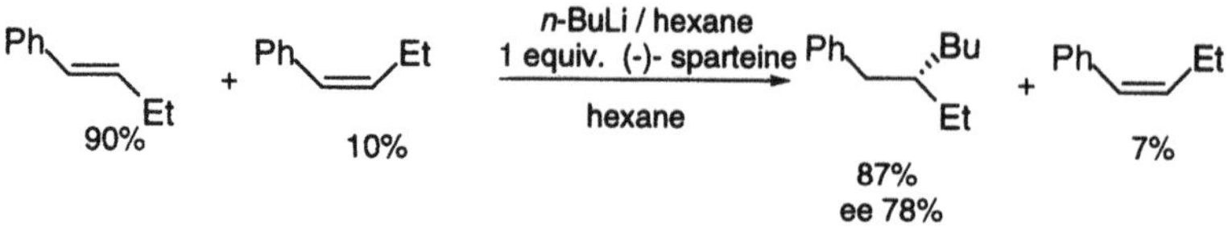

Scheme 23.

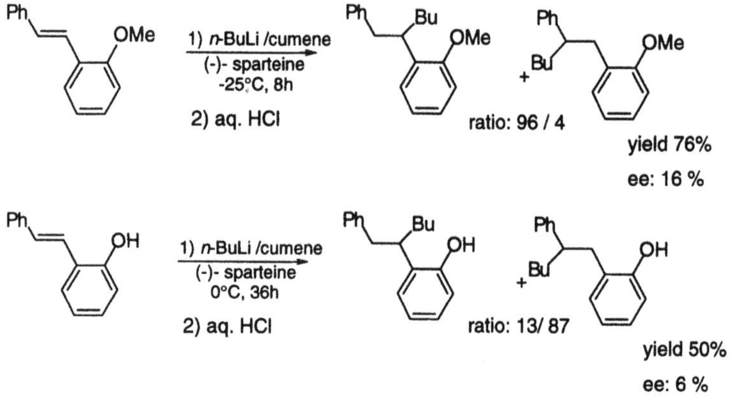

Scheme 24.

The phenyl group of cinnamyl alcohol has been replaced by a variety of groups (heteroatoms, alkynyl or alkenyl moieties...) in order to test the enantioselective carbolithiations of the resulting substrates [46]. Low enantioselectivities have been obtained, except in the case of dienic alcohols [47]. Sparteine, like TMEDA, promotes the substitution of the hydroxy group of penta-2,4-dien-1-ol by an alkyllithium reagent (via an addition-elimination-readdition process) (Scheme 25), but if the pentadienol is substituted on carbon 5, then the expected addition on carbon 2 takes place, and an allyllithium is formed. The latter is hydrolyzed non-regioselectively, but hydrogenation of the mixture of the adducts delivers a saturated alcohol in 71% yield, with an ee of 74% (Scheme 26). A δ-silylated pentadienol behaves in a similar way [47] (Scheme 27).

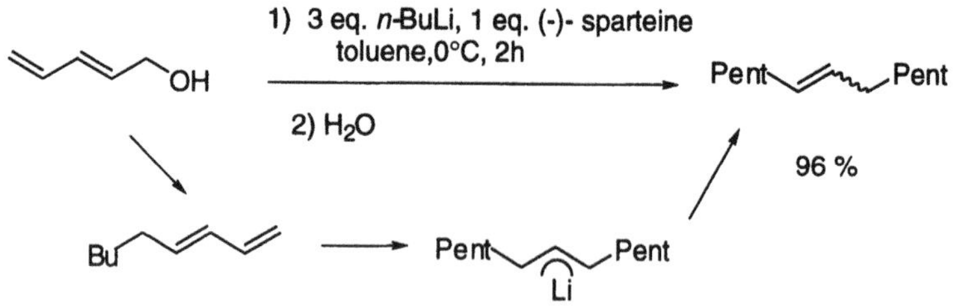

Scheme 25.

Recently, the strategy used for cyclopropanation, as described in Scheme 20 has been extended to such substrates: the mixed acetals derived from dienyl al-

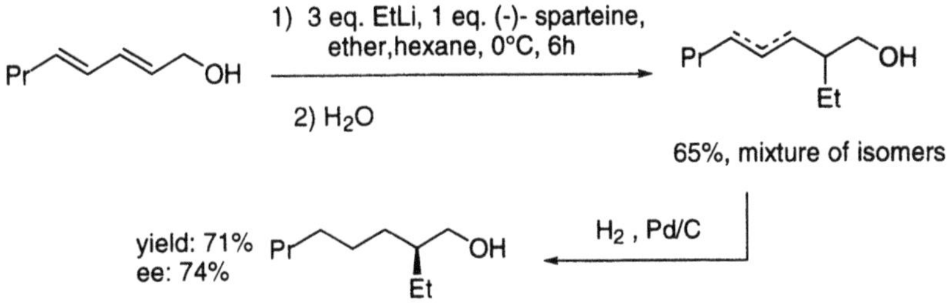

1) 3 eq. EtLi, 1 eq. (-)- sparteine,
ether,hexane, 0°C, 6h
2) H2O
Pr
OH
Et
65%, mixture of isomers
H2 , Pd/C
yield: 71%
ee: 74%

Scheme 26.

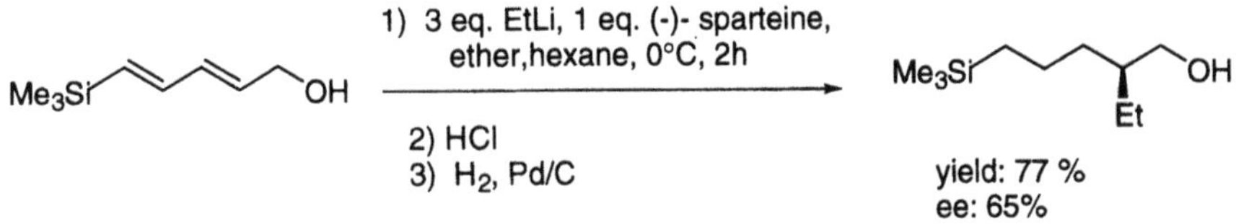

1) 3 eq. EtLi, 1 eq. (-)- sparteine,
ether,hexane, 0°C, 2h
2) HCl
3) H2, Pd/C
Me3Si
OH
Et
yield: 77 %
ee: 65%

Scheme 27.

cohols and acetone dimethyl acetal, when treated with an alkyllithium in the presence of 10% (–)-sparteine, lead to *trans*-1-alkyl-2-vinylcyclopropanes in 50–83% yield, with ees in the 50–83% range [48] (Scheme 28).

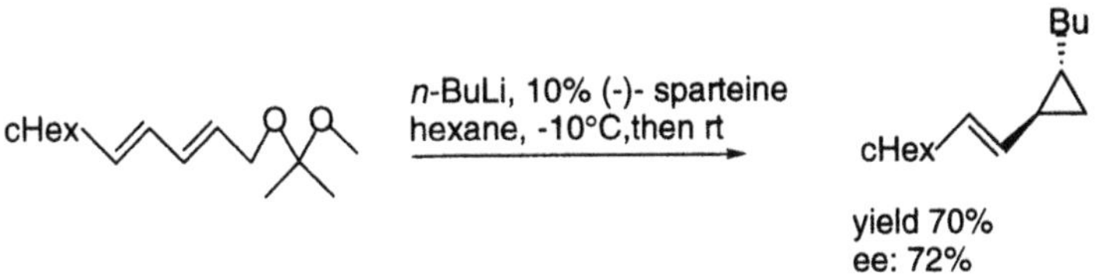

cHex
n-BuLi, 10% (-)- sparteine
hexane, -10°C,then rt
Bu
yield 70%
ee: 72%

Scheme 28.

Enantioselective ring opening of symmetrical bicyclic dihydrofurans via addition-elimination of *n*-butyllithium in the presence of (–)-sparteine has led to ring enlargement with ees up to 52%, whereas premixing of the two reagents at 0°C for 15 min, prior to the addition step (at –40°C), allowed a catalytic use of sparteine, giving comparable ees (Scheme 29) [49].

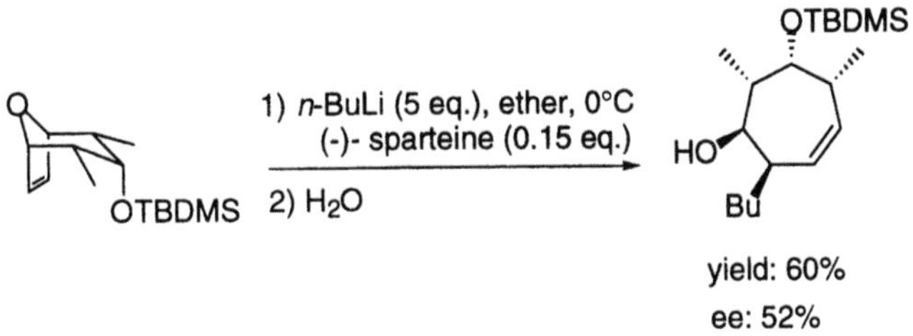

OTBDMS
1) n-BuLi (5 eq.), ether, 0°C
(-)- sparteine (0.15 eq.)
2) H2O
OTBDMS
HO
Bu
yield: 60%
ee: 52%

Scheme 29.

In the field of nucleophilic additions to arenechromium tricarbonyls [50], the chromium complex of phenyloxazoline adds various organolithium reagents regio- and enantioselectively in the presence of (–)-sparteine (or chiral 1,2-diethers). The lithium adduct thus formed has been trapped with propargyl bromide, leading to a *trans*-disubstituted cyclohexadiene of high ee [51] (Scheme 30).

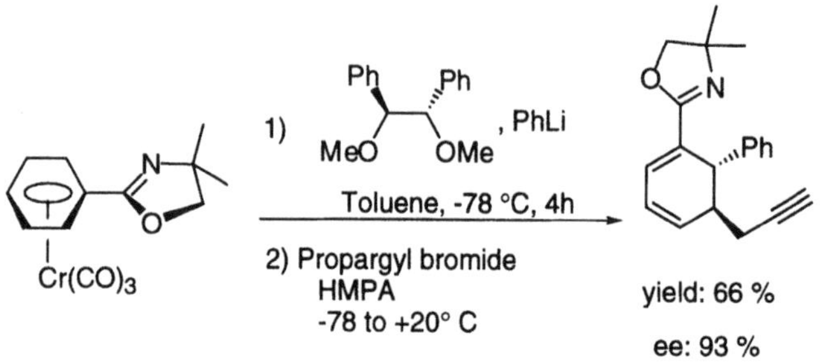

$Cr(CO)_3$

1) MeO OMe Ph Ph, PhLi

Toluene, -78 °C, 4h

2) Propargyl bromide
HMPA
-78 to +20° C

yield: 66 %

ee: 93 %

Scheme 30.

In a chiral approach to ferrocenyldiamines, various aryllithiums, in the presence of sparteine, and in toluene, were added to 6-dimethylaminofulvene, leading to chiral lithium cyclopentadienides. The latter can be transformed into ferrocenes, via an Fe(II) salt. When the aryllithium is *ortho*-substituted by a methyl, and in the presence of a chiral binaphthyl alcoholate, instead of sparteine, the ratio of chiral to *meso*-ferrocene is 88/12 (in 94% yield) and the chiral one is almost pure (ee>99%) (Scheme 31) [52].

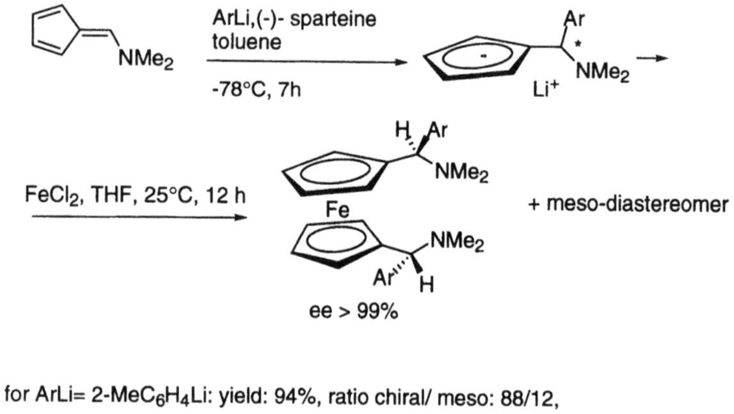

NMe_2

ArLi,(-)- sparteine
toluene

-78°C, 7h

Li^+

FeCl$_2$, THF, 25°C, 12 h

+ meso-diastereomer

ee > 99%

for ArLi= 2-MeC$_6$H$_4$Li: yield: 94%, ratio chiral/ meso: 88/12,

Scheme 31.

The presence of a chiral appendage on an unactivated ethylenic substrate may also induce a diastereoselective carbolithiation: Thus, cinnamaldehyde has been transformed into a *gem*-amino alcoholate [53], with the lithium monoamide of an enantiomerically enriched 1,2-diamine [54] (Scheme 32). This chiral substrate is able to add an alkyllithium reagent regioselectively. The resulting benzyllithium can be trapped by methyl iodide to give the corresponding dialkylated aldehyde, after hydrolysis, by a process which corresponds to an "umpolung" of the usual nucleophilic conjugate addition-electrophilic capture of the starting aldehyde. Reduction of the adduct leads to a primary alcohol with an ee of 93%. This reaction has been extended to a β-silylated acrolein [54] (Scheme 32) with a similar stereoselection.

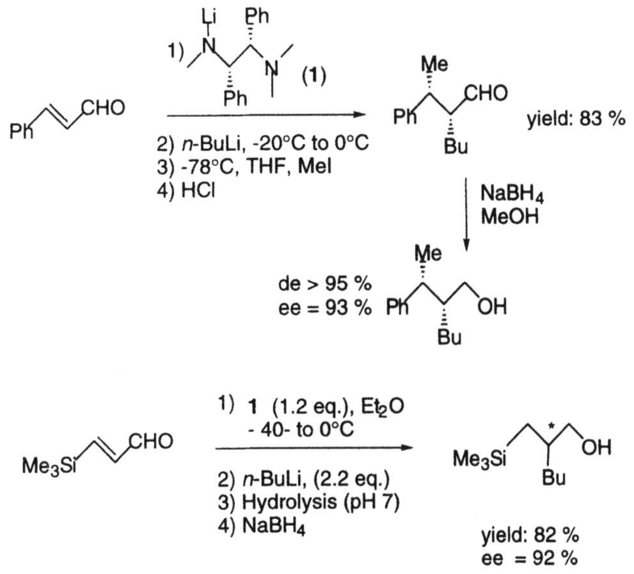

1) (1)
2) *n*-BuLi, -20°C to 0°C
3) -78°C, THF, MeI
4) HCl
yield: 83 %
$NaBH_4$
MeOH
de > 95 %
ee = 93 %

1) **1** (1.2 eq.), Et_2O
- 40- to 0°C
2) *n*-BuLi, (2.2 eq.)
3) Hydrolysis (pH 7)
4) $NaBH_4$
yield: 82 %
ee = 92 %

Scheme 32.

3
Enantioselective Intramolecular Carbolithiations

The (–)-sparteine-mediated enantioselective carbolithiation of phenyl-substituted alkenes has been performed intramolecularly to prepare enantioenriched cyclopentanes [55]. Scheme 33 shows that the (*S*)-configured (α-carbamoyloxy)-alkyllithium, made by enantioselective deprotonation, still undergoes a *syn* addition, followed first by epimerization of the intermediate benzyllithium, and then by the electrophilic attack of the latter with inversion of configuration: three stereogenic centers have been created . A similar stereoselective cyclization has been used to transform piperidine-derived substrates into indolizidines [56].The reaction proceeds equally well, if the styryl moiety, present here, is replaced by a 1-phenylbutadienyl or a 1-phenylbut-1-yn-3-enyl moiety [57].

In the preceding substrate of Scheme 33, if a phenylethynyl moiety is used instead of a styryl one, problems may arise from the acidity of the propargylic hydrogens. However, the 5-*exo-dig* cyclization takes place readily if a bulky substituent is located on the propargylic position which may be of (*R*) or (*S*) configuration. In both cases the (*S*)-configurated (α-carbamoyloxy)alkyllithium is still

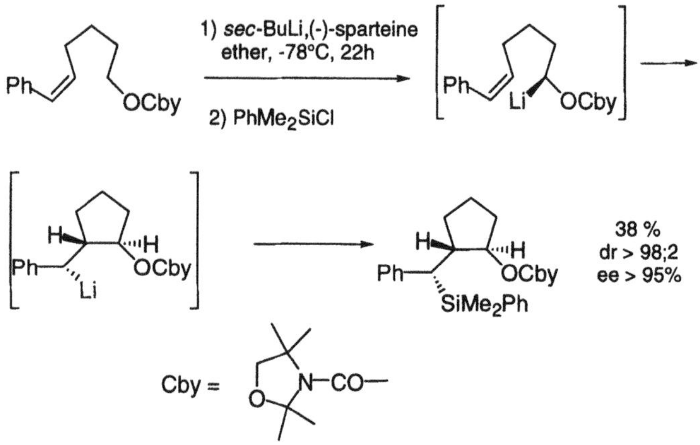

Scheme 33.

formed, and according to the stereochemistry at the propargylic center, a unique exomethylene cyclopentane, either *trans* or *cis* is formed in excellent ee [58] (Scheme 34).

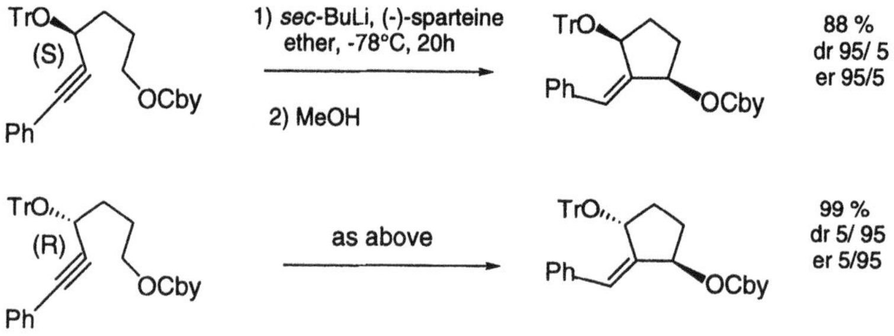

Scheme 34.

Reverting to the simple phenylated substrate of Scheme 33, if the styryl moiety is *ortho*-substituted by a phenyldimethylsilyloxy group, the 5-*exo-trig* cyclization is now followed by a retro-[1,4]-Brook rearrangement. Whatever the (*E*) or (*Z*) configuration of the styryl group, the intermediate benzyllithium epimerizes, and a pentacoordinated silicon allows transfer of the silicon moiety diastereoselectively to form a unique *C*-silylated adduct (er>98:2, dr>99:1) [59] (Scheme 35).

An interesting enantioselective cyclocarbolithiation has been disclosed by Hoppe's group, starting from a racemic carbamate (**1**) derived from indene (Scheme 36). Enantioselective deprotonation of the two enantiomers now leads to two lithio-diastereomers. A kinetic resolution takes place since the *syn* carbolithiation from (*R*,*S*)-**2** maintains the carbamate group in a pseudoequatorial position, whereas for (*S*,*S*)-**2**, this group would be forced in an *endo* position: as a result, the tricyclic adduct (–)-**3** is formed in high purity after hydrolysis, where-

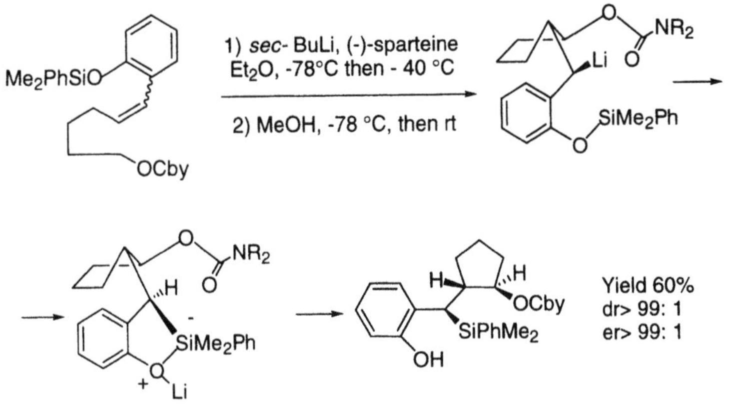

Scheme 35.

as (*S*,*S*)-**2** reverts to the optically active starting material (*S*)-(+)-**1**. If the reaction is performed with the phenylated analogue **4**, the intermediate benzhydryllithium, formed after the first cyclization, is prone to induce a subsequent cyclopropanation via a γ-elimination of the OCby group, and (–)-**5** is formed [60].

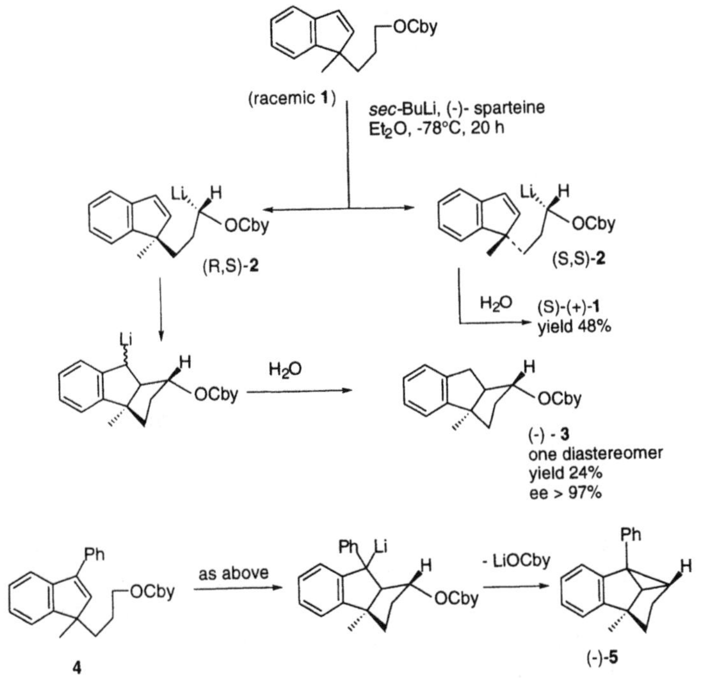

Scheme 36.

Beak's approach to enantiomerically enriched α-aminolithium reagents [61] has been used to prepare pyrrolizidines [62] from a 2-stannylated-*N*-(1-buten-

3-yl)pyrrolidine. If the tin/lithium exchange is performed at low temperatures in THF, followed by warming to 0°C, no cyclization takes place; but in pentane/ether (10/1), although the cyclization requires room temperature, the desired bicyclic (+)-pseudoheliotridane is formed in 90% yield, as a single diastereomer of 94% ee (Scheme 37). This result shows that the intermediate lithium reagent is configurationally stable ($t_{1/2}\approx 5$ h) at 23°C, and reacts with retention of configuration.

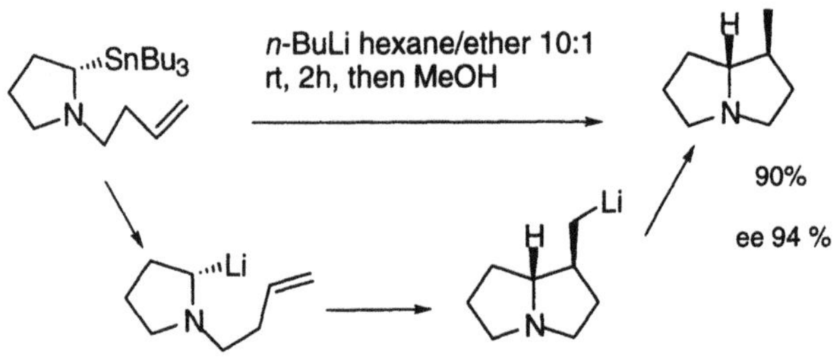

Scheme 37.

Changing the *N*-butenyl moiety of the starting reagent in Scheme 37 into an *N*-pentenyl one gives access to indolizidines via 6-*exotrig* cyclization [62b, 63] (Scheme 38). In the same low polarity solvent, at room temperature, the reaction now requires 6 h, ($t_{1/2}\approx 90$ min) and allows for epimerization of the lithium reagent which racemizes within 1 h. But if the cyclization process is boosted by the presence of a phenylthio moiety on the double bond, the cyclization operates readily ($t_{1/2}\approx 4$ min) and leads to a 70/30 mixture of the diastereomeric indolizidines with ees of 72 and 74% (Scheme 38).

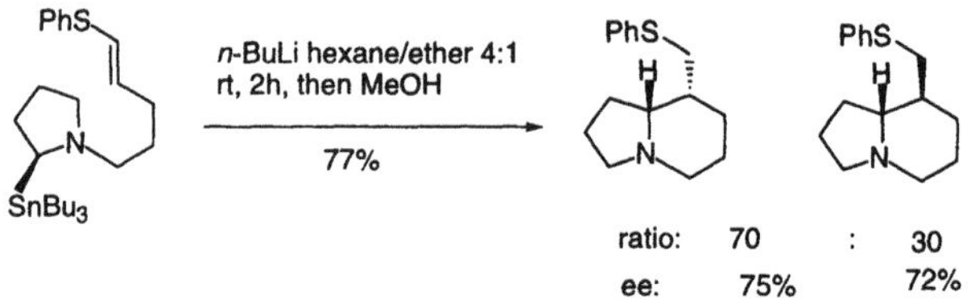

Scheme 38.

When *N*-lithiomethyl-*N*-benzylamino-1-but-3-ene is prepared by tin/lithium exchange in the presence of (–)-sparteine, the corresponding pyrrolidine is formed with a low er of 64: 36. If the benzyl group is replaced by a chiral *R*-phenethyl group, transmetallation in THF at –78°C, in the presence or the absence of sparteine, leads to a 3:1 mixture of diastereomeric pyrrolidines (Scheme 39) [64].

Access to enantiomerically enriched lithium derivatives α to oxygen, by tin/lithium exchange, with retention of configuration [65], has been addressed by Nakai and coworkers, not only to study the stereochemistry of the [1,2]- and

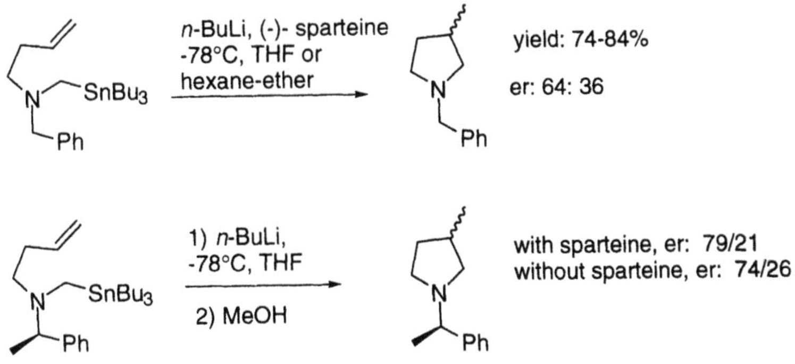

Scheme 39.

[2,3]-Wittig rearrangements [66], but also to perform intramolecular carbolithiations on a terminal C=C bond, or an internal C=C bond bearing an allylic leaving group [55d, 67] (Scheme 40). A *trans*-2,3-disubstituted tetrahydrofuran is formed with a total retention of configuration.

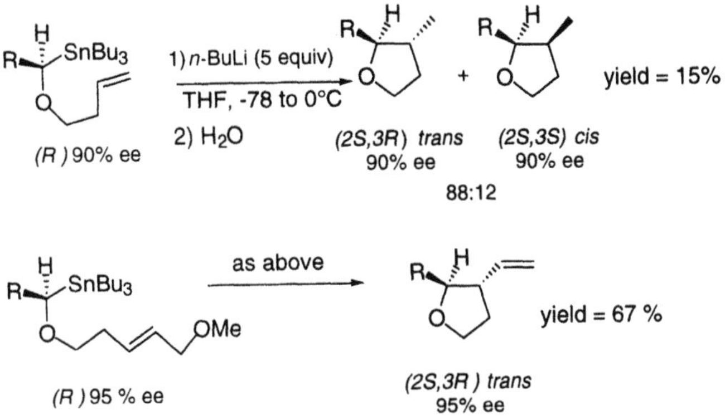

Scheme 40.

In the reduction of a chiral α-cyanoether by lithium di-*tert*-butyl-biphenylide (LDBB), the intermediate radical captures a second electron more rapidly than it cyclizes, and delivers an organolithium derivative which is able to give a 5-*exo-trig* cyclization: the resulting spiro-derivative is obtained with 42% ee [68] (Scheme 41).

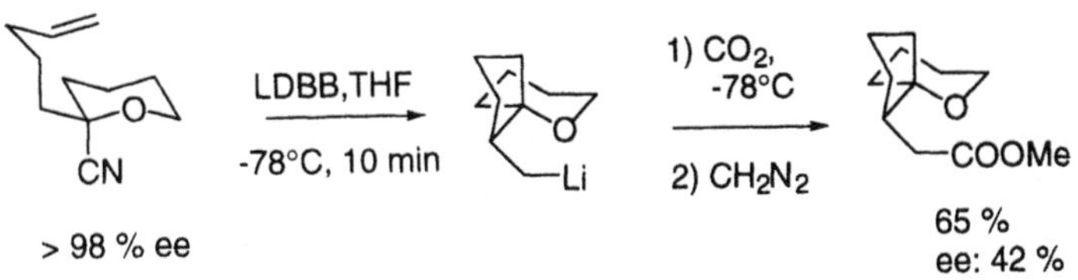

Scheme 41.

Finally, complexes of (–)-sparteine and aryllithiums can cyclize with good enantioselectivities onto a terminal C=C bond, tethered to the *ortho*-position of the aromatic ring (in the heterocyclic or homocyclic series). Thus, when the *ortho*-lithiated-*N,N*-diallylated aniline, in Scheme 42, is submitted to a Br/Li exchange at –78 °C in pentane-ether, and is allowed to warm to room temperature, after the addition of 2 equivalents of sparteine: the corresponding 2-methylindoline is formed in 69% yield, and 93/7 er. The analogous *ortho*-but-3-en-1-ylbromobenzene requires more stringent conditions (1 h at +22 °C), and furnishes the desired 3-methylindane in 73% yield with a lower er (71/29) [69].

A vinyllithium (2-lithiohepta-1,6-heptadiene) also cyclizes in the presence of sparteine at 0 °C, to form an exomethylene cyclopentane in 73% yield, but the er is also limited to 70/30 [69a] (Scheme 42).

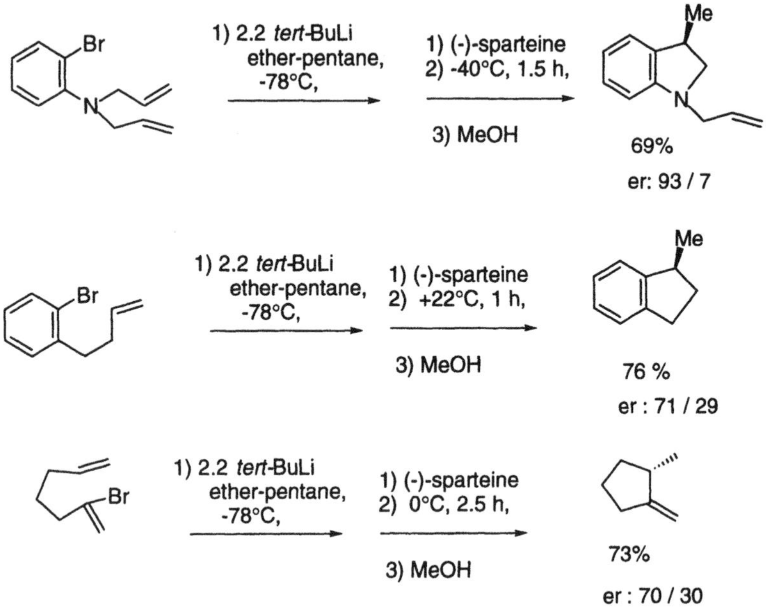

Scheme 42.

When two aryllithium moieties are part of the same ethylenic substrate, representing a *meso*-diether, as in Scheme 43, the addition-elimination process (S_N2' reaction) can be directed by chiral ligands: chiral amino alcoholates, or alcoholates derived from binaphthol promote the formation of one enantiomer of the resulting dihydrobenzofuran [70].

In the field of diastereoselective intramolecular carbolithiations, a stereogenic center, already present in the vicinity of the C=C bond, can induce a facial choice. Thus, the epimerizable tertiary benzyllithium of Scheme 44, [71] undergoes a 5-*exo-trig* cyclization under the influence of the neighboring chiral lithium alkoxide, and delivers four isomeric cyclopentanes (in a ratio of 82:8:7.5:2.5) in 90% yield.

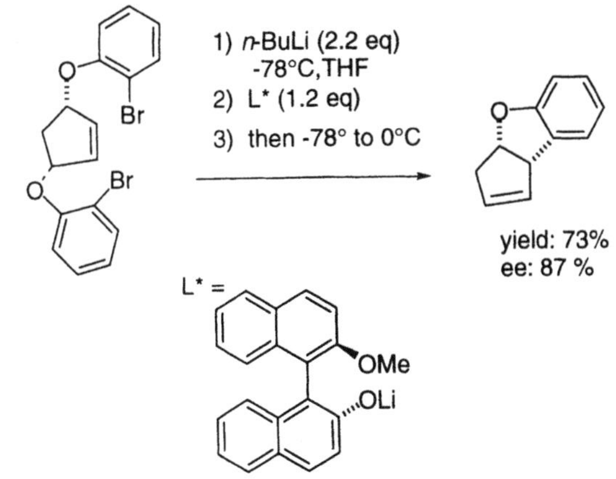

1) *n*-BuLi (2.2 eq) -78°C, THF
2) L* (1.2 eq)
3) then -78° to 0°C

yield: 73%
ee: 87 %

L* = (OMe, OLi)

Scheme 43.

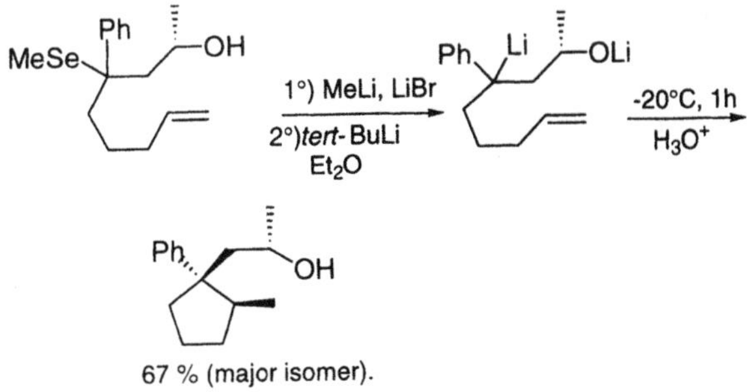

1°) MeLi, LiBr
2°) *tert*-BuLi
Et_2O

-20°C, 1h
H_3O^+

67 % (major isomer).

Scheme 44.

4 Conclusion

The enantioselective carbolithiation of unactivated C=C bonds is, so far, restricted to limited cases: In the intramolecular carbocyclization, best results are obtained in the case of five-membered rings, starting with a primary alkyllithium bearing a terminal C=C bond in the δ-position (which may be "activated" by a phenyl or a phenylthio group, or by the presence of an allylic leaving group). The intermolecular process is also allowed, in the cinnamyl or the styryl series, and it is now extended to the case of conjugated dienols. High enantiomeric excesses can be attained in the presence of sparteine, which so far, represents the best chiral ligand for such enantioselective additions, and can be used in catalytic amounts in certain cases. The presence of a low polarity solvent is crucial, as it favors the Li/C=C π-bonding, and reduces problems due to polymerization. In the intramolecular version, it is possible to start with an enantiomerically enriched secondary lithium derivative, if cyclization is fast enough, as compared to epimerization. For inter- or intramolecular carbolithiations, a chiral center, already present on the substrate can induce a highly diastereoselective addition.

Without doubt, the "state of the art" reported here, only represents a starting point for further investigations in this vast domain. Not only is a new stereogenic center created, but also a new organolithium derivative is formed, which may further react in a diastereoselective fashion.

Note added in proof. Different allyl 2-lithioaryl ethers undergo tandem carbolithiation/γ-elimination in Et_2O/TMEDA affording *o*-cyclopropyl phenol or naphtol derivatives in a diastereoselective manner; the use of (–)-sparteine as a chiral ligand instead of TMEDA allows the synthesis of cyclopropane derivatives with up to 81% ee [72].

References

1. For general reviews on carbometallation, see: (a) Knochel P (1991). In: Trost BM, Fleming I (eds) Comprehensive organic synthesis. Pergamon Press, Oxford, vol 4 p 865; (b) Yamamoto Y, Asao N (1993) Chem Rev 93:2207; (c) Knochel P (1995). In: Able EW, Stone FGA, Wilkinson G (eds) Comprehensive organometallic chemistry II. Pergamon Press, Oxford, vol 11 p 159; (d) Negishi E, Kondakov DY (1996) Chem Rev 96:417; (e) Marek I, Normant JF (1998). In: Diederich F, Stang P (eds) Metal catalysed cross-coupling reactions. Wiley, VCH, Weinheim, p 271; (f) Fallis A, Forgione P (2001) Tetrahedron 57:5899
2. For a review on enantioselective carbometallation, see: Marek I (1999) J Chem Soc Perkin Trans 1 535
3. Ziegler K, Bähr K (1928) Ber dtsch chem Ges 61:28
4. Morton M (1983) Anionic polymerization: principles and practice, Academic Press, New York
5. Bailey WF, Patricia JJ, Del Gobbo VC, Jarret RM, Okarma PJ (1985) J Org Chem 50:1999
6. Bailey WF, Khanolkar AD, Gavaskar K, Ovaska TV, Rossi K, Thiel Y, Wiberg KB (1991) J Am Chem Soc 113:5720
7. (a) Bailey WF, Khanolkar AD, Gavaskar KV (1992) J Am Chem Soc 114:8053; (b) Bailey WF (1994). In: Advances in detailed reaction mechanisms JAI Press Inc, vol 3 p 251; (c) Krief A, Barbeaux P (1991) Tetrahedron Lett 32:417
8. Bailey WF, Gavaskar KV (1994) Tetrahedron 50:5957
9. Krief A, Kenda B, Remacle B (1995) Tetrahedron Lett 36:7917
10. (a) Bailey WF, Jiang XL (1994) J Org Chem 59:6528; (b) an exception is found if the double bond belongs to a conjugated diene: Cooke Jr MP, Huang JJ (1997) Synlett 535
11. (a) Oliver JP, Smart JB, Emerson MT (1966) J Am Chem Soc 88:4101; (b) Denis JSt, Dolzine TW, Oliver JP (1972) J Am Chem Soc 94:8260; (c) Dolzine TW, Oliver JP (1974) J Organomet Chem 78:165; (d) Denis JSt, Oliver JP, Dolzine TW, Smart JB (1974) J Organomet Chem 71:315; (e) Rölle T, Hoffmann RW (1995) J Chem Soc Perkin Trans 2 1953; for theoretical aspects: Hehre WJ, Radom L, Schleyer PvR, Pople JA (1986). In: Ab initio molecular orbital theory. Wiley, New York; (f) Houk KN, Rondan NG, Schleyer PvR, Kaufmannn E, Clark T (1985) J Am Chem Soc 107:2821
12. Bailey WF, Nurmi TT, Patricia JJ, Wang W (1987) J Am Chem Soc 109:2442
13. (a) Krief A, Barbeaux P (1987) J Chem Soc Chem Commun 1214; (b) Krief A, Barbeaux P (1990) Synlett 511
14. (a) Chamberlin AR, Bloom SH (1986) Tetrahedron Lett 27:551; (b) Chamberlin AR, Bloom SH Cervini LA, Fotsch CH (1988) J Am Chem Soc 110:4788; (c) Bailey WF, Jiang XL, Mc Leod CE (1995) J Org Chem 60:7791; (d) Zhang D, Liebeskind LS (1996) J Org Chem 61:2594; (e) Bailey WF, Jiang XL (1996) J Org Chem 61:2596; (f) in the heterocyclic series: Barluenga J, Sanz R, Fananas FJ (1997) Tetrahedron Lett 38:2763
15. (a) Cheng D, Zu S, Liu X, Norton SH, Cohen T (1999) J Am Chem Soc 121:10241; (b) Cheng D, Knox KR, Cohen T (2000) J Am Chem Soc 122:412

16. (a) Coldham I, Hufton R (1996) Tetrahedron 52:12541; (b) Coldham I, Lang-Anderson MMS, Rathmell RE, Snowden DJ (1997) Tetrahedron Lett 38:7621; (c) Coldham I, Fernandez JC, Snowden DJ (1999) Tetrahedron Lett 40:1819; (d) For azabicyclic structures: Coldham I, Fernandez JC, Price KN, Snowden DJ (2000) J Org Chem 65:3788
17. For the formation of azetidines, see: Coldham I (1993) J Chem Soc Perkin Trans 1 1275
18. F. Chen F, B. Mudryk B, T. Cohen T (1994) Tetrahedron 50:12793
19. (a) Mudryk B, Cohen T (1993) J Am Chem Soc 115:3855; (b) Chen F, Mudryk B, Cohen T (1999) Tetrahedron 55:3291
20. (a) Bailey WF, Punzalan ER, Dela EW, Taylor DK (1995) J Org Chem 60:297; (b) Harvey JN, Viehe HG (1995) J Prakt Chem 337:253
21. Krief A, Kenda B, Barbeaux P, Guittet E (1994) Tetrahedron 50:7177
22. (a) Pearson WH, Lovering FE (1994) Tetrahedron Lett 35:9173; (b) Pearson WH, Postich MJ (1994) J Org Chem 59:5662; (c) Pearson WH, Lovering FE (1995) J Am Chem Soc 117:12336
23. (a) Klumpp GW, Veefkind AH, Graaf WL, Bickelhaupt F (1967) Ann Chem 706:47; (b) Veefkind AH, Bickelhaupt F, Klumpp GW (1969) Recl Trav Chim Pays-Bas 88:1058; (c) Veefkind AH, Schaaf JVD, Bickelhaupt F, Klumpp GW (1971) J Chem Soc Chem Commun 722; (d) Kool M, Klumpp GW (1978) Tetrahedron Lett 1873
24. Wittig G, Otten J (1963) Tetrahedron Lett 601
25. (a) Felkin H,. Swierczewski G, Tambute A (1969) Tetrahedron Lett 707; (b) Grandall JK, Clark AC (1969) Tetrahedron Lett 325; (c) Crandall JK, Clark AC (1972) J Org Chem 37:4236; (d) Dimmel DR, Huang SJ (1973) J Org Chem 38:2756
26. Kato T, Marumoto S, Sato T, Kuwajima I (1990) Synlett 671
27. (a) Klein S, Marek I, Normant JF (1994) J Org Chem 59:2925; (b) for a review on enantioselective carbometalation in the cinnamyl series, see: Norsikian S, Marek I, Klein S, Poisson JF, Normant JF (1999) Chem Eur J 5:2055
28. Lichtenfeld CM, Ahlbrecht H (1996) Tetrahedron 52:10025
29. Lichtenfeld CM, Ahlbrecht H (1999) Tetrahedron 55:2609
30. Wei X, Taylor RJK (1996) J Chem Soc Chem Commun 187
31. Wei X, Taylor RJK (1996) Tetrahedron Lett 37:4209
32. Wei X, Johnson P, Taylor RJK (2000) J Chem Soc Perkin Trans 1 1109
33. Wei X, Taylor RJK (1997) Tetrahedron Lett 38:6467
34. Wei X, Taylor RJK (2000) Angew Chem Int Ed 39:409
35. Superchi S, Satomayor N, Miao G, Joseph B, Campbell MG, Snieckus V (1996) Tetrahedron Lett 37:6061
36. Barluenga J, Gonzalez R, Fanonas FJ, Yus M, Foubelo F (1994) J Chem Soc Perkin Trans 1 1069
37. Pearson WH, Stevens EP (1998) J Org Chem 63:9812
38. Klein S, Marek I, Poisson JF, Normant JF (1995) J Am Chem Soc 117:8853
39. Norsikian S, Marek I, Poisson JF, Normant JF (1997) J Org Chem 62:4898
40. The enantiomeric excesses were determined according to: (a) Alexakis A, Frutos JC, Mutti S, Mangeney P (1994) J Org Chem 59:3326; (b) Alexakis A, Mutti S, Mangeney P (1992) J Org Chem 57:1224
41. The formation of a cyclopropane by 1,3-elimination between an organogembismetallic and a methoxy-methyl ether was already described; (a) Marek I, Beruben D, Normant JF (1995) Tetrahedron Lett 36:3695; (b) Beruben D, Marek I, Normant JF, Platzer N (1995) J Org Chem 60:2488; (c) Beruben D, Marek I, Normant JF, Platzer N (1993) Tetrahedron Lett 34:7575
42. For some representative formations of chiral cyclopropanes, see (a) Charette AB, Marcoux JF (1995) Synlett 1197 and references cited therein; (b) Salaün J (1989) Chem Rev 89:1247; (c) Romo D, Meyers AI (1992) J Org Chem 57:6265; (d) Davies HML, Huby NJS, Cantrell WRJr, Olive JL (1993) J Am Chem Soc 115:9468; (e) Doyle MP (1993) In: Catalytic asymmetric synthesis, Ojima I (ed) VCH Publishers, New York, Chapter 3; (f) Pfaltz A (1993) Acc Chem Res 26:339; (g) Evans DA, Woerpel KA, Scott MJ (1992) Angew Chem Int Ed Engl 31:430; (h) Denmark SE, O'Connor SP (1997) J Org Chem 62:584; (i) Doyle MP, Davies

SB, Hu W (2000) Org Lett 2:1145; (j) Charette AB, Lemay J (1997) Angew Chem Int Ed Engl 36:1090; (k) Charette AB, Molinaro C, Brochu C (2001) J Am Chem Soc 123:12168
43. For a previous report on the synthesis of chiral cyclopropane mediated by a stoichiometric amount of (–) sparteine, see Paetow M, Kotthaus M, Grehl M, Frölich R, Hoppe D (1994) Synlett 1034
44. (a) Krief A, Hobe M (1992) Tetrahedron Lett 33:6527; (b) Krief A, Hobe M, Dumont W, Badaoui E, Guittet E, Evrard G (1992) Tetrahedron Lett 33:3381
45. Norsikian S, Marek I, Normant JF (1997) Tetrahedron Lett 38:7523
46. Norsikian S (1999) PhD thesis, University P & M Curie Paris
47. Norsikian S, Baudry M, Normant JF (2000) Tetrahedron Lett 41:6575
48. Majumdar S, de Meijere A, Marek I (2002) Synlett 423
49. Lautens M, Gajda C, Chiu P (1993) J Chem Soc Chem Commun 1193
50. Semmelhack MF (1995). In: Trost BM, Fleming I (eds) Comprehensive organometallic chemistry, Pergamon Press, Oxford, vol 12 p 517
51. Amurrio D, Khan K, Kündig P (1996) J Org Chem 61:2258
52. Suzuka T, Ogasawara M, Hayashi T (2002) J Org Chem 67:3355
53. Comins DL (1992) Synlett 671
54. Brémand N, Mangeney P, Normant JF (2001) Tetrahedron Lett 42:1883
55. (a) Woltering MJ, Fröhlich R, Hoppe D (1997) Angew Chem Int Ed Engl 36:1764; (b) Hoppe D, Hense T (1997) Angew Chem Int Ed Engl 36:2282; (c) Hoppe D, Woltering MJ, Oestreich M, Fröhlich R (1999) Helv Chim Acta 82:1860; (d) Tomooka K, Komine N, Sasaki T, Shimizu H, Nakai T (1998) Tetrahedron Lett 39:9715
56. Woltering MJ, Fröhlich R, Wibbeling B, Hoppe D (1998) Synlett 797
57. Oestreich M, Hoppe D (1999) Tetrahedron Lett 40:1881
58. (a) Oestreich M, Fröhlich R, Hoppe D (1998) Tetrahedron Lett 39:1745; (b) Oestreich M, Fröhlich R, Hoppe D (1999) J Org Chem 39:8616
59. Kleinfeld SH, Wegelius E, Hoppe D (1999) Helv Chim Acta 82:2413
60. Laqua H, Fröhlich R, Wibbeling B, Hoppe D (2001) J Organomet Chem 624:96
61. (a) Kerrick ST, Beak P (1991) J Am Chem Soc 113:9708; (b) Beak P, Basu A, Gallagher DJ, Park YS, Thayumanavan S (1996) Acc Chem Res 29:552
62. (a) Coldham I, Hufton R, Snowden DJ (1996) J Am Chem Soc 118:5322; (b) Ashweek NJ, Coldham I, Snowden DJ, Vennall GP (2002) Chem Eur J 8:195; (c) for a general review on heterocycles via carbolithiation, see: Mealy MJ, Bailey WF (2002) J Organomet Chem 646:59
63. Coldham I, Vennall GP (2000) Chem Commun 1569
64. Coldham I, Hufton R, Price KN, Rathmell RE, Snowden DJ, Vennal GP (2001) Synthesis 1523
65. (a) Still WC, Sreekumar C (1980) J Am Chem Soc 102:1201; (b) Sawyer JS, Kucerovy A, Macdonald TL McGarvey GJ (1988) J Am Chem Soc 110:842
66. (a) Tomooka K, Igarashi T, Komine N, Nakai T (1992) Tetrahedron Lett 33:5795; (b) Tomooka K, Igarashi T, Nakai T (1994) Tetrahedron 50:5927
67. Tomooka K, Komine N, Nakai T (1997) Tetrahedron Lett 38:8939
68. Rychnovsky SD, Hata T, Kim AI, Buckmelter J (2001) Org Lett 3:807
69. (a) Bailey WF, Mealy MJ (2000) J Am Chem Soc 122:6787; (b) Gil GS, Groth UM (2000) J Am Chem Soc 122:6789
70. Nishiyama H, Sakata N, Motoyama Y, Wakita H, Nagase H (1997) Synlett 1147
71. Krief A, Bousbaa J (1996) Synlett 1007
72. Barluenga J, Fañanás FJ, Sanz R, Marcos C (2002) Org Lett 4:2225

Author Index Volumes 1–5

Subject Index

O

Springer

www.ingramcontent.com/pod-product-compliance
Ingram Content Group UK Ltd.
Pitfield, Milton Keynes, MK11 3LW, UK
UKHW021834190726
13853UKWH00003B/1299
* 9 7 8 3 6 6 2 1 4 6 1 1 8 *